Electron Correlation in New Materials and Nanosystems

NATO Science Series

A Series presenting the results of scientific meetings supported under the NATO Science Programme.

The Series is published by IOS Press, Amsterdam, and Springer in conjunction with the NATO Public Diplomacy Division

Sub-Series

I. Life and Behavioural Sciences	IOS Press
II. Mathematics, Physics and Chemistry	Springer
III. Computer and Systems Science	IOS Press
IV. Earth and Environmental Sciences	Springer

The NATO Science Series continues the series of books published formerly as the NATO ASI Series.

The NATO Science Programme offers support for collaboration in civil science between scientists of countries of the Euro-Atlantic Partnership Council. The types of scientific meeting generally supported are "Advanced Study Institutes" and "Advanced Research Workshops", and the NATO Science Series collects together the results of these meetings. The meetings are co-organized bij scientists from NATO countries and scientists from NATO's Partner countries – countries of the CIS and Central and Eastern Europe.

Advanced Study Institutes are high-level tutorial courses offering in-depth study of latest advances in a field.
Advanced Research Workshops are expert meetings aimed at critical assessment of a field, and identification of directions for future action.

As a consequence of the restructuring of the NATO Science Programme in 1999, the NATO Science Series was re-organised to the four sub-series noted above. Please consult the following web sites for information on previous volumes published in the Series.

http://www.nato.int/science
http://www.springer.com
http://www.iospress.nl

Series II: Mathematics, Physics and Chemistry – Vol. 241

Electron Correlation in New Materials and Nanosystems

edited by

Kurt Scharnberg
University of Hamburg, Germany

and

Sergei Kruchinin
Bogolyubov Institute for Theoretical Physics,
Kiev, Ukraine

Springer

Published in cooperation with NATO Public Diplomacy Division

Proceedings of the NATO Advanced Research Workshop on
Electron Correlation in New Materials and Nanosystems, held in
Yalta, Ukraine 19–23 September 2005.

A C.I.P. Catalogue record for this book is available from the Library of Congress.

ISBN 978-1-4020-5658-1 (PB)
ISBN 978-1-4020-5659-8 (eBook)

Published by Springer,
P.O. Box 17, 3300 AA Dordrecht, The Netherlands.

www.springer.com

Printed on acid-free paper

TABLE OF CONTENTS

II.2 Cuprate and other unconventional superconductors

PREFACE

These proceedings reflect much of the work presented and extensively discussed in a stimulating and congenial atmosphere at the NATO Advanced Research Workshop "Electron correlation new materials and nanosystems", held at the "Yalta" Hotel, Yalta, Ukraine from 19-23 September 2005. The lively discussion sessions in the evenings, unfortunately, could not be included in the proceedings but in some sense they were continued during a rigorous refereeing process which lead to substantial modifications of many contributions. This refereeing process, together with the request by the publisher "to have those articles which have been written by non-native English speakers carefully proofread and, if necessary, corrected by a native English speaker working closely with the editor", caused considerably delay in the submission of these proceedings to the publisher. On the other hand, given this extra time, several participants who in Yalta had declined to submit manuscripts, could be persuaded to present their latest research in these proceedings after all. Other authors used this opportunity to update their manuscripts. So, on average these proceedings represent state of the art research as of the summer of 2006 rather than September 2005.

Since neither of us work in an English-speaking environment, we tried to enlist the help of referees to meet the publisher's request for carefully proofreading manuscripts. We would like to use this opportunity to express our sincere thanks to the referees for the help we got.

Thanks are also due to Johny Sebastian from Springer's texsupport, who modified the original style file in accord with the editor's wishes. These changes helped in particular to squeeze many manuscripts on an even number of pages and thus reduce the physical size of this tome substantially.

The topics discussed included a wide range of novel materials with emphasis on superconductors, mesoscopic and nanostructured systems like quantum wires, quantum dots, nanotubes and various hybrid structures involving ferromagnets and superconductors or organic substances and metals. Studies of these systems were presented which addressed the problems of understanding the fundamental physical processes as well as their applications to quantum computing and spintronics. The workshop closed with a session on various types of sensors. Most contributions were presented orally, but in order not to overload the program and to leave enough time for discussion, there was also a poster session. Some of the papers presented as posters have been included in these proceedings.

This workshop addressed a range of topics rather wider than is usual. While the relaxed atmosphere provided by this Black Sea resort with its natural beauty and architectural gems, filled with captivating history, encouraged and facilitated the communication with colleagues having rather different backgrounds, the editors decided to focus the proceedings more sharply and thus not to include the sessions on sensors, for which only two manuscripts had been submitted.

We are grateful to members of the International Advisory Committee A. Balatsky and I.Yanson for their consistent help and suggestions.

We would like to thank the NATO Public Diplomacy Division, Collaborative Programmes Section, for the essential financial support without which this meeting could not have taken place. Thanks are also due to the National Academy of Science of Ukraine and the Ministry of Ukraine for Education and Science for support.

Kurt Scharnberg and Sergei Kruchinin
August 2006.

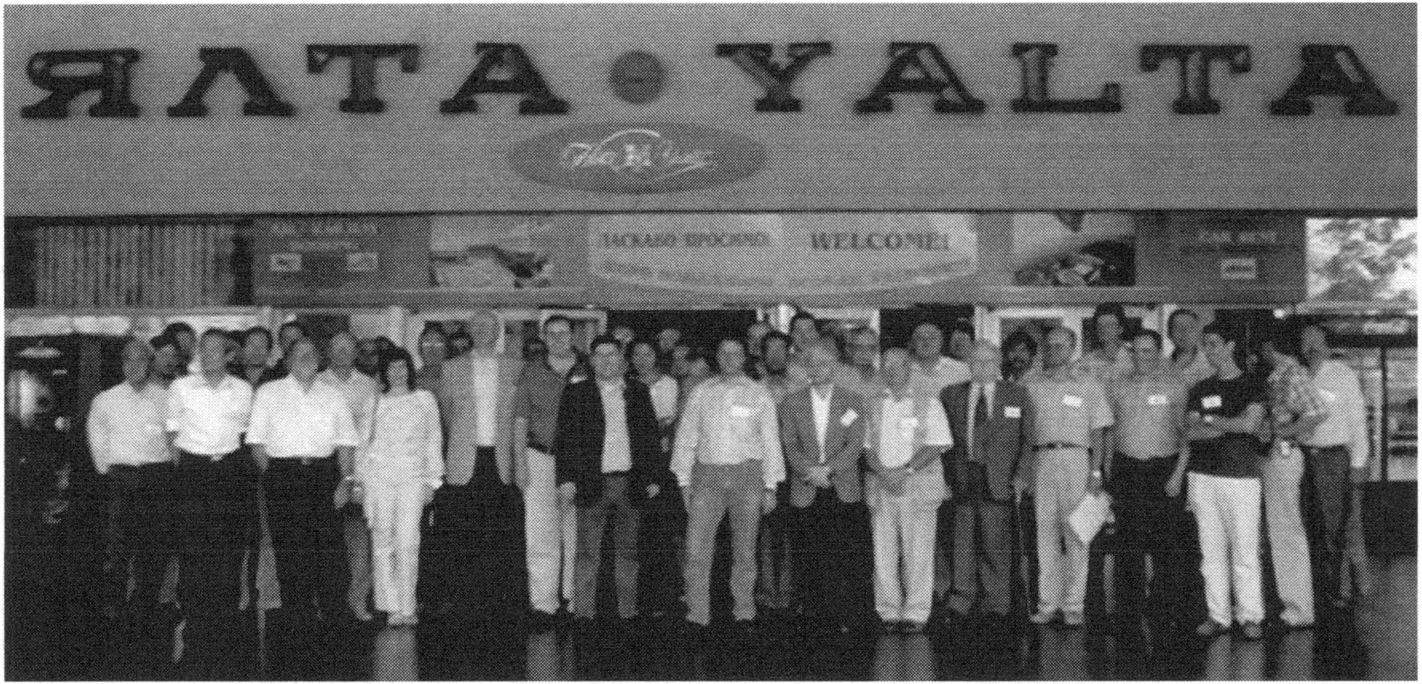
YALTA
WELCOME!

PART I

Quantum Nanodevices

TRANSPORT PROPERTIES OF FULLERENE NANODEVICES

Toward the New Research Field of Organic Electronic Devices

Akihiko Fujiwara[*], Yukitaka Matsuoka, Nobuhito Inami, Eiji Shikoh

School of Materials Science, Japan Advanced Institute of Science and Technology, 1-1 Asahidai, Tatsunokuchi, Ishikawa 923-1292, Japan, and CREST, Japan Science and Technology Agency, 4-1-8 Honchou, Kawaguchi, Saitama 332-0012, Japan

Abstract. We report transport properties of C_{60} thin film field-effect transistors (FETs) with a channel length of several-ten nanometers. Nonlinear drain current I_D versus source-drain voltage V_{DS} characteristics were observed at room temperature. We discuss this phenomenon in terms of the crossover from a diffusive conductance in bulk regime to a coherent one in the nanometer scale.

Key words: Fullerene; Nanodevice; Organic electronics; Transport properties; Field-effect transistor; Crossover

1. Introduction

The miniaturization of transistors enables us to put a billion transistors on a chip operating with the clock periods of a billionth of a second. However, as transistors get smaller in size, there are many undesirable effects, such as short-channel effects and the increase of off-currents. Moreover, quantum effects will become significant. To overcome these effects, a device based on a new principle of operation, such as a single-electron transistor (SET)[1], is required. In recent years, C_{60} has attracted considerable attention as the material for an island of the SET because it can be regarded as an ideal quantum dot by itself. C_{60} is a closed cage, nearly spherical molecule consisting of 60 carbon atoms with a diameter of about one nanometer. Its high symmetry results in a unique electronic structure, such as the three-fold degenerate lowest-unoccupied-molecular orbital (LUMO) and the five-fold degenerate highest-occupied-molecular orbital (HOMO)[2]. In addition,

---

[*] Corresponding author: Akihiko Fujiwara; e-mail: fujiwara@jaist.ac.jp

K. Scharnberg and S. Kruchinin (eds.),
Electron Correlation in New Materials and Nanosystems, 3–8.
© 2007 *Springer.*

the electronic structure of crystalline C_{60} is hardly modified from that of a free C_{60} molecule, namely, a molecular orbital, because crystalline C_{60} is a nearly ideal molecular crystal with van der Waals interaction. The quantized electronic levels are conserved even when C_{60} is in a cluster or a crystalline state.

A FET is a macroscopic system with dominant classical effects, whereas an SET is a nano-scale system with dominant quantum mechanical effects. The transport properties of C_{60} thin film FETs with a channel of several-decades of micrometers[3,4] and of the C_{60} SET with an island of several nanometers[5] have been reported. The device structures of a FET and a SET are qualitatively different in inorganic devices. However, in organic devices they are the same: the difference is only the size. This comes from two factors. One is the electronic structure. It originates from the molecular orbital even in the crystal and hardly depends on the size as discussed above. Another is the existence of the barrier at the contact between the channel area and the electrodes for the electron conduction. It acts as the tunnel barrier for an SET and as the Schottky barrier for an FET. The latter is not favorable but cannot be excluded so far. In organic devices, therefore, the marginal electronic states between a macroscopic system and a nano-scale one are expected (Fig. 1). In this work, to clarify the C_{60} device properties in this marginal area, we have investigated the transport properties of C_{60} thin film FETs with a channel length of ca. 20 nanometers.

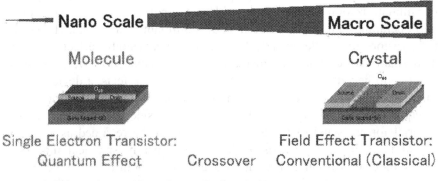

Figure 1. Schematic overview of organic electronics.

2. Experimental details

Figure 2 shows the schematic cross section of the fabricated C_{60} thin film FET with a diagram of the measurement setup. The Au source and drain electrodes with thickness of 100 nm were fabricated on the 400 nm SiO_2 layer that was made on the surface of a heavily doped n-type silicon wafer, using an electron-beam lithography method. The doped silicon layer of the wafer was used for a gate electrode. The distance between source and drain electrodes, *i.e.* the channel length of the fabricated C_{60} thin film FETs was approximately 20 nm.

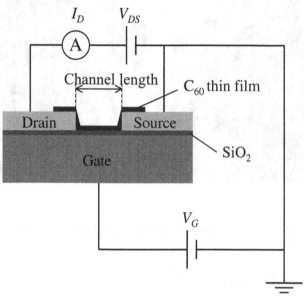

Figure 2. Schematic cross section of the fabricated C60 thin film FET (700 nm channel length) with a diagram of the measurement setup.

A typical scanning electron microscope (SEM) image of fabricated source and drain electrode is shown in Fig. 3. Commercially available C_{60} (99.98 %) was used for the formation of the thin films channel layer. A C_{60} thin film of 150 nm thickness was formed on the SiO_2 layer using vacuum ($< 10^{-4}$ Pa) vapor deposition at the deposition rate of 0.01 nm/s.

It is well known that the *n*-type organic semiconductor is very sensitive to chemically and physically adsorbed O_2 and/or H_2O molecules, which can generate traps of electrons and suppress carrier transport[6-8]. Therefore, before measurements, the samples were annealed at 120 °C under 10^{-3} Pa for a few days. The drain and gate electrodes were biased with dc voltage sources and the source electrode was grounded. The transport properties of C_{60} FETs were measured at room temperature under 10^{-4} Pa without exposure to air after annealing.

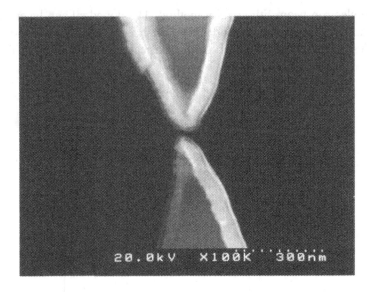

Figure 3. Typical scanning electron microscope (SEM) image of fabricated source and drain electrode.

3. Results and discussion

Figure 4 shows the source-drain voltage V_{DS} dependence of the drain current I_D. I_D increases nonlinearly with increasing V_{DS} and is enhanced by V_G: the I_D versus V_{DS} curves are almost symmetrical. The symmetrical characteristics can be related to those observed in the SET operation rather than the FET operation in which a pronounced asymmetric I_D versus V_{DS} response is observed. As for the V_G dependence, an enhancement of I_D is similar to the FET operation.

It is worth noting that the device structures of the C_{60} FET[3,4,9,10] and the C_{60} SET[5] are the same in principle. Weak contact between the inorganic metal electrodes and organic semiconductor, acting as tunnel barrier in the SET, exists even in the FET as Schottky barrier, although no such an obstacle exists in the inorganic FETs. Therefore, a reduction (an expansion) in size of the FET (SET) will lead to the appearance of the SET (FET) mode of operation in organic devices. The device size shown in Fig. 3 is about 20 nm and is of the same order of characteristic size in which the quantum effect is observed. In addition, the device operation is, in part, similar to that observed in both the SET and FET. On the other hand, the devices with the channel length of about 50 - 700 nm operate as a FET[10,11]. Considering the characteristics of devices and their size-dependence it is plausible that the crossover from the FET character to that of the SET takes place around a channel length of 20 nm. More detailed and systematic experiments on the crossover from the macroscopic behavior (the FET operation) to the microscopic quantum behavior (the SET operation) will clarify the mechanism of electron transport in organic materials.

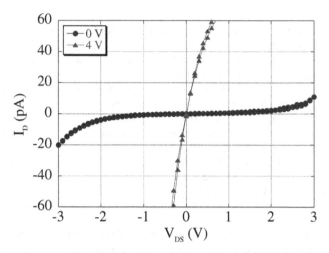

Figure 4. I_D versus V_{DS} plots for $V_G = 0$ V (circle) and 4 V (triangle).

4. Conclusion

We have investigated the transport properties in short-channel C_{60} thin film FETs. The I_D versus V_{DS} plots showed symmetric nonlinear characteristics. This phenomenon can be interpreted as the crossover between a diffusive conductance in bulk regime and a coherent one in the nanometer scale. The marginal area is estimated to fall around 20 nm.

Acknowledgements

The authors are grateful to Dr. M. Akabori, Professor S. Yamada, and the technical staffs of the Center for Nano-materials and Technology at the Japan Advanced Institute of Science and Technology for use of the electron-beam lithography system and other facilities in the clean rooms, as well as for technical support. This work is supported in part by the JAIST International Joint Research Grant, the Grant-in-Aid for Scientific Research (Grant Nos. 16206001, 1731005917, 17540322) from the Ministry of Education, Culture, Sports, Science and Technology (MEXT) of Japan, and the NEDO Grant (Grant No. 04IT5) form the New Energy and Industrial Technology Development Organization (NEDO), and the Kurata Memorial

Hitachi Science and Technology Foundation, the Support for International Technical Exchange from TEPCO Research Foundation.

References

1. H. Grabert, M. H. Devoret, Single Charge Tunneling, NATO ASI Series vol. 294, Plenum Press, New York, 1992.
2. M. S. Dresselhaus, G. Dresselhaus, P. C. Eklund, *Science of Fullerenes and Carbon Nanotubes* (Academic Press, New York, 1996).
3. R. C. Haddon, A. S. Perel, R. C. Morris, T. T. M. Palstra, A. F. Hebard, R. M. Fleming, C_{60} thin film transistors, *Appl. Phys. Lett.* **67**, 121-123 (1995).
4. K. Horiuchi, K. Nakada, S. Uchino, S. Hashii, A. Hashimoto, N. Aoki, Y. Ochiai, M. Shimizu, Passivation effects of alumina insulating layer on C_{60} thin-film field-effect transistors, *Appl. Phys. Lett.* **81**, 1911-1912 (2002).
5. H. Park, J. Park, A. K. L. Lim, E. H. Anderson, A. P. Alivisatos, P. L. McEuen, Nanomechanical oscillations in a single-C_{60} transistor, *Nature* **407**, 57-60 (2000).
6. A. Hamed, Y. Y. Sun, Y. K. Tao, R. L. Meng, P. H. Hor, Effects of oxygen and illumination on the *in situ* conductivity of C_{60} thin films, *Phys. Rev. B* **47**, 10873-10880 (1993).
7. B. Pevzner, A. F. Hebard, M. S. Dresselhaus, Role of molecular oxygen and other impurities in the electrical transport and dielectric properties of C_{60} films, *Phys. Rev. B* **55**, 16439-16449 (1997).
8. R. Könenkamp, G. Priebe, B. Pietzak, Carrier mobilities and influence of oxygen in C_{60} films, *Phys. Rev. B* **60**, 11804-11808 (1999).
9. S. Kobayashi, T. Takenobu, S. Mori, A. Fujiwara, Y. Iwasa, Fabrication and characterization of C_{60} thin-film transistors with high field-effect mobility, *Appl. Phys. Lett.* **82**, 4581-4583 (2003).
10. Y. Matsuoka, N. Inami, E. Shikoh, A. Fujiwara, Transport properties of C_{60} thin film FETs with a channel of several-hundred nanometers, *Sci. Technol. Adv. Mater.* **6**, 427-430 (2005).
11. Y. Matsuoka, N. Inami, E. Shikoh, A. Fujiwara, unpublished.

NANOSCALE STUDIES ON METAL-ORGANIC INTERFACES

N. Chandrasekhar (n-chandra@imre.a-star.edu.sg)

Institute of Materials Research and Engineering, 3 Research Link, Singapore 117602

Abstract. We report ballistic electron emission microscopy (BEEM) studies on two metal organic interfaces, Ag-polyparaphenylene (PPP) and Ag- Poly-1-methoxy-4-(2-ethylhexyloxy)-p-phenylenevinylene (MEHPPV), and a metal-molecule interface Ag-terthiophene-Au, which are evaporated, spin-coated, and self assembled on an Au film respectively. All systems show spatially non-uniform carrier injection. Physical origins of the non-uniform carrier injection and its implications are discussed. The observed injection barriers are smaller than expected. We explain these using a model of metal induced gap states. For the metal-molecule system, a WKB calculation is carried out and compared with the experimental data. The results indicate that molecular levels are being accessed in the BEEM experiment, since the measured currents are larger than a purely tunneling contribution. Our results are consistent with previously published results on a similar molecule. Implications for device applications are briefly discussed.

Key words: BEEM, Interfaces, Electronic transport, Manoscience

1. Introduction

Metal-organic (MO) interfaces have traditionally been investigated by current-voltage (I-V), capacitance-voltage (C-V) and ultraviolet (UV) spectroscopy, all of which average over macroscopic areas [1]. In contrast, prototype devices incorporating molecules as active components are sub-micron [2-4]. Recent work [4] has shown that nanoscale conductance inhomogeneities can exist at MO interfaces. The physical origin of these conducting filaments remains obscure. Filament growth and dissolution has been identified as being responsible for the switching behavior in other systems as well [5,6]. Lau et al. [4] report the observation of a single switching center, and suggest a runaway process of filament growth driven by increasing current density and/or electric field. Memory effects observed in inorganic semiconductors [7] have been invoked to explain the behavior of some organic devices [8,9]. Organic device configurations that have been investigated to date are either Langmuir Blodgett (LB) films [4] or self-assembled monolayers (SAM) [2,3]. At the present time, it is unclear whether the inhomogeneities originate from microstructural perturbations such as asperities at the interfaces with the contacting electrodes, or whether they are an inherent electronic property of MO interfaces. SAM and LB films are not rigid, and despite the implementation of precautionary

K. Scharnberg and S. Kruchinin (eds.),
Electron Correlation in New Materials and Nanosystems, 9–21.
© 2007 *Springer.*

measures, it is uncertain whether the integrity of the organic is maintained after deposition of the metal film [10]. For instance, in metal-inorganic semiconductor (MIS) interfaces, unless the semiconductor surface is prepared with care and the metal is chosen so that it is lattice matched, the metal film is polycrystalline, causing significant variations in the electronic transparency of the interface [11,12].

In this paper, we use ballistic electron emission microscopy and spectroscopy to study charge transport across Ag-polyparaphenylene oligomer (PPP), Ag-Poly-1-methoxy-4-(2-ethylhexyloxy)-p-phenylene-vinylene (MEHPPV) and Ag-terthiophene (T3C4SH)-on-Au interfaces. This technique allows us to determine the distribution of Schottky barrier (SB) values with nanometer scale spatial resolution, unlike conventional spectroscopy and current-voltage measurements that average over millimeter areas. For the molecule, experimental results are compared with a Wentzel Kramers Brillouin (WKB) calculation to discuss the tunneling contribution in the measurement.

2. Ballistic electron emission microscopy (BEEM)

2.1. PRINCIPLE

A device configuration and schematic for BEEM, is shown in Fig. 1. An organic semiconductor is overlaid with a thin metal film (typically < 10 nm, termed the base), with an ohmic contact on the opposite side (termed the collector). The top metal film is grounded, and carriers are injected into it using a scanning tunneling microscope (STM) tip. These carriers are injected at energies sufficiently high above the metal's Fermi energy, so that they propagate ballistically before impinging on the interface. There is spreading of carriers in the metal film due to mutual Coulomb repulsion as

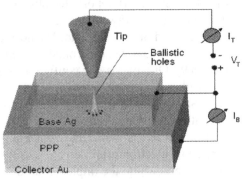

Figure 1. Schematic for a ballistic emission emission microscopy experiment. The buried Ag-PPP interface is studied.

well as some scattering by imperfections. When the energy of the carriers exceeds the Schottky/injection barrier, they propagate into the semiconductor and can be collected at the contact on the bottom. Typically the tunneling current is attenuated by a factor of 1000, so that collector currents are in the picoampere range for tunneling currents of few nA. Spectroscopy and imaging can be done on this structure, by monitoring the collector current as a function of STM tip bias voltage at a fixed location, or as a function of tip position at a fixed STM tip bias. One of the fundamental advantages of BEEM is the ability to investigate transport properties of hot electrons with high lateral resolution, typically at the nanometer scale.

2.2. EXPERIMENTAL

Choice of the base depends on the injection barrier that is to be measured. We have chosen Ag, since most organics are hole transport materials and the Fermi energy of Ag is favorably aligned with the highest occupied molecular orbital (HOMO). The Ag film is nominally 10 nm thick. The experiments were done at 77K in a home-assembled STM system. Sample preparation and characterization have been discussed in one of our earlier papers [13]. The current noise of the setup is typically 1 pA. Ag has been shown to yield "injection limited" contacts for hole injection into the polyparaphenylene/vinylene (PPV) family of organics [14]. It is important to ensure that the Ag film is reasonably flat, since the BEEM actually grounds the area of the metal investigated by the tip. Unless this requirement is met, attempts to tunnel into patches of the metal film, which are poorly connected, can lead to tip crashes.

3. Results

Figure 2a shows a raw current-voltage (I-V) spectroscopy over a 0 to 2 V range for the Ag-PPP interface. Metal organic interfaces likely have a high density of trap sites, and noise can be caused by trapping and release of charge from these sites. Repeated acquisition of spectra at the same point were found to damage the sample, as evidenced by instability of the spectrum. Qualitatively, this curve is similar to BEEM spectra seen for MIS interfaces. The Schottky or injection barrier is usually taken to be the point where the collector current begins to deviate from zero.

Extraction of the SB from BEEM data, such as that shown in Fig. 3 requires modeling of the spectral shape. Bell and Kaiser [15] used a planar tunneling formalism for determining the shape. The solid line is a Kaiser-Bell fit to the raw data (dots), and has the form:

$$I_b = A (V-V_o)^n \qquad (1)$$

where I_b is the collector current, and V_o is the injection barrier. We find that the value of V_o ranges from 0.3 to 0.5. This should be contrasted with the injection barrier determined by the Schottky-Mott (SM) rule, which yields a value of 0.9 V assuming alignment of vacuum levels for the metal and organic. The exponent n varies from 2.76 to 3.13. This is substantially higher than 2, which is commonly used to fit BEEM data for MIS interfaces. However, this is not surprising, since the exponent n is influenced by scattering at the interface. There will be more scattering at the MO interface due to lattice mismatch, and non-conservation of momentum vector (for MIS interfaces, conservation of k is usually implicit). The V_o values, extracted in this manner, are shown as a histogram in Fig. 3(c). Substantial deviation of the barrier from the SM rule, and its distribution are noteworthy and will be discussed later.

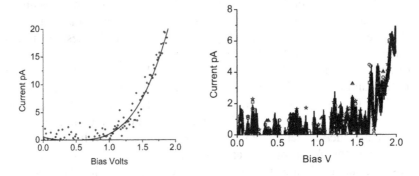

Figure 2. BEEM spectrum of (a) Ag-PPP interface, with Kaiser-Bell fit shown as the solid line and (b) for Ag-MEHPPV interface. Each symbol represents one raw spectrum. Only few points of the spectra for both organics are shown for clarity. The solid line for MEHPPV is an average of over 20 individual spectra. See text

An STM image of the top Ag film, at 0.5 V and 1 nA is shown in Fig. 3(a). The I-V and dI/dV enable choice of imaging conditions suitable to the interface. For instance, based on the spectroscopy data, it is possible to determine that bias voltage of 1 V should yield a measurable collector current. Plots of the collector current as a function of the STM tip position are images of electronic transparency of the interface. Such an image is shown in Fig. 3(b). The image clearly indicates non-uniform transparency of the interface over the region scanned by the STM. The bright spots indicate transparent regions. The size of such regions appears to be 10 nanometers. BEEM studies of MIS interfaces often show a correlation between the STM, STM derivative and BEEM images. This is due to lateral variation of the surface density of states [11,12]. In this work, the electronic transparency of the interface and the surface morphology of the Ag film have little correlation. It is possible to estimate the electric field at the interface, assuming appropriate physical parameter values for PPP, using standard equations from semiconductor physics [16]. For a dielectric constant of 3, and a carrier concentration of $10^{13}/cm^3$ we obtain a field of 10^5 V/m. We note that these fields are at least two orders of magnitude lower than the fields typically applied to organic devices during I-V spectroscopy or operation.

Figure 2(b) shows raw and averaged BEEM spectra obtained from an Ag-MEHPPV interface. When compared to spectra for the Ag-PPP interface, two noteworthy differences are readily apparent. First, the noise in the raw spectra is higher; and second, the current at 2V is much smaller. The current is expected to be smaller, since the SB is 0.1 eV higher for Ag-MEHPPV as compared to Ag-PPP. The higher noise is not surprising, since a spin coated organic interface will be more disordered than an evaporated organic interface. Increased disorder would imply a larger density of trapping sites, and more noise. The quality of the spectral data on Ag-MEHPPV precludes an analysis of the kind done above for Ag-PPP. The phenyl rings in the spin coated MEHPPV are expected to lie in the plane of the substrate, unlike those of PPP, where they are expected to lie perpendicular to the plane of the substrate. This variation in geometry has implications for charge transfer. The latter geometry is more conducive to charge transfer, as shown by first principles theoretical calculations.

Figures 4(a) and (b) show STM and BEEM images of an approximately 150 nm square area for the Ag-MEHPPV system. The BEEM image of Ag-MEHPPV also indicates nonuniform transparency of the interface. The bright spots are the more transparent regions. The size of such regions is typically a few nanometers. Due to coulomb repulsion, the injected charges spread out in the metal base to as much as 5 nm. Further spreading results when the charges cross over into the organic. The lateral resolution of BEEM is determined by these factors. For MO interfaces, due to the lack of k-conservation, the precise resolution of BEEM is difficult to determine. Keeping this in mind, it is intriguing that isolated bright spots of lateral extent less than 2 nm are seen in the BEEM current images. These likely arise from interfacial defects which provide excess electronic states for the charges. To summarize, spatial nonuni-formity of injection is observed for both Ag-PPP and Ag-MEHPPV interfaces.

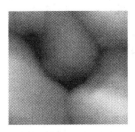  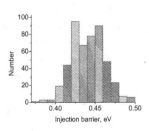

Figure 3. (a) STM topography of 10 nm Ag film on PPP. The height variation is 1.2 nm. (b) corresponding BEEM current image, with full scale of 3.5 pA, at 0.8 V. Scan size is 50 nm for both images. (c) Observed distribution of Schottky barrier values.

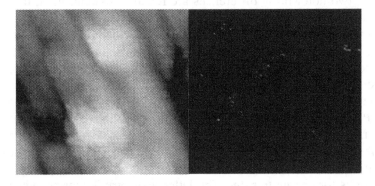

Figure 4. (a) STM topography of 10 nm Ag film on MEHPPV. The height variation is 2 nm. (b) corresponding BEEM current image, with full scale of 5 pA, at 1.5 V. Scan size is 150 nm for both images.

We now discuss results on the terthiophene molecule. The terthio-phene with the alkanethiol segment was deposited from solution (1 mM in ethanol) and was immobilized onto a template-stripped gold surface, prepared by the procedure of Wagner et al [17]. The Ag was evaporated through a mechanical mask (1x2 mm). The film was deposited at a rate of 0.1 Å/s and has a thickness of 8 nm. The STM in Fig. 4(a) image shows the granular structure of Ag deposited onto the molecule at 77 K. The BEEM current image in Fig. 4(b) again shows non-uniform transparency of the interface with bright spots that are a few nanometres in size. BEEM spectra from the bright and dark regions show significant differences, and we this below.

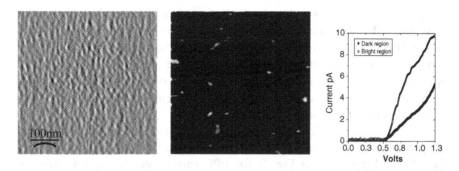

Figure 5. (a) STM topography of 8 nm Ag film on T3C4SH. (b) BEEM current image, with full scale of 3 pA, at 0.6 V, over an area 50 nm square. (c) BEEM spectra over bright and dark regions showing difference in transport. The open circles (top) correspond to the bright regions and the filled squares (bottom) correspond to the dark regions of the BEEM image.

The BEEM threshold voltage deduced from spectra over the bright regions indicates Schottky barrier of 0.4 V, whereas the spectra over the dark regions yields a SB of 0.5 V, as shown in Fig. 4(c). We conservatively estimate that an area of approximately 5 nm square is being probed at the interface. Since the molecule is estimated to be 1.2 nm long, direct tunneling from the Ag to the Au may contribute to the observed BEEM current. Therefore, a calculation within the framework of the WKB approximation for tunneling from silver through a potential barrier to gold was done to compare with the BEEM results [18]. We have assumed that T3C4SH is a 1.5 nm long tunneling barrier, the work functions for Ag and Au are 4.3 and 5.1 eV respectively. The calculated I-V for a 10 nm square region is shown in Fig. 6. The tunneling current contribution is three orders of magnitude smaller than the BEEM current. Therefore we can be quite confident that molecular levels are being accessed.

4. Discussion

4.1. TRANSMISSION AND GAP STATES

The BEEM process can be divided into three distinct steps. The first is the tunneling of the charges from the STM tip to the metal overlayer (the base). The second is the propagation through this metal layer, and last is the transmission across the interface [11,12]. Data for MIS interfaces has been analyzed with considerable success in this manner. Each of these processes is a function of the energy of the charges (electrons/holes). The product or convolution of these three functions yields the derivative of the BEEM spectrum, the dI/dV. The first step requires knowledge of spatial and energetic distribution of the electron current at the metal surface after tunneling. The functional dependence of this current on energy is well known [11]. The next step is attenuation/propagation in the metal film. This depends on the electron path length through the metal film and is energy dependent. Its functional form is an exponential decay [11]. The third step is the transmission across the interface, and depends on the energy and direction of the electrons.

Retaining the functional forms of the tunneling current distribution, and the propagation through the metal film, taking their product, and dividing the dI/dV obtained from Fig. 2 for PPP earlier, yields the transmission function shown in Fig. 6a by the solid line. This compares well with published literature, as we discuss below.

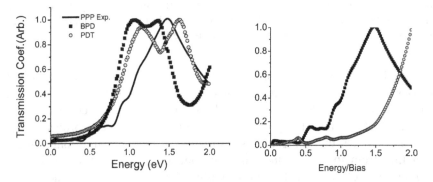

Figure 6. (a) Transmission for Ag-PPP interface and comparison with theory. Solid line is obtained from our data. The filled square and open circle are calculations by Xue and Ratner [19]. (b) Calculated LDOS (open circles) at Ag-PPP interface shown in red. Filled squares indicate the transmission function shown in Fig. 6a.

Xue and Ratner [19] have studied the transmission across an Au-phenyl dithiol (PDT) and Au-bi-phenyl dithiol (BPD) structure. Most of the potential drop occurs at the Au-phenyl ring interface. Once charge is transferred from Au to the phenyl ring, further transport does not change the potential, although it is accompanied by changes in the electron density on the molecules. In the case of the Ag –PPP interface, one can assume that the primary injection process is the transfer of charge from the Ag to the first phenyl ring of the PPP, in the absence of any evidence to the contrary. Therefore, comparing the normalized trans-mission function determined from our data with the calculation of Xue and Ratner [19], we find two common features: (a) both have a curvature that is concave upwards, i.e. they scale with a power of energy which is greater than one, (b) both peak towards the HOMO levels of the organic. Given the crude approximations that have been made, the agreement between theory and ex-periment is noteworthy. In contrast, the transmission function for a MIS inter-face is the available density of states in the semiconductor and scales with $E^{1/2}$.

Metal induced gap states, or metal wave functions tailing into the gap of inorganic semiconductors are well established [20,21]. Recently, there has been an appreciation of the importance of such states in MO contacts. Theoretical calculations by Xue and Ratner [19], have shown that MIGS can arise from proximity of metal to the organic.electrochemically gated transistor was used to probe the density of states (DOS). The important findings in this work are : significant tailing of the metal DOS into the gap, the assignment of the HOMO (determined by cyclic voltammetry) in the literature is not exact, and the tail of the DOS in the gap has quite a complex structure.

We have discussed the separation of the BEEM process into three distinct steps. The geometry of a BEEM device is such that it can also be analyzed as a resonant diode. It is therefore appropriate to use the relation

$$I = (2e/h) \, S \, \mu \qquad\qquad (2)$$

for a resonant diode to determine the LDOS (local density of states) from the transmission function. S is the LDOS, I the current, and μ the energy. The LDOS, which can be determined from the transmission function for PPP is shown in Fig. 6(b). Consistent with the observations of Muelenkamp et al. [22], we observe significant tailing of electronic states into the gap. In addition, we also find that the HOMO as determined by the LDOS does not peak at the transmission. This is acceptable, since the HOMO and LUMO values used in this work are also determined by cyclic voltammetry, and therefore could be inaccurate, as in the case of PPV.

4.2. INHOMOGENOUS CHARGE INJECTION

All the MO interfaces studied in this work show inhomogeneous charge transfer or nanoscale SB patches. We now determine the current that goes through these patches, based on the observed SB distribution. Tung [23] was the first to point out the physical significance of inhomogeneous Schottky barriers, i.e. patches of low barrier height embedded in an otherwise uniform background. We follow the treatment of Tung [23] to evaluate the effect of the observed barrier height distribution in our Ag-PPP diodes. Our BEEM current images (Fig. 3b) for Ag-PPP diodes show regions of low Schottky barrier which are typically 10 nm in diameter, as evidenced by a local increase of BEEM current. The distribution of

Schottky barrier values (Fig. 3c) indicates a 40% reduction of the barrier over these regions, compared to the uniform back-ground of approximately 0.5 eV. The following are the equations for a circular patch, the appropriate geometry for the observations reported in this work.

$$V = V_{bb}(1\text{-}z/W)^2 - \Delta \, R_o^2/2z^2 + V_n \tag{3}$$

V is the potential at zero applied bias, the condition appropriate for BEEM, and z is the distance. In this equation, V_{bb} or band bending potential is defined as $\Phi_B^0 - V_n$, where Φ_B^0 is the average SB height, and V_n is the separation between the conduction band minimum and the Fermi level in the semiconductor far from the interface. W is the depletion width, Δ is the local SB deviation from the average SB. R_o is the low SB patch. The current through such a low SB patch is given as:

$$I_{patch} = A^* \, A_{eff} \, T^2 \, \exp \, \{-\beta\Phi_B^0 + (\beta\gamma \, V_{bb}^{1/3}/ \, \eta^{1/3})\} \tag{4}$$

for zero applied bias. Here A* is the effective Richardson's constant, A_{eff} is the effective area of the patch, $\beta=1/kT$. γ and η are given as:

$$\gamma = 3(\Delta \, R_o^2/4)^{1/3} \tag{5}$$
$$\eta = \varepsilon_s/qN_D \tag{6}$$

where ε_s is the permittivity of the organic, q is the elementary charge, and N is the carrier concentration in the organic.

In Fig. 7(a), we show the potential in the vicinity of low SB patch as function of position. In agreement with experimental observations, the patch diameter is taken to be 10 nm, and the magnitude of the SB is lowered by 40%, as shown by the measured barrier height distribution for the Ag-PPP interface. Figure 7(b) shows the current that passes through the patch and its vicinity when a finite bias voltage, comparable to that encountered during device operation, is applied. The resulting electric field is of the order of 10^7 V/cm. It is clear that the current through the low SB patch is 1000 times larger than that in the surrounding region. Clearly, this would lead to high current density filaments which can cause ionization damage to the organic and electromigration of the metal.

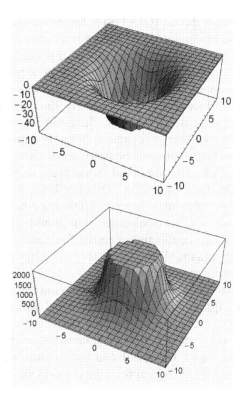

Figure 7. (a) Potential profile across a low Schottky barrier patch, 10 nm in radius, with a deviation of 40% from the background, and (b) current through the patch.

5.   Concluding remarks

We have successfully applied the BEEM technique to study charge transport across metal-organic and metal-molecule interfaces with nanometer scale spatial resolution. Inhomogeneous charge injection has been observed for all interfaces investigated in this work. Inhomogeneous charge injection does not preclude device applications, since Schottky barrier inhomogeneities are also seen with inorganic semiconductors such as Silicon Carbide [24]. We have also calculated the transmission function for the Ag-PPP interface, and found qualitative agreement with theory. Metal induced gap states are found to influence the charge transport and modify the injection barriers. Low Schottky barrier patches cause large current densities which can results in damage to devices. We discuss two examples below.

Organic light emitting diodes (OLEDS) suffer from finite lifetimes and non-uniformities in light emission, usually termed dark spots [24]. Our work has shown that non-uniform charge injection is common to optically active organics such as PPP and MEHPPV. An excess population of one type of carrier (for example, holes) caused by low Schottky barrier patches such as those found in this work, would locally limit the recombination efficiency, and thereby cause dark spots. In addition, the high carrier density in the low Schottky barrier regions is likely to damage the metal or the organic or the interface. This will be detrimental to the lifetime of the device. Recent work with self assembled monolayers of molecules which exhibit negative differential resistance (NDR) has been controversial [25]. It is now acknowledges that the observed NDR may not be due to the electrical characteristics of the molecules, but could arise due to the creation and destruction of conducting metallic filaments. We have shown that such filaments arise naturally at low Schotty barrier patches. Similar results have been reported in other systems [9].

Roughness, microstructure of the metal, topography of the organic and microstructure of the organic film are all factors which influence the BEEM current image. The connection between local defects/microstructure and electronic properties is by now generally acknowledged [25]. Therefore, controlling microstructure of the metal and the organic can increase the uniformity of charge injection and improve the performance of devices.

Acknowledgements

The molecules were provided by Prof. P. Bauerle. Discussions with Professors W. Knoll, C. Joachim and A. Dodabalapur are gratefully acknowledged.

References

1. Ishii, H.; Sugiyama, K.; Ito, E.; Seki, K., Energy Level Alignment and Interfacial Electronic Structures at Organic/Metal and Organic/Organic Interfaces, Adv. Mat. 1999, 11, 605.
2. Gittins, D. I.; Bethell, D.; Schiffrin, D. J.; Nichols, R. J., A nanometre-scale electronic switch consisting of a metal cluster and redox-addressable groups, Nature 2000, 408, 67.
3. Donhauser, Z. J.; Mantooth, B. A.; Kelly, K. F.; Bumm, L. A.; Monnell, J. D.; Stapleton, J. J.; Price, D. W.; Rawlett, A. M.; Allara, D. L.; Tour, J. M.; Weiss, P. S., Conductance switching in single molecules through conformational changes, Science 2001, 292, 2303.
4. Lau C. N.; Stewart D. R.; Williams R. S.; and Bockrath M., Direct observation of nanoscale switching centres in metal/molecule/metal structures, Nano Lett. 2004, 4, 569.

5. Sakamoto, T.; Sunamura, H.; Kawaura, H.; Hasegawa, T., Nanoscale switches using copper sulfide, Appl. Phys. Lett. 2003, 82, 3032.
6. Terabe, K.; Nakayama, T.; Hasegawa, T. and Aono, M., Formation and disappearance of a nanoscale silver cluster realized by solid electrochemical reaction, J. Appl. Phys. 2002, 91, 10110.
7. Simmons, J.G.; and Verderber, R.R., New conduction and reversible memory phenomena in thin insulating films, Proc. Roy. Soc. A. 1967, 301, 77.
8. Ma, L. P.; Liu, J.; and Yang, Y., Organic electrical bistable devices and rewritable memory cells, Appl. Phys. Lett. 2002, 80, 2997.
9. Bozano, L. P.; Kean, B.W.; Deline, V. R.; Salem, J. R.; Scott, J. C.; Mechanism for bistability in organic memory elements, Appl. Phys. Lett. 2004, 84, 607.
10. De Boer, B.; Frank, M. M.; Chabal, Y. J.; Jiang, W.; Garfunkel, E.; Bao, Z., Metallic contact formation for molecular electronics: Interactions between vapor-deposited metals and self-assembled monolayers of conjugated mono- and dithiols, Langmuir 2004, 20, 1539.
11. Narayanamurti, V.; and Kozhevnikov, M.; BEEM imaging and spectroscopy of buried structures in semiconductors, Phys. Rep. 2001, 349, 447.
12. Prietsch, M.; Ballistic-electron emission microscopy (BEEM): Studies of metal/semiconductor interfaces with nanometer resolution, Physics Reports 1995, 253, 163.
13. Troadec, C., Kunardi, L., and Chandrasekhar, N., Ballistic emission spectroscopy and imaging of a buried metal/organic interface, Appl. Phys. Lett,, 2005, 86, 072101, Metal-organic interfaces at the nanoscale, Nanotechnology, 2004, 15, 1818, Switching in organic devices caused by nanoscale Schottky barrier patches, J. Chem. Phys., 2005, 122, 204702.
14. Baldo, M. A.; and Forrest, S. R., Interface limited injection in amorphous organic semiconductors, Phys. Rev. B 2001, 64, 085201.
15. (15 Bell, L. D.; and. Kaiser, W. J., Ballistic electron emission microscopy : a nanometer scale probe of interfaces and carrier transport, Ann. Rev. Mat. Sci. 1996, 26, 189.
16. Sze, S. M., Physics of Semiconductor devices, Wiley, NY 1990.
17. Wagner, P., Hegner, M., Guntherodt. H. J., and Semenza, G., Formation and in Situ modification of monolayers chemisorbed on ultraflat template-stripped gold surfaces Langmuir, 1995, 11, 3867
18. Simmons, J. G., Generalized thermal J-V characteristic for the electric tunnel effect, J. Appl. Phys., 1963, 35, 2655.
19. Xue, Y.; and Ratner, M., Microscopic study of electrical transport through individual molecules with metallic contacts. I. Band lineup, voltage drop, and high-field transport, Phys. Rev. B 2003, 68, 115406.
20. Heine, V.; Theory of surface states, Phys. Rev. A 1965, 138, 1689.
21. Monch, W., Electronic Properties of Semiconductor Interfaces, Springer, Berlin, 2004.
22. Hulea, I. N.; Brom, H.B.; Houtepen, A. J.; Vanmaekelbergh, A.; Kelly, J.J.; Meulenkamp, E. A.; Wide energy-window view on the Density of States and hole mobility in Poly(p-Phenylene Vinylene) Phys. Rev. Lett. 2004, 93, 166601.
23. Sullivan J.P.; Tung, R.T.; Pinto, M.R.; Graham, W. R.; Electron transport of inhomogeneous Schottky barriers : A numerical study, J. Appl. Phys. 1991, 70, 7403.
24. Liew, Y. F., Aziz, H., Hu, N-X., Chan, H. S-O., Xu, G., and Popovic, Z., Investigation of the sites of dark spots in organic light-emitting devices, Appl. Phys. Lett. 2000, 77, 2650.
25. Service, R. F.; Next generation technology hits an early mid-life crisis, Science, 2003, 302, 556.

ELECTRON-ELECTRON INTERACTION IN CARBON NANOSTRUCTURES

A.I. Romanenko (air@che.nsk.su), O.B. Anikeeva, T.I. Buryakov,
E.N. Tkachev, A.V. Okotrub
*Nikolaev Institute of Inorganic Chemistry, Lavrentieva 3, Novosibirsk, 630090
Russia; Novosibirsk State University, Lavrentieva 14, Novosibirsk, 630090
Russia*

V.L. Kuznetsov, A.N. Usoltseva
Boreskov Catalysis, Lavrentieva 5, Novosibirsk, 630090 Russia

A.S. Kotosonov
Institute of Carbon, Moscow, Russia

Abstract. The electron-electron interaction in carbon nanostructures was studied. A new method which allows to determine the electron-electron interaction constant λ_c from the analysis of quantum correction to the magnetic susceptibility and the magnetoresistance was developed. Three types of carbon materials: arc-produced multiwalled carbon nanotubes (arc-MWNTs), CVD-produced catalytic multiwalled carbon nanotubes (c-MWNTs) and pyrolytic carbon were used for investigation. We found that $\lambda_c = 0.2$ for arc-MWNTs (before and after bromination treatment); $\lambda_c = 0.1$ for pyrolytic graphite; $\lambda_c > 0$ for c-MWNTs. We conclude that the curvature of graphene layers in carbon nanostructures leads to the increase of the electron-electron interaction constant λ_c.

Key words: Electron-electron interaction; Nanostructures; Electronic transport; Galvanomagnetic effects; Quantum localization

1. Introduction

The carbon nanostructures are formed of graphene layers which always have some curvature. As a result, these materials are characterized by new properties which are not present in graphite consists of plane graphene layers. The curvature of the graphene layers influences the electron-electron interaction in these systems. The most interesting consequence of the graphene layers' curvature is the existence of a superconducting state in bundles of single-walled carbon nanotubes with diameters of 10 Å at temperatures below $T_c \approx 1$ K (Kociak et al., 2001) as well as the onset of superconductivity at

K. Scharnberg and S. Kruchinin (eds.),
Electron Correlation in New Materials and Nanosystems, 23–35.
© 2007 *Springer.*

$T_c \approx 16$ K in nanotubes with diameters of $4\,\text{Å}$ (Tang et al., 2003) and at $T_c \approx 12$ K in entirely end-bonded multiwalled carbon nanotubes (Takesue et al., 2006). In contrast, in graphite no superconducting state is observed. Y. Kopelevich et al. (Kopelevich et al., 2000) proposed that superconductivity may appear in ideal graphite and that the absence of superconductivity in real samples is related to the defects always present in graphite. According to theoretical predictions of (Gonzalez et al., 2001) topological disorder can lead to an increase in the density of states at the Fermi surface and to an instability of an electronic subsystem. These changes in the electronic system could lead to a superconducting state. However, such topological disorder leads to a curvature of initially flat graphene layers. We assume, therefore, that in carbon nanostructures the superconducting state is related to the curvature of graphene layers. The curvature of surfaces is always present in the crystal structure of nanocrystallites. As a result, in such structures the electron-electron interaction should be modified. This paper is devoted to the analysis of experimental data which allows to determine the electron-electron interaction constant λ_c in carbon nanostructures formed by curved graphene layers.

2. Experimental methods

The method of our investigation of the electron-electron interaction constant is based on the joint analysis of quantum corrections to the electrical conductance, magnetoconductance and magnetic susceptibility. For all nanostructures formed by graphene layers the presence of structural defects leads to the diffusive motion of charge carriers. As a result, at low temperatures, quantum corrections to the electronic kinetic and thermodynamic quantities are observed. For the one-particle processes (weak localization - WL (Kawabata, 1980; Lee and Ramakrishnan, 1985)) these corrections arise from an interference of electron wave functions propagating along closed trajectories in opposite directions, provided the lengths l of these trajectories are less then the phase coherence lengths $L_\varphi(T) = (D\tau_\varphi)^{1/2}$ (D is the diffusion constant and $\tau_\varphi = T^{-p}$ is the characteristic time for the loss of phase coherence with an exponent $p = 1 \div 2$). As a result, the total conductance of the system is decreased. $L_\varphi(T)$ increases with decreasing temperature which, in turn, leads to the decrease of the total conductance. In a magnetic field there is an additional contribution to the electronic phase, which has an opposite sign for opposite directions of propagation along the closed trajectory. As a result, the phase coherence length is suppressed: $L_B = (\hbar c/2eB)^{1/2} < L_\varphi$. Here $L_B = (\hbar c/2eB)^{1/2}$ the magnetic length, c is the light velocity, e - the electron charge, B - the magnetic field. This leads to negative magnetoresistance, i.e. to an increase of conductance in a magnetic field. Quantum corrections also arise from the interaction between electrons (interaction effects - IE (Al'tshuler et al., 1983)). These corrections arise due to the phase memory between two consecutive events of electron-electron scattering. If the second scattering event happens at a distance shorter than the coherence length, $L_{IE} = (D\hbar/k_B T)^{1/2}$, from the first one ($L_{IE}$ being the length on which the information about the changes of the electronic phases due to the first scattering event is not yet lost), the second scattering will depend on the first one. As a result the effective density of states on the Fermi surface v_F is renormalized. Interaction effects contribute not only to electrical conductance, but also to thermodynamic quantities depending on v_F - magnetic susceptibility χ and heat capacity C.

3. Results and discussion

3.1. ARC-PRODUCED MULTIWALLED CARBON NANOTUBES

A characteristic peculiarity of our arc-MWNTs (Okotrub et al., 2001; Romanenko et al., PSS, 2002) is the preferential orientation of the bundles of nanotubes in the plane perpendicular to the electrical arc axis. The volume samples of our arc-MWNTs show anisotropy in their electrical conductivity $\sigma_{II}/\sigma_{\perp} \approx 100$ (Okotrub et al., 2001; Romanenko et al., PSS, 2002). σ_{II} is the electrical conductivity in the plane of preferential orientation of the bundles of nanotubes, $\sigma_{\perp}$ is the conductance perpendicular to this plane. The average diameter of individual nanotubes is $d_{MWNT} \approx 140\,\text{Å}$. According to the electron paramagnetic resonance data, the concentration of paramagnetic impurities in our samples is less than 10^{-6}. This excludes a substantial contribution of the impurities to the susceptibility. The MWNTs brominated at room temperature in bromine vapour (Romanenko et al., PSS, 2002) have a composition of $CBr_{0.06}$. The addition of bromine leads to an increase of the conductivity, which can be attributed to an increase in the concentration of hole current carriers.

According to experimental and theoretical data, the basic contribution in χ of quasi-two-dimensional graphite (QTDG), including MWNTs, gives orbital magnetic susceptibility χ_{or} connected with extrinsic carriers (EC). Figures 1(a) and 2(a) present the magnetic susceptibility χ of arc-MWNTs samples before bromination and after bromination as a function of temperature respectively. Available models well reproduce the temperature dependence of magnetics susceptibility for MWNTs only at T > 50 K. In the low-temperature region the experimental data deviate from the theoretical ones. According to theoretical consideration the magnetic susceptibility χ

of quasi-two-dimensional graphite (QTDG) is generally contributed by the component χ_D (Kotosonov et al., 1997)

$$\chi_{or}(T) = -\frac{5.45 \times 10^{-3}\gamma_0^2}{(T + \delta)[2 + exp(\eta) + exp(-\eta)]}, \tag{1}$$

where γ_0 is the band parameter for two-dimensional case, δ is the additional temperature formally taking into account "smearing" the density of states due to electron nonthermal scattering by structure defects, $\eta = E_F/k_B$ $(T + \delta)$ represents reduced Fermi level (E_F), k_B is the Boltzmann constant. Using an electrical neutrality equation in the 2D graphite model (Kotosonov et al., 1997) η can be derived by $\eta = \text{sgn}(\eta_0)[0.006\eta_0^4 - 0.0958\eta_0^3 + 0.532\eta_0^2 - 0.08\eta_0]$ with an accuracy no less then 1%. The η_0 is determined by $\eta_0 = T_0/(T + \delta)$, where T_0 being degeneracy temperature of extrinsic carriers (EC) depends on its concentration n_0 only. The value of δ can be estimated independently as $\delta = \hbar/\pi k_B\tau_0$, where $\hbar$ is the Planck constant, τ_0 is a relaxation time of the carrier nonthermally scattered by defects. Generally, the number of EC in QTDG is equal to that of scattering centers and δ depends only on T_0, i.e. $\delta = T_0/r$, where r is determined by scattering efficiency. These parameters were chosen to give the best fit of the experimental data.

According to theoretical calculations (Al'tshuler et al., 1983), the correction $\Delta\chi_{or}$ to the orbital susceptibility χ_{or} in the Cooper channel dominates the quantum correction to the magnetic susceptibility $\chi(T, B)$ in magnetic fields smaller than $B_c = (\pi k_B T/g\mu_B)$ ($B_c = 9.8$ T at 4.2 K). These corrections are determined by the value and the sign of the electron-electron interaction constant λ_c and are proportional to the diamagnetic susceptibility of electrons χ_{or}. In graphite and MWNTs the diamagnetic susceptibility is greater than in any other diamagnetic material (excluding the superconductors), and the correction to χ_{or} should also be large. $\Delta\chi(T)_{or} = \chi(T)_{or}^{exp} - \chi(T)_{or}$ was found by (Lee and Ramakrishnan, 1985; Al'tshuler et al., 1983):

$$\frac{\Delta\chi_{or}(T)}{\chi_{or}(T)} = -\frac{\frac{4}{3}(\frac{l_{el}}{\hbar})ln[ln(\frac{T_c}{T})]}{ln(\frac{k_B T_c\tau_{el}}{\hbar})}, (d = 2), \tag{2}$$

$$\frac{\Delta\chi_{or}(T)}{\chi_{or}(T)} = -\frac{2(\frac{\pi}{6})\xi(\frac{1}{2})(\frac{k_B T\tau_{el}}{\hbar})^{1/2}}{ln(\frac{T_c}{T})}, (d = 3), \tag{3}$$

where $\chi(T)_{or}^{exp}$ are the experimental data, $\chi(T)_{or}$ is the result of an approximation of the experimental data in an interval of temperatures 50 - 400 K by the theoretically predicted dependence (1) for quasi-two-dimensional graphite (Kotosonov et al., 1997); value of $\xi(\frac{1}{2}) \sim 1$, l_{el} is the electron mean free path; τ_{el} represents the elastic relaxation time which is about 10^{-13} sec for

MWNT (Baxendale et al., 1997); h is the thickness of graphene layers packet; d denotes the dimensionality of the system; $T_c = \theta_D exp(\lambda_c^{-1})$, where θ_D is the Debye temperature, λ_c is the constant which describes the electron-electron interaction in the Cooper channel ($\lambda_c > 0$ in a case of electron repulsion). The dependence in Eq. (2) is determined by $ln[ln(\frac{T_c}{T})]$ term because, at low temperatures, in the disordered systems, τ_{el} is temperature independent while all other terms are constants. The dependence in Eq. (3) is governed by $T^{1/2}$ term as $T_c \gg T$ and, therefore, $ln(\frac{T_c}{T})$ can be considered as a constant relative to $T^{1/2}$. The temperature dependence of the magnetic susceptibility $\chi(T)$ is shown in Figure 1 for arc-MWNTs before bromination, in Figure 2 for arc-MWNTs after bromination, and in Figure 3 for crystal graphite. Below 50 K the deviation of experimental data from the theoretical curve is observed (Romanenko et al., SSC, 2002; Romanenko et al., 2003). The additional contribution to $\chi(T)$ is presented in Figs. 1(b), 2(b), 3(b); and Figs. 1(c), 2(c), 3(c) as a function of $ln[ln(\frac{T_c}{T})]$ and $T^{1/2}$ respectively. The $\Delta\chi_{or}(T)/\chi_{or}(T)$ clearly shows the dependence given by Eq. (2) at low magnetic field and one given by Eq. (3) at high magnetic field, while at $B = 0.5$ T the temperature dependence of $\Delta\chi_{or}(T)/\chi_{or}(T)$ differs from those two limits. As seen from Fig. 1, at all magnetic fields applied to the arc-MWNTs before bromination, the absolute value of $\Delta\chi_{or}(T)/\chi_{or}(T)$ increases with decreasing temperature as has been predicted for IE in the systems characterized by electron-electron repulsion (Lee and Ramakrishnan, 1985; Al'tshuler et al., 1983). Hence, at $B = 5.5$ T a crossover from the two-dimensional IE correction to the three-dimensional one takes place. At lower magnetic field the interaction length $L_{IE}(T)$ is much shorter than the magnetic length L_B, which in turn becomes dominant at high field. An estimation of the characteristic lengths gives respectively the value of $L_{IE}(4.2K) = 130$ Å (taking into account that the diffusion constant $D = 1$ cm^2/s (Baxendale et al., 1997)) and the value of $L_B = 100$ Å at $B = 5.5$ T.

A similar dependence of $\Delta\chi_{or}(T)/\chi_{or}(T)$ was observed for arc-MWNTs after bromination (Fig. 2).

The dependence of $\Delta\chi_{or}(T)/\chi_{or}(T)$ we investigated for crystals of graphite (Fig. 3). However, only the three-dimensional dependence $\Delta\chi_{or}(T)/\chi_{or}(T) \approx T^{1/2}$ was found for graphite.

Approximation of the abnormal part of the magnetic susceptibility by theoretically predicted functions has revealed three features:

I. A crossover from the two-dimensional quantum corrections to $\chi(T)$ in fields $B = 0.01$ T to the three-dimensional quantum corrections in field $B = 5.5$ T is observed as up to bromination of arc-MWNTs, so after bromination of its when the magnetic field increases. For graphite, in all fields, the three-dimensional corrections to $\chi(T)$ are observed. This is related to the fact that magnetic length $L_B = 77$ Å in fields of $B = 5.5$ T becomes comparable to the

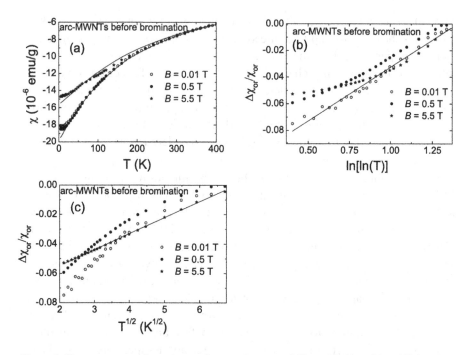

Figure 1. The temperature dependence of magnetic susceptibility $\chi(T)$ (**a**) and $\Delta\chi_{or}(T)/\chi_{or}(T)$ = $[\chi(T) - \chi_{or}(T)]/\chi_{or}(T)$ [(**b**) and (**c**)] for arc-produced MWNTs sample before bromination. The solid lines are fits: for (**a**) by Eq. (1) in interval 50 - 400 K with parameters; for curve (○) , $\gamma_0 = 1.6$ eV, $T_0 = 215$ K, $\delta = 159$ K; for (●) , $\gamma_0 = 1.6$ eV, $T_0 = 215$ K, $\delta = 159$ K; for (★) , γ_0 = 1.7 eV, $T_0 = 327$ K, $\delta = 210$ K; by Eq. (2) and Eq. (3) for (**b**) and (**c**) respectively in interval 4.5 - 45 K with parameters $T_c = 10000$ K , $l_{el}/a = 0.15$.

thickness of the graphene layers h_{MWNT} which form the MWNT. On the other hand, in fields of $B = 0.01$ T, $L_B = 1800$ Å is much longer than h_{MWNT} but it is shorter than the length of a tube l_{MWNT} ($l_{MWNT} \approx 1\mu$) and is comparable to the circumference of a nanotube. In graphite, the thickness of a package of the graphene layers is always macroscopic and it exceeds all other characteristic lengths.

II. The bromination of arc-MWNTs has led to an increase of their conductance by one order of magnitude (from 500 Ω^{-1}cm^{-1} in arc-MWNTs before bromination up to 5000 Ω^{-1}cm^{-1} after bromination). However, the relative correction to the magnetic susceptibility $\Delta\chi(T)_{or}/\chi(T)_{or}$ which determines the value of λ_c, remained constant. Thus, the bromination does not change the electron-electron interaction constant λ_c in the arc-produced multiwalled carbon nanotubes.

III. The constant of electron-electron interaction λ_c for arc-MWNTs before and after bromination has the magnitude about 0.2 (Romanenko et al.,

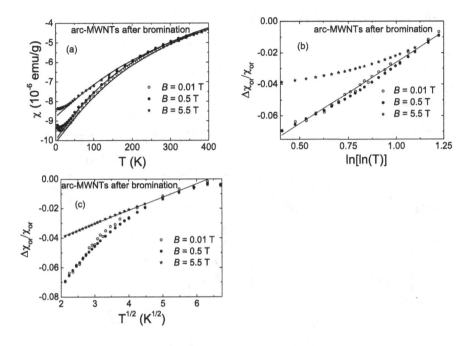

Figure 2. The temperature dependence of magnetic susceptibility $\chi(T)$ (**a**) and $\Delta\chi_{or}(T)/\chi_{or}(T)$ = $[\chi(T) - \chi_{or}(T)]/\chi_{or}(T)$ [(**b**) and (**c**)] for arc-produced MWNTs sample after bromination. The solid lines are fits: for (**a**) by Eq. (1) in interval 50 - 400 K with parameters; for curve (○) , γ_0 = 1.4 eV, T_0 = 340 K, δ = 252 K; for (●) , γ_0 = 1.4 eV, T_0 = 300 K, δ = 273 K; for (★) , γ_0 = 1.5 eV, T_0 = 435 K, δ = 325 K; by Eq. (2) and Eq. (3) for (**b**) and (**c**) respectively in interval 4.5 - 45 K with parameters T_c = 10000 K , l_{el}/a = 0.15.

SSC, 2002) which is greater than that of graphite $\lambda_c \approx$ 0.1 (Romanenko et al., 2003), i.e. the curvature of the graphene layers in MWNTs leads to the increase of λ_c.

The temperature dependence of the electrical conductivity $\sigma(T)$ of arc-MWNTs indicates the presence of quantum corrections also (Fig. 4(a) and Fig. 5(a)). At low temperatures, the temperature dependence of these quantum corrections is characteristic for the two-dimensional case (Fig. 4(b) and Fig. 5(b)):

$$\Delta\sigma(T) = \Delta\sigma_{WL}(T) + \Delta\sigma_{IE}(T).\qquad(4)$$

Here $\Delta\sigma_{WL}(T) \approx ln(L_\varphi/l_{el})$ is the correction associated with the quantum interference of noninteracting electrons in two-dimensional systems (WL) (Lee and Ramakrishnan, 1985; Kawabata, 1980) while $\Delta\sigma_{IE}(T) \approx ln(L_{IE}/l_{el})$ is the correction associated with the quantum interference of interacting electrons (IE) in such systems (Lee and Ramakrishnan, 1985; Al'tshuler et al.,

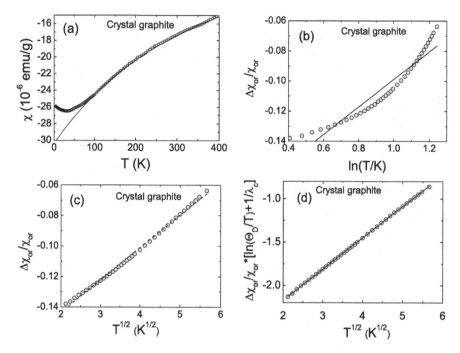

Figure 3. Temperature dependences of a magnetic susceptibility $\chi(T)$ for graphite measured in a magnetic field = 0.01 T. Continuous lines show: the regular parts $\chi(T)$ (a); two-dimensional quantum corrections to $\chi(T)$ (b); three-dimensional quantum corrections to $\chi(T)$ (c), (d). The solid lines are fits for (d) by Eq. (3) in interval 4.5 - 45 K with parameters $\theta_D = 1000$ K , $\lambda_c = 0.1$.

1983). The contribution of quantum corrections to the electrical conductivity should be accompanied by corrections to magnetoconductivity $\Delta\sigma(B) = 1/\rho(B)$ in low magnetic fields (Kawabata, 1980; Al'tshuler et al., 1981):

$$\Delta\sigma(B) = \Delta\sigma_{WL}(B) + \Delta\sigma_{IE}(B). \tag{5}$$

Here $\Delta\sigma_{WL}(B)$ is the quantum correction to magnetoconductance for non-interacting electrons; $\Delta\sigma_{IE}(B)$ - the quantum correction to the magnetoconductance for interacting electrons. Both corrections have the logarithmic asymptotic in high magnetic fields ($\Delta\sigma_{WL}(B) \approx ln(L_\varphi/L_B)$; $\Delta\sigma_{Int}(B) \approx ln(L_{IE}/L_B)$ at L_φ/l_B; $L_{IE}/L_B >> 1$), and the quadratic asymptotic in low magnetic fields ($\Delta\sigma_{WL}(B) \approx B^2$; $\Delta\sigma_{IE}(B) \approx B^2$ when L_φ/L_B; $L_{IE}/L_B << 1$). The quantum corrections to magnetoconductance become essential in low magnetic fields when the magnetic length is $L_B < L_\varphi$. In this case the phase of an electron is lost at distances $\approx L_B$, and quantum corrections to conductance are partially suppressed. This leads to positive magnetoconductance (negative magnetoresistance).

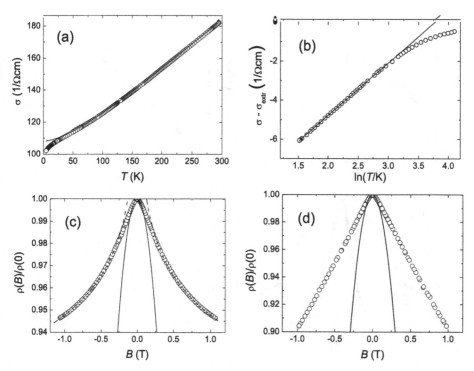

Figure 4. Data for arc-produced MWNTs (a, b, c) and for pyrolytic graphite (d). Temperature dependence of conductivity $\sigma(T)$ (**a**), anomaly part of conductivity $\Delta\sigma(T) = \sigma(T) - \sigma(T)_{ext}$ (**b**), and the relative magnetoconductivity $\sigma(B)/\sigma(0)$ from magnetic field B measured at $T = 4.2$ K (**c, d**). Continuous lines show: $\sigma(T)_{ext}$ receiving by extrapolation of approximation curve from $T \geq 50$ K to $T \leq 50$ K (**a**); two-dimensional quantum corrections to $\sigma(T)$ (**b**), asymptotic of quadratic approximation $\sigma(B)/\sigma(0) \approx B^2$ at $B \leq 0.05$ T to B up to 0.3 T (**c, d**). Dashed lines on (**c**) show the logarithmic asymptotic $\sigma(B)/\sigma(0) \approx ln(T)$ from high field to low field.

It is difficult to divide the contribution of IE in the Cooper channel (which is determined by the amplitude and sign of λ_c), WL and IE in the diffusion channel. Field dependences of magnetoresistance (Fig. 4(c)) in all intervals of the measured fields (0 - 1 T) show negative magnetoresistance, related to WL. Similar dependences are observed in graphite (Fig. 4(d)). All three mechanisms give quantum corrections to temperature dependence of conductance $\sigma(T)$. In order to observe the contribution of IE in the superconducting channel it is necessary to exclude the contribution of WL. We achieved this in catalytic carbon multiwalled nanotubes which were synthesized by a technology which prevents the formation of other phases of carbon.

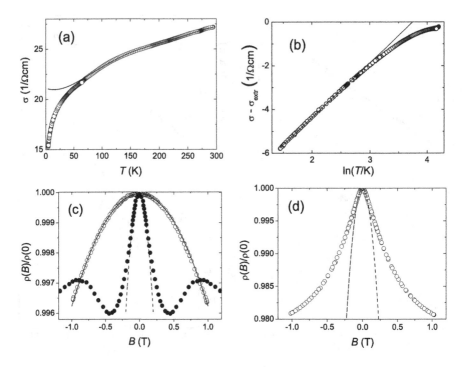

Figure 5. Data for: o - catalytic MWNTs without impurities of another forms of carbon (a, b, c); • - MWNTs prepared by a usual catalytic method(c), and Δ - soot (d). Temperature dependence of conductivity $\sigma(T)$ (a), anomaly part of conductivity $\Delta\sigma(T) = \sigma(T) - \sigma(T)_{ext}$ (b), and the relative magnetoconductivity $\sigma(B)/\sigma(0)$ from magnetic field B measured at T = 4.2 K (c, d). Continuous lines show: $\sigma(T)_{ext}$ receiving by extrapolation of approximation curve from $T \geq 50$ K to $T \leq 50$ K (a), two-dimensional quantum corrections to $\sigma(T)$ (b), asymptotic of quadratic approximation $\sigma(B)/\sigma(0) \approx B^2$ from $B \leq 0.05$ T to B up to 0.3 T for (curve • on figure c) and soot (d). Dashed line: on (c) - show the quadratic approximation $\sigma(B)/\sigma(0) \approx B^2$, on (d) - show the logarithmic asymptotic $\sigma(B)/\sigma(0) \approx ln(T)$ from high field to low field.

3.2. CATALYTIC MULTIWALLED CARBON NANOTUBES

There are always paramagnetic impurities present c-MWNTs because of the ferromagnetic metals used as catalytic agents. It is not possible to use the magnetic susceptibility in order to obtain information on $\Delta\chi_{or}(T)$ of c-MWNT, because the contribution of paramagnetic impurities dominates at low temperatures. We carried out an analysis of the conductivity data for which the contribution of paramagnetic impurities is negligible. In figure 5(c) the dependence of the relative magnetoconductivity $\rho(B)/\rho(0)$ on the magnetic field B for c-MWNTs is shown.

The closed circles - c-MWNTs prepared by a usual catalytic method (Kudashov et al., 2002); the open circles - with use of the special procedure (Couteau et al., 2003) which allows to prepare c-MWNTs practically without inclusions of the another form of carbon. Comparing these curves it is possible to see, that in the refined samples in the low magnetic fields (at $B \leq 0.2$ T) the contribution to the negative magnetoconductivity, related to WL, we not observe. $\sigma(B)/\sigma(0)$ for refined c-MWNTs is described by quadratic dependence $\sigma(B)/\sigma(0) \approx B^2$ for all values of field B. $\sigma(B)/\sigma(0)$ for c-MWNT prepared by a usual method are described by quadratic dependence only in low magnetic fields ($B < 0.1$ T). In figure 5(d) the dependence of the relative magnetoconductivity $\sigma(B)/\sigma(0)$ on the magnetic field B is shown for soot. As can be seen from figure 5 the curves for soot and c-MWNTs prepared by a usual catalytic method are very similar at low fields. We suggest that this fact is connected to the presence of impurities of soot in c-MWNTs prepared by a usual catalytic method. Investigation of $\sigma(T)$ of the c-MWNTs shows the presence of quantum corrections $\sigma(T) \approx ln(T)$ and suggests the two-dimensional character of these corrections (Kawabata, 1980; Lee and Ramakrishnan, 1985; Al'tshuler et al., 1983). The magnitude of the magnetic field which suppresses the temperature correction ($\delta\sigma(4.2$ K$)/\sigma(4.2$ K$) \approx 2.7\%$) is estimated as $B \approx 8.5$ T. This corresponds to a quite reasonable magnitude of magnetic length $L_B \approx 60$ Å. Negative magnetoconductivity for IE is also in agreement with the conclusion that $\lambda_c > 0$. However this result has been already obtained from the resistance data.

4. Conclusion

Analysis of the anomalous part of the magnetic susceptibility χ at temperatures below 50 K has allowed us to estimate the electron-electron interaction constant, λ_c, in the arc-produced MWNTs ($\lambda_c \approx 0.2$) and in graphite ($\lambda_c \approx 0.1$). We found that λ_c does not change in arc-produced MWNTs when the concentration of current carriers is modified by bromination. The analysis of the anomalous part of the conductance and the positive magnetoconductivity demonstrate a dominating of contribution of weak localization effects. In pure catalytic MWNTs we found positive magnetoconductivity related only to interaction effects which points to the positive sign of λ_c in these nanotubes. Thus, the analysis of temperature and field dependences of the magnetic susceptibility, conductivity and magnetoconductivity allows us to estimate the electron-electron interaction constant, λ_c, and to determine the effective dimensionality of the current carriers in inhomogeneous systems. On the base of our results we can conclude that the curvature of graphine layers in carbon nanostructures is responsible for the change of constant electron-electron interaction λ_c.

Acknowledgements

The work was supported by Russian Foundation of Basic Research (Grants No: 05-03-32901, 06-02-16005, 06-02-16433); Russian Ministry of Education and Sciences (Grant РНП.2.1.1.1604); Joint Grant CRDF, and Russian Ministry of Education and Sciences (NO-008-X1).

References

Al'tshuler, B. L., Aronov, A. G., and Zyuzin A. Yu. (1983) Thermodynamic properties of disordered conductors, *Sov. Phys. JETP* **57**, 889–895.

Al'tshuler, B. L., Aronov, A.G., Larkin, A.I., and Khmel'nitski, D.E. (1981) Anomalous magnetoresistance in semiconductors, *Sov. Phys. JETP* **54**, 411–419.

Baxendale, M., Mordkovich, V.Z., Yoshimura, S., and Chang, R.P.H. (1997) Magnetotransport in bundles of intercalated carbon nanotubes, *Phys. Rev. B* **56**, 2161–2165.

Couteau, E., Hernadi, K., Seo, J.W., Thien-Nga, L., Miko, C., Gaal, R., Forro, L. (2003) CVD synthesis of high-purity multiwalled carbon nanotubes using CaCO3 catalyst support for large-scale production, *Chem. Phys. Lett.* **378**, 9-17.

Gonzalez, J., Guinea, F., and Vozmediano, M. A. H. (2001) Electron-electron interactions in graphene sheets, *Phys. Rev. B* **63**, 134421-1 –134421-8.

Kawabata A. (1980) Theory of negative magnetoresistance in three-dimensional systems, *Solid State Commun.* **34**, 431–432.

Kociak, M., Kasumov, A.Yu., Guron, S., Reulet, B., Khodos, I. I., Gorbatov, Yu. B., Volkov, V. T., Vaccarini, L., and Bouchiat, H. (2001) Superconductivity in Ropes of Single-Walled Carbon Nanotubes, *Phys. Rev. Lett.* **86**, 2416–2419.

Kopelevich, Y., Esquinazi, P., Torres,J. H. S., and Moehlecke, S. (2000) Ferromagnetic- and Superconducting-Like Behavior of Graphite, *Journal of Low Temperature Physics* **119**, 5–6.

Kotosonov, A. S., Kuvshinnikov, S. V. (1997) Diamagnetism of some quasi-two-dimensional graphites and multiwall carbon nanotubes, *Phys. Lett. A* **229**, 377–380

Kudashov, A. G., Okotrub, A. V., Yudanov, N. F., Romanenko, A. I., Bulusheva, L. G., Abrosimov, O. G., Chuvilin, A. L., Pazhetov, E. M., Boronin, A. I. (2002) Gas-phase synthesis of nitrogen-containing carbon nanotubes and their electronic properties, *Physics of Solid State* **44**, 652–655.

Kudashov, A. G., Abrosimov, O. G., Gorbachev, R. G., Okotrub, A. V., Yudanova L. I., Chuvilin, A. L., Romanenko, A. I. (2004) Comparison of structure and conductivity of multiwall carbon nanotubes obtained over Ni and Ni/Fe catalysts, *Fullerenes, Nanotubes, and Carbon Nanostructures* **12**, 93–97.

Lee P A, Ramakrishnan T. V. (1985) Disordered electronic systems, *Rev. Modern Phys.* **57**, 287–337.

Mott, N. F. (1979) Electron Processes in Noncrystalline Materrials, *Oxford, Clarendon Press*, 350 p.

Okotrub, A.V., Bulusheva, L.G., Romanenko, A.I., Chuvilin, A.L., Rudina, N.A., Shubin, Y.V., Yudanov, N.F., Gusel'nikov, A.V. (2001) Anisotropic properties of carbonaceous material produced in arc discharge, *Appl. Phys. A* **71** 481-486.

Okotrub, A. V., Bulusheva, L. G., Romanenko, A. I., Kuznetsov, V.L., Butenko, Yu.V., Dong, C., Ni, Y., Heggic, M.I. (2001) Probing the electronic state of onion-like carbon. in Electronic Properties of Molecular Nanostructures (AIP Conference Proceedings, New York, 2001) 591 p. 349–352.

Romanenko, A. I., Anikeeva, O. B., Okotrub, A. V., Bulusheva, L. G., Yudanov, N. F., Dong, C., and Ni, Y. (2002) Transport and Magnetic Properties of Multiwall Carbon Nanotubes before and after Bromination, *Physics of Solid State* **44**, 659–662.

Romanenko, A. I., Okotrub, A. V., Anikeeva, O. B., Bulusheva, L. G., Yudanov, N. F., Dong, C., Ni, Y. (2002) Electron-electron interaction in multiwall carbon nanotubes, *Solid State Commun.* **121**, 149–153.

Romanenko, A. I., Anikeeva, O. B., Okotrub, A. V., Kuznetsov, V.L., Butenko, Yu.V., Chuvilin, A.L., Dong, C., Ni, Y. (2002) Diamond nanocomposites and onion-like carbon in Nanophase and nanocomposite materials, vol. 703, 258, Eds. S. Komarneni, J.-I. Matsushita, G.Q. Lu, J.C. Parker, R.A. Vaia, Material Research Society, (Pittsburgh 2002) p. 259–264 Temperature dependence of electroresistivity, negative and positive magnetoresistivity of graphite.

Romanenko, A. I., Okotrub, A. V., Bulusheva, L. G., Anikeeva, O. B., Yudanov, N. F., Dong, C., Ni, Y. (2003) Impossibility of superconducting state in multiwall carbon nanotubes and single crystal graphite, *Physica C* 388-389, 622–623.

Romanenko, Romanenko, Anikeeva, O. B., Zhmurikov, E. I., Gubin, K. V., Logachev, P.V., Dong, C., Ni, Y. (2004) Influence of the structural defects on the electrophysical and magnetic properties of carbon nanostructures, *in The Progresses In Function Materials (11th APAM Conference proceedings*, Ningbo, P. R. China, 2004) 59–61.

Takesue, I., Haruyama, J., Kobayashi, N., Chiashi, S., Maruyama, S., Sugai, T., Shinohara, H., (2006) Superconductivity in Entirely End-Bonded Multiwalled Carbon Nanotubes, *Phys. Rev. Lett.* **96**, 057001-1–057001-4.

Tang, Z.K., Zhang, L.Y., Wang, N., Zhang, X.X., Wang, J.N., Li, G.D., Li, Z.M., Wen, G.H., Chan, C.T., Sheng, P. (2003) Ultra-small single-walled carbon nanotubes and their superconductivity properties, *Synthetic metals* **133-134**, 689–693.

SINGLE-LEVEL MOLECULAR RECTIFIER

E.G. Petrov (epetrov@bitp.kiev.ua)
*Bogolyubov Institute for Theoretical Physics, Ukrainian National Acad. Sci.,
14-b Metrologichna str., 03143 Kiev, Ukraine*

Abstract. Based on the nonequilibrium density matrix approach, a unified description of
an electron transmission through a molecule embedded between electrodes is carried out.
The analytic expressions for the combined sequential and tunnel current through the isolated
reaction molecular level are derived with Coulomb interaction as well as an electron–vibration
coupling taken into account. The combined hopping–tunnel model of the single–level molecular rectification is proposed and the physical conditions for the exsistance of a molecular
rectifier are formulated.

Key words: Electron transmission / Coulomb interaction / Molecular rectifier

PACS number(s): 05.20.Dd; 05.60.+w

1. Introduction

The idea of creating molecular computers with an enormous memory advanced in the 1970-1980s (Aviram and Ratner, 1974; Carter, 1984) is no
longer a fantasy. Recent experimental results on molecular conductivity (Reed
et al., 1997; Tian et al., 1988; Metzger, 1999; Kergueris et al., 1999; Moresco
et al., 2001; Weber et al., 2002; Chen and Reed, 2002; Chen et al., 2003;
Kubatkin et al., 2003; Kushmerik et al., 2004; Kagan and Ratner, 2004)
show that single molecules and molecular wires are able to rectify microcur-
rents and to act as transistors. The current experimental problem is to man-
ufacture the nanomolecules with well defined conductive properties. Molecular junctions are factors that dramatically influence the conducting properties of the "left electrode–molecule–right electrode" (LMR) device (Kergueris et al., 1999; Kushmerik et al., 2004; Datta, 1995; Nitzan, 2001; Nitzan
and Ratner, 2003; Zahid et al., 2003; Kuznetsov and Ulstrup, 2002; Cuevas
et al., 2003; Selzer et al., 2004; Petrov, 2005). Conventional theoretical approaches based on Landauer–Bütteker formalism allow the description of
molecular current–voltage ($I - V$) characteristics if the nonequilibrium molecular Green's function is known (Di Ventra et al., 2000; Ghosh et al., 2004;
Xue et al., 2002; Galperin and Nitzan, 2003; Peccia and Di Carlo, 2004).

K. Scharnberg and S. Kruchinin (eds.),
Electron Correlation in New Materials and Nanosystems, 37–57.
© 2007 *Springer.*

The problem is then reduced to quantum–mechanical calculations of molecular energies with molecule–electrode interactions taken into account. The calculation scheme assumes that electron–vibrational molecular states act as virtual states and hence are not populated by the transferred electrons in the course of transmission. this means that the respective current can be referred to as tunnel current even at a resonant transmision.

However, the electrons are transmitted from one electrode to another not only along the tunnel route. Inelastic hopping electron transfer processes also participate in the current formation (Petrov et al., 1995; Kergueris et al., 1999; Petrov and Hänggi, 2001; Segal and Nitzan, 2002; Petrov et al., 2004; Petrov et al., 2005). The main difference of the hopping route from the tunnel route is due to the hoppings the transferred electrons really occupy the molecule. These electrons are present in excess for the molecule and thus they recharge the molecule. Since the tunnel route is associated with the definite molecular charge state, the inelastic hopping processes are able to control the tunnel mechanism of current formation via the molecular recharge. The goal of the present communication is to formulate the mechanism of such control.

2. Model and basic equations

To clarify the physics of current formation, we consider a specific case of a molecule with a single reaction level. It allows us to derive analytic expressions for both the sequential and the tunnel current components and to analyze the molecular $I - V$ characteristics at the arbitrary values and polarity of the applied voltage. Besides, Coulomb interaction between the electrons captured by a molecule in the course of transmission is also taken into consideration. It has been shown more than 10 years ago (Häusler et al., 1991; Yeyati et al., 1993) that Coulomb interaction between the transferred electrons occupying the isolated level in mesoscopic two–barrier structures, influences essentially the single–level electron transmission. In the next section, we will see that the same interaction controls the formation of the specific transmission channels associated with the molecular recharge. [In real molecules, a single–level charge transmission becomes possible if e.g. the molecule contains a metallic atom with an active isolated level.]

To provide the correct description of the transfer process in the LMR device (cf. Fig. 1) we consider the molecule as an open quantum system coupled to macroscopic electrodes and a heat bath. In this model, it becomes possible to employ the method of nonequilibrium density matrix (Blum, 1996) and thus to derive the master equation for the multi–electron distribution function $P_a(t)$. Following the earlier proposed approach (Petrov et al., 2004), one can

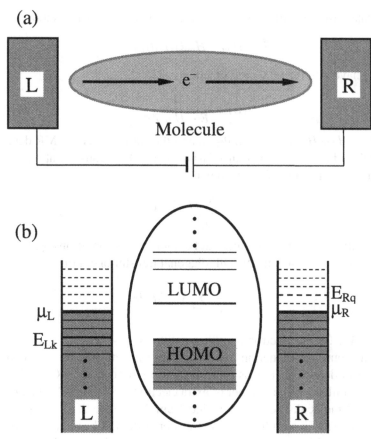

Figure 1. A molecule as a mediator of electron transfer between the electrodes (a). Single—electron scheme of energy levels in the "electrode L–molecule–electrode R" (LMR) device (b). LUMO and HOMO levels are well separated from the other molecular levels. Electron (hole) transmission occurs through the LUMO (HOMO) level depending on the level position with respect to electrochemical potentials μ_L and μ_R of the left and right electrodes, respectively.

show that

$$\dot{P}_a(t) = -\sum_b [K_{a\to b} P_a(t) - K_{b\to a} P_b(t)]. \qquad (1)$$

In this balance–like master equation, the transfer rates are determined through the expression

$$K_{a\to b} = \frac{2}{\hbar} \operatorname{Re} \int_0^\infty d\tau\, T_{ab}(\tau)\hat{T}_{ab}^+(0). \qquad (2)$$

Matrix element

$$T_{ba}(\tau) = \langle b| \exp(-iH_0\tau/\hbar)\, \hat{T} \exp(iH_0^+\tau/\hbar)|a\rangle \qquad (3)$$

is calculated using the operator for the transition on the energy shell

$$\hat{T} = H_{int} + H_{int}\hat{G}(E)H_{int} .\qquad (4)$$

Here, we have introduced Green function

$$\hat{G}(E) = \frac{1}{E - H + i0^+}\qquad (5)$$

where $H = H_0 + H_{int}$ is the Hamiltonian of the entire system "LMR device + heat bath" and $E = E_{tr}$ is the transmission energy. Electron–vibrational states $|a\rangle$ are the proper states of Hamiltonian

$$H_0 = \sum_a E_a |a\rangle\langle a|\qquad (6)$$

and thus satisfy the orthonormality and completeness conditions, $\langle a|b\rangle = \delta_{a,b}$ and $\sum_a |a\rangle\langle a| = 1$, respectively. Off–diagonal transfer operator

$$H_{int} = \frac{1}{2} \sum_{a,b} V_{a,b} |a\rangle\langle b|\qquad (7)$$

is responsible for the electron hoppings within the LMR.

   For the purpose of definity, we consider the transmission through the LUMO level denoted below as l. [Transmission through the HOMO level is described similarly.] The number of extra electrons $N = N_{l\uparrow} + N_{l\downarrow}$ occupying the isolated molecular level l is varied from $N = 0$ to $N = 2$. Since we study the transfer in a multi–electron transfer system, it becomes convenient to determine multi–electron states $|a\rangle$ as well as Hamiltonian H in the occupation number representation (Davydov, 1976). In this representation, each multi–electron state $|a\rangle$ is specified by the number of electrons occupying the single–electron states. Let $N_{j\sigma_j}(= 0, 1)$ be the number of electrons with spin projection $\sigma_j(=\uparrow, \downarrow)$ occupying the jth single–electron state. In our case, single–electron states are either the wave vectors $L\mathbf{k}(R\mathbf{q})$ of an electron in the conductive band of left(right) electrode or the MO l. We assume that the exchange interaction between an electron occupying the molecule and an electron belonging to the electrode is too small to give a noticable contribution in a total electronic energy of the LMR system. Neglecting this interaction, one can represent a multi–electron state as the product

$$|a\rangle = \prod_{\mathbf{k}\sigma_{\mathbf{k}}} |N_{L\mathbf{k}\sigma_{\mathbf{k}}}\rangle \prod_{\mathbf{q}\sigma_{\mathbf{q}}} |N_{R\mathbf{q}\sigma_{\mathbf{q}}}\rangle |M\alpha\rangle .\qquad (8)$$

Here, the molecular electron–vibrational state appears as

$$|M\alpha\rangle = |M\rangle|\chi_{N\alpha}\rangle\qquad (9)$$

where $|M\rangle = |N_{l\uparrow}, N_{l\downarrow}\rangle$ is the electronic molecular state while $|\chi_{N\alpha}\rangle$ is the vibrational state that depends on the number of extra electrons $N = N_{l\uparrow} + N_{l\downarrow}$ occupying the lth MO. In the absence of magnetic field, the states $|1_{l\uparrow}, 0_{l\downarrow}\rangle$ and $|0_{l\uparrow}, 1_{l\downarrow}\rangle$ are energetically equivalent. In what follows the energetically different molecular electronic states are denoted via $M = 0, 1, 2$, in complete correspondence with the number of excess electrons N on the molecule. Symbol $\alpha = 0, 1, 2, ...$ indicates the vibrational level at a given N (only a single reaction mode is assumed to be involved in the transfer process). Energy E_a that corresponds to the state (8), is given by expression

$$E_a = \sum_{k\sigma_k} E_{Lk} N_{Lk\sigma_k} + \sum_{q\sigma_q} E_{Rq} N_{Rq\sigma_q} + \varepsilon_M + \hbar\omega_M (\alpha + 1/2) \qquad (10)$$

In Eq. (10), $\varepsilon_M = \varepsilon(N_{l\uparrow}, N_{l\downarrow})$ is the molecular electronic energy and ω_M is the frequency of reactive mode depending on the molecular charge. To obtain the analytic results we employ the Hubbard model (Hubbard, 1963) where

$$\varepsilon_M = \varepsilon_0 + \varepsilon_l(V) (N_{l\uparrow} + N_{l\downarrow}) + U N_{l\uparrow} N_{l\downarrow} . \qquad (11)$$

Here, ε_0 is the energy of a ground molecular state. This state is assumed to be charge neutral. Therefore, it is not affected by the applied voltage V. Quantity $\varepsilon_l(V) = \varepsilon_l(0) + \eta eV$ is the energy of an extra electron occupying the lth level. The voltage division factor η defines the shift of the "center of gravity" of this level against the V (Tian et al., 1988; Nitzan, 2001; Petrov, 2005). The parameter U characterizes Coulomb repulsion among the extra (transferred) electrons occupying the lth MO. The transition matrix element V_{ab} that characterizes electron hopping between the states a and b, reads ($r = L, R$)

$$V_{ab} = \sum_{rk\sigma} \sum_{\alpha\alpha'} [V_{0rk}\langle\chi_{0\alpha}|\chi_{1\alpha'}\rangle |1_{rk\sigma}\rangle|0_{l\sigma}, 0_{l-\sigma}\rangle|\chi_{0\alpha}\rangle\langle\chi_{1\alpha'}| \langle 0_{l-\sigma}, 1_{l\sigma}|\langle 0_{rk\sigma}|$$
$$+ V_{1rk}\langle\chi_{1\alpha}|\chi_{2\alpha'}\rangle |1_{rk\sigma}\rangle|0_{l\sigma}, 1_{l-\sigma}\rangle|\chi_{1\alpha}\rangle\langle\chi_{2\alpha'}| \langle 1_{l-\sigma}, 1_{l\sigma}|\langle 0_{rk\sigma}| + h.c.] . \qquad (12)$$

In Eq. (12), the coupling V_{Mrk} characterizes an electron hopping from the kth level of the rth electrode on the lth MO only if the MO does ($M = 1$) or does not ($M = 0$) contain an extra electron; $\langle\chi_{M\alpha}|\chi_{M+1\alpha'}\rangle$ is the overlap integral between the corresponding vibrational molecular states.

The current through a molecule can be obtained from the general expression

$$I = e\dot{N}_L , \qquad (13)$$

where $\dot{N}_L$ is the time–derivative of the electron number of the left electrode. [Note that $\dot{N}_L = -\dot{N}_R$.] Quantity

$$N_L(t) = \sum_{k\sigma_k} P_{Lk\sigma_k}(t) \qquad (14)$$

is determined through the sum of band–level occupancies $P_{Lk\sigma_k}(t)$. To derive kinetic equation for this quantity let us note that $(r = L, R)$

$$P_{rk\sigma_k}(t) = \sum_{N_{rk\sigma_k}=0,1} N_{rk\sigma_k} P(N_{rk\sigma_k}, t), \tag{15}$$

where $P(N_{rk\sigma_k}, t)$ is the probability of an electron with spin projection σ_k occupying $(N_{rk\sigma_k}=1)$ or not occupying $(N_{rk\sigma_k}=0)$ the $r$kth band level. Kinetic equation for the $P(N_{rk\sigma_k}, t)$ follows directly from the master equation (1) if one notes that $(r = L)$

$$P(N_{Lk\sigma_k}, t) = \sum_{\{N_{Lk'\sigma_k'}\}\neq N_{Lk\sigma_k}} \sum_{\{N_{Rq\sigma_q}\}} \sum_{M\alpha} P_a(t). \tag{16}$$

In Eq. (16), the summation covers all possible sets of occupation numbers for both electrodes except the precise occupation number $N_{Lk\sigma_k}$. Besides, the summation involves the electron–vibrational molecular states $M\alpha$. Based on the Eqs. (1) and (15), one obtains

$$\dot{P}_{Lk\sigma_k}(t) = - \sum_{N_{Lk\sigma_k}} N_{Lk\sigma_k}$$

$$\times \sum_{\{N_{Lk'\sigma_k'}\}\neq N_{Lk\sigma_k}} \sum_{\{N_{Rq\sigma_q}\}} \sum_{M\alpha} [K_{a\to b} P_a(t) - K_{b\to a} P_b(t)]. \tag{17}$$

Similarly, one derives a kinetic equation for the molecular occupancy $P(M, t)$. The equation reads

$$\dot{P}(M, t) = - \sum_{\{N_{Lk\sigma_k}\}} \sum_{\{N_{Rq\sigma_q}\}} \sum_{\alpha} [K_{a\to b} P_a(t) - K_{b\to a} P_b(t)]. \tag{18}$$

The precise form of kinetic equations (17) and (18) can be obtained if one specifies the form of the multi–electron distribution function $P_a(t)$. In line with the structure of the multi–electron state (8) we employ the ansatz

$$P_a(t) = \prod_{k\sigma_k} P(N_{Lk\sigma_k}, t) P(M, t) \prod_{q\sigma_q} P(N_{Rq\sigma_q}, t) W_{M\alpha}. \tag{19}$$

Here,

$$W_{M\alpha} = \frac{\exp\left[-\hbar\omega_M (\alpha + 1/2)/k_B T\right]}{\sum_\alpha \exp\left[-\hbar\omega_M (\alpha + 1/2)/k_B T\right]} \tag{20}$$

is the bath equilibrium density matrix (vibrational distribution function). Using the form (19) and the normalization conditions

$$\sum_{N_{rk\sigma_k}=0,1} P(N_{rk\sigma_k}, t) = 1, \quad \sum_M P(M, t) = 1, \tag{21}$$

one is able to reduce the kinetic equations (17) and (18) to the form suitable for further analytic solution.

The inspection shows that kinetic equations for the band–level occupancies $P_{rk\sigma_k}(t)$ contain two types of transfer rates characterizing an electron transfer associated with a single–electron state $rk\sigma_k$. The first type refers to an inelastic electron hopping from the kth band level of the rth electrode to a charge–neutral molecule ($M = N = 0$) or to a singly reduced molecule ($M = N = 1$). The corrsponding transition matrix element T_{ab}, Eq. (3) is calculated with the first term of transition operator $\hat{T}$, Eq. (4), i.e. with operator H_{int}, Eq. (12). This leads to the following expression for the first type of transfer rate

$$K_{rkM \to M+1} = \frac{2\pi}{\hbar} |V_{M\,rk}|^2 \sum_{\alpha,\alpha'} W_{M\alpha} \langle \chi_{M\alpha} | \chi_{M+1\alpha'} \rangle^2$$
$$\times \delta\big(E_{rk} + \varepsilon_M - \varepsilon_{M+1} + \hbar\omega_0(\alpha - \alpha')\big). \qquad (22)$$

[The form (23) is independent on spin projection of the transferred electron.] It is important to stress that the overlap integral $\langle \chi_{M\alpha} | \chi_{M+1\alpha'} \rangle$ is calculated between the eigen vibrational functions of the Mth and $(M + 1)$th electronic terms so that $\varepsilon_M + \hbar\omega_0\alpha$ is the eigen electron–vibrational energy of the N–multiply charged molecule. The complete forward (electrode–molecule) electron transfer rate reads

$$\chi_r(M) = \sum_k f_r(E_{rk}) K_{rkM \to M+1} \qquad (23)$$

The backward (molecule–electrode) transfer rate is expressed via the forward one as

$$\chi_{-r}(M) = \sum_k (1 - f_r(E_{Lk})) K_{M+1 \to rkM} = e^{\Delta E_{rM}(V)/k_B T} \chi_r(M). \qquad (24)$$

For the purpose of definity, we consider the case when the left electrode is held at zero voltage ($\mu_L = E_F, \mu_R = E_F + eV$). It allows to present the gaps $\Delta E_{rM}(V) = \varepsilon_{M+1} - \varepsilon_M - \mu_r$ as ($M = N = 0, 1$)

$$\Delta E_{LM}(V) = \Delta E_M(0) + \eta eV, \quad \Delta E_{RM}(V) = \Delta E_M(0) - (1 - \eta)eV \qquad (25)$$

where $\Delta E_M(0) = \Delta E^{(0)} + N U$ and $\Delta E^{(0)} = \varepsilon_l - E_F$. The relative position of the electronic gaps (25) is represented in Fig. 2. To simplify the form of transfer rates, the Holstein model (Holstein, 1959) is employed for a heat bath. In the framework of this model, the vibrational frequency ω_M is independent of the molecular charge state N, e.g. $\omega_0 = \omega_1 = \omega_2$. The following considerarion is carried out under the conditions when the frequency of a single reaction

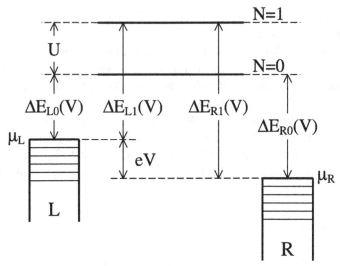

Figure 2. Relative position of the electronic gaps (25) that appear during the electron transmission through a charge–neutral molecule ($N = 0$) and a singly reduced molecule ($N = 1$).

mode, ω_0 is rather large, so that $\exp(-\hbar\omega_0/k_BT) \ll 1$. It means that in Eq. (23) one can set $W_{M\alpha} \approx \delta_{\alpha,0}$.

Now, let us introduce the quantity

$$\Gamma_{rM}(E) = 2\pi \sum_{\mathbf{k}} |V_{M\,r\mathbf{k}}|^2 \delta(E - E_{r\mathbf{k}}) \tag{26}$$

that defines the broadening of the molecular level l if the latter contains one ($M + 1 = 1$) or two ($M + 1 = 2$) captured electrons. This broadening is formed by the off–diagonal electronic interaction of the molecule with the rth macroscopic electrode. In the wide–band limit valid for the most noble metals, the broadening is independent of the transmission energy E (Tian et al., 1988; Datta, 1995; Nitzan, 2001). Thus, setting $\Gamma_{rM}(E) \approx \Gamma_{rM}$ one comes to the following expression for the electrode–molecule transfer rate

$$\chi_r(M) \approx \frac{1}{\hbar} \Gamma_{rM} F_{rM}, \quad F_{rM} = 2\pi \sum_{\alpha} \langle \chi_{M0} | \chi_{M+1\alpha} \rangle^2 n(\Delta E_{rM}(\alpha; V)). \tag{27}$$

Here, the distribution function $n(x) = [\exp(x/k_BT) + 1]^{-1}$ is specified by the vibration–dependent gap

$$\Delta E_{rM}(\alpha; V) = \Delta E_{rM}(V) + \hbar\omega_0\alpha. \tag{28}$$

The second type of transfer rate characterizes the direct unistep electron hopping between the $L\mathbf{k}$th and the $R\mathbf{q}$th band states. The corresponding matrix element T_{ab} is calculated with the second term, $H_{int}\hat{G}(E)H_{int}$ of the transition operator (4). It has been shown (Petrov, 2005) that in the case of macroscopic electrodes, the LMR Hamiltonian $H = H_0 + H_{int}$ can be reduced to the effecive diagonal Hamiltonian $H^{(eff)} = H_{electrode} + H_M^{(eff)}$, so that the Green's function (cf. Eqs. (4) and (5)) appears as

$$\hat{G}(E) \approx \frac{1}{E - H^{(eff)}}.$$ (29)

In the occupation number space, the Hamiltonian associated with the electrode's electrons, reads

$$H_{electrode} = \sum_{r=L,R} \sum_{k\sigma} \sum_{N_{k\sigma}} E_{rk} N_{k\sigma} |N_{k\sigma}\rangle\langle N_{k\sigma}|$$ (30)

while the effective molecular Hamiltonian is

$$H_{mol}^{(eff)} = \sum_{M\alpha} E_{M\alpha} |M\rangle|\chi_{M\alpha}\rangle\langle\chi_{M\alpha}|\langle M|.$$ (31)

Molecular electron–vibrational energy

$$E_{M\alpha} = \epsilon_M + \hbar\omega_0 (\alpha + 1/2) - i\Gamma_M/2$$ (32)

contains the renormalized pure electronic energy $\epsilon_M = \varepsilon_M - \text{Re}\,\Sigma_M$ as well as the broadening $\Gamma_M = 2\text{Im}\,\Sigma_M$ where $\Sigma_M = \Sigma_{LM} + \Sigma_{RM}$ is the self–energy resulted from an off–diagonal molecular–electrode interaction. In the wide-band limit under consideration, the energy addition $\text{Re}\,\Sigma_M$ and the broadening $\Gamma_M = \Gamma_{LM} + \Gamma_{RM}$ are assumed to be independent of the transmission energy E. In what follows, we will suppose that the addition $\text{Re}\,\Sigma_M$ is already included in the molecular energy ε_M.

The calculation of the direct transfer rate under the condition $W_{M\alpha} \approx \delta_{\alpha,0}$ yields

$$K_{Lk\to Rq}^{(M)} = \frac{2\pi}{\hbar} \sum_{\alpha} |T_{Lk0,Rq\alpha}^{(M)}|^2 (M + 1) \delta(E_{Lk} - E_{Rq} - \hbar\omega_0\alpha).$$ (33)

The result is independent of the spin projection of the transferred electron. In Eq. (33), a distant electrode–electrode coupling

$$T_{Lk0,Rq\alpha}^{(M)} = \sum_{\tilde{\alpha}} \frac{V_{MLk}V_{MRq}^*\langle\chi_{M0}|\tilde{\chi}_{M+1\tilde{\alpha}}\rangle\langle\tilde{\chi}_{M+1\tilde{\alpha}}|\chi_{M\alpha}\rangle}{E - \tilde{\varepsilon}_{M+1} - \hbar\omega_0\tilde{\alpha} + i\Gamma_{M+1}/2}$$ (34)

is calculated at the transmission energy $E = E_{Lk} + \varepsilon_M$. To estimate the coupling (34), one has to take into account the following. In Eq. (34), the summation is performed over the electron–vibrational molecular states $|M + 1\rangle\tilde{\chi}_{M\tilde{\alpha}}\rangle$. Since only a single MO participates in the transmission, a molecular charge can not exceed 2. Therefore, only molecular states $M = 0$ and $M = 1$ mediate a direct (unistep) inter–electrode electron hopping. Energy conservation law $E_{Lk} = E_{Rq} + \hbar\omega_0\alpha$ does not include the intermediate energy $\tilde{\varepsilon}_{M+1} + \hbar\omega_0\tilde{\alpha}$. It means that the respective intermediate electron–vibrational states $|M + 1\rangle|\tilde{\chi}_{M+1\tilde{\alpha}}\rangle\rangle$ have to be refered to as virtual states with the corresponding broadened energies. Thus, the problem appears where the electronic energy $\tilde{\varepsilon}_{M+1}$ of virtual state is positioned with respect to the proper electronic energy ε_{M+1}. Let Z_N be the equilibrium reaction nuclear coordinate for the N–multiply charged molecule. In the adiabatic approximation, each electronic transition $M \to M + 1$ occurs at the fixed nuclear coordinates. Therefore, an electronic energy in the new charged state $N + 1$ (but at the same equilibrium nuclear coordinate Z_N) reads $\varepsilon_{M+1}^{(0)} = \varepsilon_{M+1}(Z_N)$. However, after the molecular recharge has happened, the nuclei shift their equilibrium position. The Z_N is then replaced by the Z_{N+1}. Nuclear displacement $\Delta Z_N = Z_{N+1} - Z_N$ is accompanied by the reduction of electronic energy from its unrelaxed value $\varepsilon_{M+1}^{(0)}$ to the proper (relaxed) energy $\varepsilon_{M+1} = \varepsilon_{M+1}^{(0)} - E_{M\to M+1}^{(reorg)}$ where $E_{M\to M+1}^{(reorg)} = (1/2)\hbar\omega_0\zeta_N^2$ is the energy of nuclear reorganization during the $M \to M + 1$ electronic transition. The value of the reorganization energy depends on the vibrational frequency ω_0 as well as the dimensionless nuclear displacement $\zeta_N = \Delta Z_N \sqrt{m_Z\omega_0/\hbar}$ (m_Z is the effecive mass associated with the Zth normal coordinate). Here, we are not concerned with the relaxation problem related to formation of the reorganization energy and consider only two limiting cases. In the first case, we assume that virtual electronic energy $\tilde{\varepsilon}_{M+1}$ coincides with the proper (relaxed) energy ε_{M+1}. It means that the terms M and $M + 1$ are shifted along normal coordinate Z by the value ΔZ_N and thus the overlap integral $\langle\chi_{M0}|\tilde{\chi}_{M+1\tilde{\alpha}}\rangle$ does not vanish at $\tilde{\alpha} \neq 0$. Just this happens at inelastic molecule–electrode and electrode–molecule electron hopping. In the second case, a virtual electronic energy $\tilde{\varepsilon}_{M+1}$ coincides with the unrelaxed energy $\varepsilon_{M+1}^{(0)} = \varepsilon_{M+1} + E_{M\to M+1}^{(reorg)}$. Now a nuclear displacement between the relaxed term M and the unrelaxed term $M + 1$ is absent and thus the overlap inegral is reduced to the Kronecker symbol $\delta_{\tilde{\alpha},0}$.

Integral transfer rate characterizing the direct electron flow from the left electrode to the right electrode reads

$$K_{L\to R}^{(M)} = \sum_{kq} K_{Lk\to Rq}^{(M)} f_L(E_{Lk})(1 - f_R(E_{Rk} - eV)). \qquad (35)$$

Based on expression (35), one can derive the net electron flow from the left electrode to the right electrode, $Q_{L\to R}^{(M)} = K_{L\to R}^{(M)} - K_{R\to L}^{(M)}$. Below we give the form valid at isoenergetic (pure elastic, $E_{Lk} = E_{Rq}$) transmission,

$$Q_{L \to R}^{(M)} = -Q_{R \to L}^{(M)} = \frac{2}{\hbar} \frac{\Gamma_{LM} \Gamma_{RM}}{\Gamma_{LM} + \Gamma_{RM}} \sum_{\alpha} \langle \chi_{M0} | \tilde{\chi}_{M+1\alpha} \rangle^4 \Phi_M(\alpha). \tag{36}$$

Here, the function

$$\Phi_M(\alpha) = \arctan \left[\frac{2\Delta \tilde{E}_{RM}(\alpha; V)}{\Gamma_{LM} + \Gamma_{RM}} \right] - \arctan \left[\frac{2\Delta \tilde{E}_{LM}(\alpha; V)}{\Gamma_{LM} + \Gamma_{RM}} \right] \tag{37}$$

defines the voltage behavior of the net elastic flow. If the relaxed (proper) virtual molecular terms mediate a flow formation, then the gap $\Delta \tilde{E}_{rM}(\alpha; V)$ coincides with the one from Eq. (28). When the mediation is associated with unrelaxed electronic terms, then $\Delta \tilde{E}_{rM}(\alpha; V) = \Delta E_{rM}(\alpha; V) + E_{M \to M+1}^{(reorg)}$.

The integral transfer rates $\chi_{-r}(M)$ and $\chi_r(M)$ (cf. Eqs. (24) and (27)) as well as the net electron flow $Q_{L \to R}^{(M)}$ specify the time–derivative ($M = 0, 1$),

$$\dot{N}_L = -2 \sum_M [\chi_L(M) P(M, t) - \chi_{-L}(M) P(M+1, t) + (M+1) Q_{L \to R}^{(M)} P(M, t)]. \tag{38}$$

To obtain the expression for occupancies $P(M)$ one has to solve the set of respective kinetic equations. These equations follow from Eq. (18) and read

$$\dot{P}(M, t) = -2B(0) P(0, t) + 2A(0) P(1, t),$$

$$\dot{P}(1, t) = -(A(0) + B(1)) P(1, t) + B(0) P(0, t) + A(1) P(2, t),$$

$$\dot{P}(2, t) = -2A(1) P(2, t) + 2B(1) P(1, t), \tag{39}$$

where we have introduced the notations ($M = 0, 1$)

$$A(M) \equiv \chi_{-L}(M) + \chi_{-R}(M), \quad B(M) \equiv \chi_L(M) + \chi_R(M). \tag{40}$$

3. Results and discussion

Below we consider the formation of a stationary current. Solving the set (39) at stationary condition $\dot{P}(M, t) = 0$ and taking into consideration the normalization condition $P(0, t) + 2P(1, t) + P(2, t) = 1$, one obtains

$$P(0) = A(0)A(1)/D, \quad P(1) = B(0)A(1)/D, \quad P(2) = B(0)B(1)/D,$$

$$D = B(0)B(1) + (A(0) + 2B(0))A(1). \tag{41}$$

The stationary molecular occupancies $P(M)$ specify the stationary time–derivative $\dot{N}_L$ and thus the stationary current (cf. Eqs. (13) and (38)). The latter appears as a sum of two components,

$$I = I_{hop} + I_{dir}. \tag{42}$$

The first one,

$$I_{hop} = \sum_{M=0,1} I_{hop}^{(M)}, \quad I_{hop}^{(M)} = I_0 h \left[\chi_L(M) P(M) - \chi_{-L}(M) P(M+1) \right], \quad (43)$$

is associated with an exclusively inelastic hopping transmission through the isolated molecular level l. The second current component,

$$I_{dir} = \sum_{M=0,1} I_{dir}^{(M)}, \quad I_{dir}^{(M)} = I_0 h (M+1) Q_{L \to R}^{(M)} P(M), \quad (44)$$

characterizes a direct (tunnel) electron transmission between the electrodes. [We have introduced the elementary current unit $I_0 = 2|e|/h \approx 80 \mu A$. Therefore, all energy characteristics specifying the transfer rates as well as the flow have to be expressed in eV.] Both contributions reflect the fact that current formation through a single level is controlled by two transmission channels. The first channel involves the transmission through a charge–neutral molecule ($M = N = 0$) while the second channel is associated with an electron transmission through a singly charged molecule ($M = N = 1$). The opening of each channel is defined by the probability $P(M)$ to find a molecule in the ($M = N$)–multiply charged state. In turn, the value of the noted probability depends strongly on the relation between the hopping transfer rates characterizing exclusively inelastic kinetic process (cf. Eqs. (40), (41) and Fig. 3). Thus, the inelastic processes govern the direct (tunnel) electron transmission.

Firstly, we demonstrate this important conclusion by analyzing the current formation under large Coulomb repulsion between the extra electrons occupying the molecule in the course of transmission. The calculations are based on general expressions (42)–(44). The overlap integrals $\langle \chi_{M0} | \chi_{M+1\alpha} \rangle$ in expressions (24), (27), and (36) have been estimated at $\zeta_0 = \zeta_1 \equiv \zeta$. If the overlapping occurs between the vibrational functions related to proper electron–vibrational states, then

$$\langle \chi_{M0} | \chi_{M+1\alpha} \rangle = (\zeta^2/2)^{\alpha/2} (1/\sqrt{\alpha!}) \exp(-\zeta^2/4). \quad (45)$$

At the same time, one has to set

$$\langle \chi_{M0} | \tilde{\chi}_{M+1\alpha} \rangle = \delta_{0,\alpha} \quad (46)$$

if only electron–vibrational functions $|\chi_{M0}\rangle$ and $|\tilde{\chi}_{M+1\alpha}\rangle$ belong to the proper and unrelaxed molecular terms, respectively. Fig. 4 depicts the symmetric molecular $I - V$ characteristics in the case of small and large nuclear displacement. The comparision of the conventional Landauer–Bütteker (L–B) model with the model proposed in the present communication is also given.

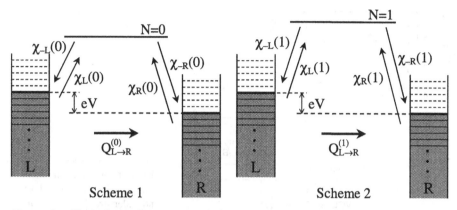

Figure 3. Single level electron transmission routes through a charge neutral molecule ($M = N = 0$, Scheme 1) and a singly reduced molecule ($M = N = 1$, Scheme 2). Sequential (hopping) route is characterized by the transfer rates $\chi_{L(R)}(M)$ and $\chi_{-L(-R)}(M)$ while a direct electrode–electrode route is connected with the tunnel flows $Q_{L \to R}^{(M)} = -Q_{R \to L}^{(M)}$. Note that the gaps $\Delta E_{L1}(V)$ and $\Delta E_{R1}(V)$ associated with the resonant switching on of the second transmission channel ($M = N = 1$) exceed the respective gaps $\Delta E_{L0}(V)$ and $\Delta E_{L1}(V)$ associated with the resonant switching on the first transmission channel ($M = N = 0$), cf. Eq. (25) and Fig. 2.

In the case of a single–level transmission and under large Coulomb repulsion between the extra electrons occupying the isolated molecular level, the L–B model leads to expression

$$I_{dir}^{(L-B)} = I_0\, h\, Q_{L \to R}^{(0)}, \tag{47}$$

It differs from Eq. (44) derived in the framework of unified description (based on the density matrix approach). The latter assumes an important role of inelastic processes (via the molecular occupancies $P(M)$) in formation of the total current components. In the case of a single–level transmission, the L–B model leads to the featureless current behavior. The proposed model of unified description suggests the appearance of sequential (hopping) and direct (tunnel) current components. The hopping current component exibits a stepwise behavior that completely correlates with the behavior of the $P(M)$. In particular, a resonant inelastic transmission as well as a drop in quantity $P(0)$ occurs at the same resonant voltages $V = V_{L0} = 0.6$ eV (region $V > 0$) and $|V| = V_{R0} = 0.6$ eV (region $V < 0$) where

$$V_{L0} = \frac{1}{\eta |e|} \Delta E^{(0)}, \quad V_{R0} = \frac{1}{(1-\eta)|e|} \Delta E^{(0)}. \tag{48}$$

The modification of a direct current is caused by the occupation of a molecular level by an extra electrons that are captured (kinetically) by a molecule in the course of electron transmission. It has been already noted above that

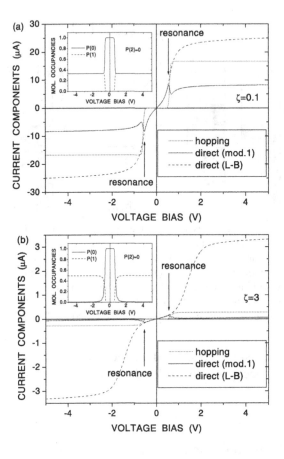

Figure 4. Symmetric $I - V$ characteristics of the molecule with a single reaction level and under strong Coulomb interaction (U = 10 eV) between the extra electrons. Calculations are based on Eqs. (43) and (44) as well as on Eqs. (36), (37), and (41) at $k_B T$ = 0.012 eV, $\hbar\omega_0$ = 0.1 eV, $\Delta E^{(0)}$ = 0.3 eV, ζ = 0.1, η = 0.5, $\Gamma_{L0} = \Gamma_{R0}$ = 0.1 eV, $\Gamma_{L1} = \Gamma_{R1}$ = 0.3 eV. Notations (L–B) and (mod.1) refer, respectively, to the Landauer–Bütteker model and the proposed model where a tunneling occurs through the proper virtual molecular terms.

this fact is reflected in Eq. (44) via the factor $P(M)$. Under large Coulomb repulsion (i.e. at large Hubbard parameter U, Eq. (11)) the gaps $\Delta E_{L1}(V)$ and $\Delta E_{R1}(V)$ associated with a resonant switching on of the second transmission channel ($M = N = 1$) exceed strongly the respective gaps $\Delta E_{L0}(V)$ and $\Delta E_{R0}(V)$ associated with the first transmission channel ($M = N = 0$), cf. Fig. 2 and Eq. (25). Since the $\Delta E_{L1}(V)$ and $\Delta E_{R1}(V)$ are positive at any given V, the hopping electrode–molecule transfer rates $\chi_L(1)$ and $\chi_R(1)$ (cf. Eq. (24) and the scheme 2 in Fig. 3) are determined by the thermally activated process and thus are too small to create a two electron population of the molecular level. Setting $\chi_L(1) \approx \chi_R(1) \approx 0$ and using Eq. (41), one derives

$$P(0) \approx \frac{A(0)}{A(0) + 2B(0)}, \quad P(1) \approx \frac{B(0)}{A(0) + 2B(0)}, \quad P(2) \approx 0. \quad (49)$$

Under large Coulomb repulsion, just these molecular occupanies determine the weight of current components (43) and (44) in a total current (42). At

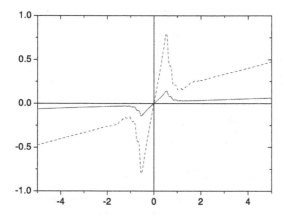

Figure 5. Formation of direct current components under strong Coulomb repulsion between the extra electrons occupying the reaction molecular level. At large electron–vibration coupling, the direct electron tunneling between the electrodes becomes much more effective if it is mediated by unrelaxed molecular terms (mod. 2) rather than by proper (relaxed) molecular terms (mod. 1) (a). Total current (curve 3) as the sum of the hopping (curve 1) and the direct (curve 2) components associated with the tunneling mediated by unrelaxed molecular terms (b). The peaks appear at $V = V_{L0}$ ($V > 0$) and $|V| = V_{R0}$ ($V < 0$). The vibrational structure in $I - V$ characteristics of the direct component is determined by the inelastic processes forming the molecular occupancy $P(0)$ (cf. Eq. (44)). Calculations according to Eqs. (36), (37), (41), (43), and (44). The parameters are the same as in Fig. 4 except $\zeta = 3$

$\chi_L(1) \approx \chi_R(1) \approx 0$ and $P(2) \approx 0$ the general expressions (43) and (44) appear in more simple form

$$I_{hop} \approx I_{hop}^{(0)} = I_0 \, h \, [\chi_L(0) \, P(0) - \chi_{-L}(0) \, P(1)] \qquad (50)$$

and

$$I_{dir} \approx I_{dir}^{(0)} = I_0 \, h \, Q_{L \to R}^{(0)} \, P(0) \,. \qquad (51)$$

Eq. (51) contains the controlling factor $P(0)$. It makes the Eq. (51) differs the one given by L–B model, Eq. (47).

In the framework of unified description, the direct current component exibits strongly nonlinear behavior, in particular the $I - V$ characteritic exibits the peak. The formation of the peak is associated with the sudden drop in the occupancy $P(0)$ at $V = V_{L0}$ ($|V| = V_{R0}$). Comparision of Fig. 4a with Fig. 4b shows that the current components decrease if only a nuclear displacement ζ increases. In Fig. 5a, two versions of the formation of the modified direct current components are represented. It can be seen clearly that the tunneling with participation of the unrelaxed virtual molecular electronic terms becomes more efficient in comparision with the tunneling mediated by the

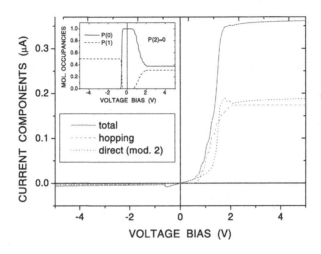

Figure 6. Molecular single–level rectification effect under strong Coulomb interaction. The current is formed by the sum of hopping and direct electron transmission. The latter occurs with participation of unrelaxed molecular electronic terms. The rectification is associated with the asymmetric molecular occupancy $P(0)$ (see the insertion). The second transmission channel gives a small contribution to the current. Calculations according to Eqs. (36), (37), (41), (43), and (44). The parameters are the same as in Fig. 4 except $\zeta = 3$, $\Gamma_{L0} = 0.001$ eV, $\Gamma_{R0} = 0.1$ eV, $\Gamma_{L1} = 0.003$ eV, $\Gamma_{R1} = 0.3$ eV.

proper (relaxed) molecular electronic terms. In particular, the direct (tunnel) current contribution is able to exceed the inelastic (hopping) current contribution, Fig. 5b only if the unrelaxed virtual molecular terms are involved in the tunnel transmission.

Now we are able to discuss molecular rectification properties caused by asymmetric molecule–electrodes kinetic processes at $V > 0$ and $V < 0$. To look at these properties in a clear form, let us consider a symmetric disposition of a molecule with respect to the electrodes ($\eta = 0.5$). At such arrangement, the level shift against the V is independent of the voltage polarity and thus both resonant voltages (48) have identical value. It excludes the appearance of asymmetry in the $I-V$ characteristics caused by the floating the molecular level. In line with Eq. (43), the asymmetry in the hopping current component is possible only if molecular occupancies $P(M)$ become asymmetric with respect to the applied positive and negative voltages. The same is true for the direct current component (51). [Note that the flow $Q_{L \to R}^{(M)} = -Q_{R \to L}^{(M)}$ is symmetric at $\eta = 0.5$.]

Figure 6 illustrates the current asymmetry at $\eta = 0.5$. For the purpose of definity, electron coupling between the molecule and the right electrode is assumed to be much stronger than similar coupling for the molecule and the left electrode. Physically, such situation can appear if the left and right

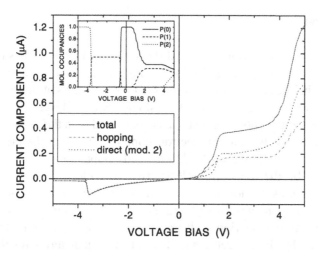

Figure 7. A molecular single–level rectification effect under middle Coulomb interaction. The current is formed by the sum of hopping and direct electron transmission. The latter occurs with participation of unrelaxed molecular electronic terms. The rectification is associated with the asymmetric molecular occupancies $P(0)$ and $P(1)$ (see the insertion). The second transmission channel gives a notable contribution in the current starting from $V \geq V_{L1}$ and $V \leq -V_{R1}$. Calculations according to Eqs. (36), (37), (41), (43), and (44). The parameters are the same as in Fig. 5 except $U = 1.5$ eV.

end groups of the molecule are chemically different. It is seen that a sudden drop in the $P(0)$ from 1 to 0 (actually to 10^{-2}) observed at $V \leq -V_{R0}$, is the physical reason that an electron transfer through the first tramsmission channel ($M = N = 0$) is nearly blocked. At the same time, only a partial decrease of the $P(0)$ (from 1 to 0.36) happens at $V \geq V_{L0}$ and thus electron transfer through the same transmission channel occurs with a rather large efficiency. If Coulomb interaction between extra electrons decreases, then molecular rectification properties are declined (compare Fig. 6 and Fig. 7 at $V = \pm 2$ eV). It is due to the second resonant transmission being switched on

$$V_{L1} = \frac{1}{\eta |e|}(\Delta E^{(0)} + U), \quad V_{R1} = \frac{1}{(1 - \eta)|e|}(\Delta E^{(0)} + U). \quad (52)$$

Therefore, already under middle Coulomb interaction, the second transmission channel ($M = N = 1$) contributies significantly to the direct (tunnel) current component even in the preresonance region $V < V_{L0}$ ($V > -V_{R0}$). At $U = 1.5$ eV, the resonant switching on of the second transmission channel occurs at $V = V_{L1} = 3.6$ eV and $V = -V_{R1} = -3.6$ eV. It leads to the appearance of a pronounced two electron population of the reaction molecular level. The simplified expressions (49)–(51) are no longer applicable and more general forms (41)–(44) needs to be employed.

4. Conclusion

In the present communication, a unified desription of electron transmission through a molecule has been performed using the nonequilibrium density matrix approach. The analytic expressions (42)–(44) for the stationary current have been derived for a molecule with a single reaction level. The current consists of two components. The first one is associated with the inelastic electron transfer process. It includes electron hoppings between the electrodes and the molecule. The hoppings lead to a steady population of the reaction molecular level by the transferred electrons. These electrons are extra for the molecule and thus the hoppings lead to a kinetic molecular recharge. The second current component is formed by a direct unistep electron transmission between the electrodes. During such transmission, the transferred electrons are not captured by the molecule and thus all molecular terms involved in formation of the direct electrode–electrode coupling refer to virtual terms. Accordingly, the direct current component is defined by the tunnel electron transmission. However, this transmission depends strongly on a charge state of the molecule. In the case of a single reaction level, only a charge neutral and a singly reduced molecular states are able to mediate a direct as well as a sequential electron transfer. Therefore, only two tunnel transmission channels ($M = N = 0$ and $M = N = 1$) participate in the current formation. The probability to switch on of the Mth channel is defined by molecular occupancy $P(M)$ (cf. Eqs. (43) and (44)). Since each occupancy is formed by the hopping processes exclusively (cf. Eqs. (40) and (41)), just inelastic electron transfer routes depicted in Fig. 3 via transfer rates $\chi_{L(R)}(M)$ and $\chi_{-L(-R)}(M)$, are responsible for controlling the sequential and the direct current components. The effect is especially significant during the resonant transmission. In this case, a stationary occupation of the molecule by a single electron (under large Coulomb repulsion between the transferred electrons occupying the reaction molecular level) or by two extra electrons (under small and middle Coulomb repulsion) becomes notable. Note that the controlling effect is absent in the conventional model of the tunnel current formation if floating of a single reaction level is independent of the polarity of the applied voltages, i.e. at $\eta = 0.5$. It follows from Eq. (47) where the direct current is defined only by the net tunnel flow $Q^{(0)}_{L \to R}$. If $\eta = 0.5$, the absolute values of the flows $Q^{(0)}_{R \to L}$ and $Q^{(0)}_{L \to R}$ coincide at identical absolute values V. Thus, the molecular rectification effect is absent even though $\Gamma_{L0} \neq \Gamma_{R0}$. At the same time, a modified model of a tunnel transmission based on the combined description of elastic and inelastic transfer processes predicts a specific kinetic rectification effect at $\eta = 0.5$ and $\Gamma_{L0} \neq \Gamma_{R0}$, cf. Eq. (51) and Fig. 6. It is completely determined by the appearance of factor $P(0)$. Physically, the occupancy $P(0)$ determines the probability to observe the molecule without excess electrons. If $\Gamma_{L0} \neq \Gamma_{R0}$ then the $P(0)$ depends strongly on the voltage polarity. Thus, the asymmetry in the occupancy $P(0)$ appears as the factor that governs the asymmetry of the direct current. The effect becomes more significant at strong Coulomb repulsion between the extra electrons. Therefore, the proposed model of combined single–level hopping–tunnel electron transmission predicts a strong rectification effect in those LMR devices where: (a) the reaction molecular level is positioned not far from the Fermi–level (to be involved in a resonant transmission at the real applied voltages); (b) the noted molecular level has to be not only well separated from the other molecular levels but to appear as a strongly localized level (to provide a strong Coulomb repulsion between the extra electrons occupying this level in the course of transmission); (c) the strong asymmetry in electron coupling between the molecule and each separate electrode is presented (to reach the asymmetry in electron hoppings between the molecule and separate electrodes); (d) the coupling between the electronic and the vibrational degrees of freedom is rather strong (to provide the asymmetric molecular recharge in the vicinity of resonant voltages $V = V_{L0}$ and $V = -V_{R0}$, cf. Fig. 6). The molecules with a single active metal atom or with a single double (triple) bond are probably those candidates that can be employed as the molecular rectifiers.

Acknowledgements

The author gratefully acknowledges the support of this work by the project No. M/230.

References

Aviram, A. and Ratner, M. (1974) Molecular Rectifiers, *Chem. Phys. Lett.* **29**, 277–279.

Blum, K. (1996) *Density Matrix Theory and Application, 2-nd ed.*, Plenum Press, N.Y.

Carter, F. L. (1984) Molecular Electronic Devices, Today's Dream–Tomorrow's Reality, *Presented at BIOTECH'84 USA: Online Publications*, Pinner, UK, pp. 127-138.

Chen, J. and Reed, M. F. (2002) Electronic Transport of Molecular Systems, *Chem. Phys.* **281**, 127-145.

Chen, H., Lu, J. Q., Wu, J., Note, R., Mizuseki, H., and Kawazoe, Y. (2003) Control of Substituent Ligand over Current through Molecular Devices: An ab initio Molecular Orbital Theory, *Phys. Rev. B* **67**, 113408(1-4).

Cuevas, J. C., Heurich, J., Pauly, F., Wenzel, W., and Schön, G. (2003) Theoretical Description of the Electrical Conduction in Atomic and Molecular Junctions, *Nanotechnology* **14**, R29-R38.

Datta, S. (1995) *Electron Transfer in Mesoscopic Systems*, University Press, Cambridge.

Davydov, A. S. (1976) *Quantum Mechanics, 2-nd ed.*, Pergamon Press, Oxford.

Di Ventra, M., Pantelides, S. T., and Lang, N. D. (2000) First–Principles Calculation of Transport Properties of a Molecular Device, *Phys. Rev. Lett.* **84**, 979-982.

Galperin, M. and Nitzan, A. (2003) NEGF–HF Method in Molecular Junction Property Calculations, *Ann. N.Y. Acad. Sci.* **1006**, 48-67.

Ghosh, A. W., Damle, P., Datta, S., and Nitzan, A. (2004) Molecular Electronics: Theory and Device Prospects, *MRS Bulletin/June*, 391-395.

Häusler, W., Kramer, B., and Masek, J. (1991) The Influence of Coulomb Interaction on Transport through Mesoscopic Two–Barrier Structures, *Z. Phys. B* **85**, 435-442.

Holstein, T. (1959) Studies of Polaron Motion: Part I. The Molecular–Crystal Model, *Ann. Phys. (New York)*, **8**, 325-342.

Hubbard, J. (1963) Electron Correlations in Narrow Energy Bands, *Proc. R. Soc. London A*, **276**, 238-257.

Kagan, C. R. and Ratner, M. A. (2004) Molecular Transport Junctions: An Introduction, *MRS Bulletin/June* 376-384.

Kergueris, C., Bourgoin, J.-P., Palacin, S., Esteve, D., Urbana, C., Magoga, M., and Joachim, C. (1999) Electron Transport through a Metal–Molecule–Metal Junction, *Phys. Rev. B* **59**, 12505-12513.

Kubatkin, S., Danilov, A., Hjort, M., Cornill, J., Bredas, J.-L., Stuhr-Hansen, N., Per Hedegard, and Bjornholm, T. (2003) Single Electron Transistor of Single Organic Molecule with Accecess to Several Redox States, *Nature* **425**, 698-701.

Kushmerik, J. G., Whitaker, C. M., Pollack, S. K., and Schull, T. L. (2004) Tuning Current Rectification Across Molecular Junctions, *Nanotechnology* **15**, S489-S493.

Kuznetsov, A. M. and Ulstrup, J. (2002) Mechanisms of Molecular Electronic Rectification through Electronic Levels with Strong Vibrational Coupling, *J. Chem. Phys.* **116**, 2149-2165.

Metzger, R. M. (1999) Electrical Rectification by a Molecule: the Advent of Unimolecular Electronic Devices, *Acc. Chem. Res.* **32**, 950-957.

Moresco, F., Meyer, G., and Rieder, K.-H. (2001) Conformational Changes of Single Molecules Induced by Scanning Tunneling Microscopy Manipulation: A Route to Molecular Switching, *Phys. Rev. Lett.* **86**, 672-675.

Nitzan, A. (2001) Electron Transmission through Molecules and Molecular Interfaces, *Annu. Rev. Phys. Chem.* **52**, 681-750.

Nitzan, A. and Ratner, M. A. (2003) Electron Transport in Molecular Wirew Junctions, *Science* **300**, 1384-1389.

Pecchia, A. and Di Carlo, A. (2004) Atomistic Theory of Transport in Organic and Inorganic Nanostructures, *Rep. Prog. Phys.* **67**, 1497-1515.

Petrov, E. G., Tolokh I. S., Demidenko, A. A., and Gorbach, V.V. (1995) Electron–Transfer Properties of Quantum Molecular Wires, *Chem. Phys.* **193**, 237-253.

Petrov, E. G. and Hänggi, P. (2001) Nonlinear Electron Current through a Short Molecular Wire, *Phys. Rev. Lett.* **86**, 2862-2865.

Petrov, E. G., May, V., and Hänggi, P. (2004) Spin–Boson Description of Electron Transmission through a Molecular Wire, *Chem. Phys.* **296**, 251-266.

Petrov, E. G. (2005) Transmission of Electrons through a Linear Molecule: Role of Delocalized and Localized Electronic States in Current Formation, *Low Temp. Phys.* **31**, 338-351.

Petrov, E. G., May, V., and Hänggi, P. (2005) Kinetic Theory for Electron Transmission through a Molecular Wire, *Chem. Phys.* **319**, 380-408.

Reed M. A., Zhou, C., Miller, C. J., Burgin, T. P., and Tour, J. M. (1997) Conductance of a Molecular Junction, *Science* **278**, 252-256.

Segal, D. and Nitzan, A. (2002) Conduction in Molecular Junctions: Inelastic Effects, *Chem. Phys.* **281**, 235-256.

Selzer, Y., Cabassi, M. A., Mayer, T. S., and Allara, D. L. (2004) Temperature Effects on Conduction through a Molecular Junction, *Nanotechnology* **15**, S483-S488.

Tian, W., Datta, S., Hong, S., Reifenberger, R., Henderson, J. I., and Kubiak, C. I. (1988) Conductance Spectra of Molecular Wires, *J. Chem. Phys.* **109**, 2874-2882.

Weber, H. B., Reichert, J., Weigent, F., Ochs, R., Beckmann, D., Mayor, M., Ahlrichs, R., and v. Löhneysen, H. (2002) Electronic Transport through Single Conjugated Molecules, *Chem. Phys.* **281**, 113-125.

Xue, Y., Datta, S., and Ratner, M. A. (2002) First–Principles Based Matrix Green's Function Approach to Molecular Electronic Devices: General Formalism, *Chem. Phys.* **281**, 151-170.

Yeyati, A. L., Martin–Rodero, A., and Flores, F. (1993) Electron Correlation Resonances in the Transport through a Single Quantum Level, *Phys. Rev. Lett.* **71**, 2991-2994.

Zahid, F., Paulsson, V., and Datta, S. (2003) Electrical Conduction through Molecules, In H. Morkos (ed.), *Advanced Semiconductors and Organic Nano-techniques*, Chapter 2, 41, Acad. Press, N.Y.

MAGNETIC UNIPOLAR FEATURES IN CONDUCTIVITY OF POINT CONTACTS BETWEEN NORMAL AND FERROMAGNETIC D-METALS (CO, NI, FE)

Yu. G. Naidyuk (naidyuk@ilt.kharkov.ua), I. K. Yanson, D. L. Bashlakov, V. V. Fisun, O. P. Balkashin, L. Y. Triputen'
B. Verkin Institute for Low Temperature Physics and Engineering, National Academy of Sciences of Ukraine, 47 Lenin Ave., 61103, Kharkiv, Ukraine

A. Konovalenko, V. Korenivski
Nanostructure Physics, Royal Institute of Technology (KTH), Stockholm, Sweden

R. I. Shekhter
Department of Physics, Göteborg University, SE–412 96 Göteborg, Sweden

Abstract. In nanocontacts between normal and ferromagnetic metals (N–F) abrupt changes of the order of 1 % are detected in differential resistance, $dV/dI(V)$, versus bias voltage, V, on achieving of high current densities, 10^9 A/cm^2. These features in $dV/dI(V)$ are observed when the electron flow is directed from the nonmagnetic metal into the ferromagnet and connected with magnetization excitations in the ferromagnet induced by the current. Applying an external magnetic field leads to a shift of the observed features to higher biasing current, confirming the magnetic nature of the effect. Such effects are observed for the non-ballistic (not spectral) regime of current flow in the nanocontacts. Thus, the current induced magneto-conductance effects in multilayered N–F structures (nanopillars) extensively studied in the recent literature have much more general character and can be stimulated by elastic electron scattering at single N–F interfaces.

Key words: Point-contact spectroscopy, Electron-phonon interaction, Magnetotransport phenomena, Spin transfer torque

1. Introduction

Nonlinearity of the current-voltage characteristics (IVC) of point contacts (PC) contains direct information about interaction of the conduction electrons with various quasiparticle excitations in metallic solids (Naidyuk and Yanson, 2005). It turns out that PC spectra (the second derivative of IVC) are proportional to the spectral function of electron-quasiparticle interactions. A

K. Scharnberg and S. Kruchinin (eds.),
Electron Correlation in New Materials and Nanosystems, 59–69.
© 2007 Springer.

necessary condition for obtaining such spectra is a small PC size – the size of the PC d should be smaller than the inelastic electron relaxation length Λ_ε. Additionally, to achieve the necessary spectral resolution the measurement temperature should be much lower than the characteristic energy of the relevant quasiparticle excitations. Studies of PC spectra for both usual (normal) and ferromagnetic metals have revealed features of the electron-phonon interaction (EPI) function (Naidyuk and Yanson, 2005; Khotkevich and Yanson, 1995) and the electron-magnon interaction function (Naidyuk and Yanson, 2005).

In the present paper we investigate new features of non-spectral character in $dV/dI(V)$ and $d^2V/dI^2(V)$ of heterocontacts between normal metals (N=Cu, Ag) and ferromagnets (F=Co, Ni, or Fe). These features are observed most distinctly in an external magnetic field and are incompatible with the spectral EPI features. They originate from disturbances of the magnetic order at the ferromagnetic surface, under the influence of extremely high current density of electrons incident from the normal metal. Similar features were observed earlier in lithographically prepared ferromagnetic nanoconstrictions (Ralph, 1994; Myers et al., 1999) or in PCs to magnetic multilayers (Tsoi et al., 1998; Tsoi et al., 2000). Theory (Slonczewski, 1996; Berger, 1996) connects these features in multilayers with transfer of spin moment between the magnetic layers mediated by polarized current of sufficiently high density.

2. Experimental technique

We have measured IVC's, their first $dV/dI(V)$ and the second $d^2V/dI^2(V)$ derivatives for heterocontacts between normal metals (Cu, Ag) and ferromagnetic Co films or bulk ferromagnetic Co, Ni, Fe. 10 or 100 nm thick Co films were deposited onto 100 nm thick bottom Cu electrodes prepared on oxidized silicon substrates. The top Co surface was contacted mechanically by a Cu or Ag needle at low temperature. External magnetic field was applied both perpendicular and parallel to the film plane. The majority of the experiments were carried out on a polycrystalline bulk Co in the form of a parallelepiped 1x1.5x5 mm, using a sharpened silver wire (0.15 mm in diameter) as the needle. Before the measurements the Co surface was electro-polished in a mix of a hydrochloric acid and alcohol (1:1) at current densities of about 2.5 A/cm^2 and voltage 8 V. The surface of the Co films was not exposed to any processing. All measurements were done at temperature 4.2 K, and contacts were created directly in liquid helium.

Not all contacts showed PC spectra suitable for further study. Apparently, this is due to the presence of a layer of natural oxide or other imperfections on the surface. So, for example, lightly touching the Co films by the Ag

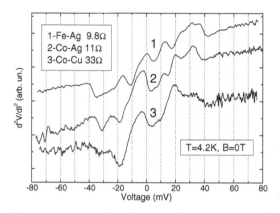

Figure 1. Spectra of PCs between bulk Fe or Co and Ag or Cu with distinct phonon maxima of Co or Fe at $|V| > 15$ mV. The Ag phonon maximum around ± 12 mV is seen in curves 1 and 2. The N-feature at $V=0$ is typical for magnetic impurities (Naidyuk and Yanson, 2005) (likely Fe or Co atoms) in Ag or Cu (so-called Kondo anomaly).

needle could yield contact resistances R_0 of several kΩ, and the elasticity of the thin Ag wire was frequently insufficient to further reduce the PC resistance. In such cases low resistance PCs, $R_0 = 1$–10Ω, were often obtained by switching on/off one of the devices in the measurement circuit, which caused a breakdown of the surface insulating layer. The technique of creating the mechanical contact may also cause significant pressure in the contact area, which can influence the ferromagnetic order at the surface. Nevertheless, we have obtained hundreds of PC spectra, which showed interesting spectral and magnetic features.

3. Spectra of bulk ferromagnets

Figure 1 shows PC spectra for heterocontacts on bulk Fe and Co in zero magnetic field with extrema symmetrically located relative to $V=0$, which correspond to maxima of the EPI function in Fe or Co and Cu or Ag (see for comparison Fig. 5.9 in (Naidyuk and Yanson, 2005)). The intensity of the phonon maxima varies, which is likely connected with an asymmetrical position of the boundary between the metals relative to the narrowest part of the contact. The spectra with prevailing Co EPI maxima at $V=19$ and 33 mV (see two bottom curves in Fig. 1) were observed more frequently. Apparently, such a spectrum corresponds to the N/F boundary being close to the PC center, since the intensity of EPI in Co and Fe are higher than those in noble metals (Naidyuk and Yanson, 2005), and the Fermi velocity is lower. It is known (Naidyuk and Yanson, 2005) that in heterocontacts the intensity of the PC spectrum is inversely proportional to the Fermi velocity. Occasionally spectra with clear contributions from the noble metals were observed (see two upper curves in Fig. 1), likely corresponding to deeper penetration of the

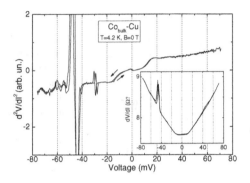

Figure 2. dV/dI (inset) and d^2V/dI^2 (main panel) for PC bulk Co–Cu, R_0=7.5Ω, B=0, T=4.2 K. Compare the phonon features in d^2V/dI^2 at about 20 mV with the maxima in the two bottom curves of Fig. 1. The weak phonon peaks indicate a deviation from the spectral (ballistic) regime of current flow. The maximum in dV/dI at negative bias (inset) corresponds to an N-shaped feature in d^2V/dI^2 and is connected with a magnetic excitation in the ferromagnetic layer (see text). dV/dI and d^2V/dI^2 were recorded by sweeping the current forth and back, which resulted in a slight horizontal shift due to a relatively large time constant of the lock-in amplifier.

respective noble metal into the ferromagnet. Thus using the EPI spectrum it is possible to infer the microstructure of the PCs under study.

Figure 2 shows the PC spectrum of another contact between bulk Co and Cu. A maximum in dV/dI is seen at negative bias, which often observed both in zero external field and in magnetic field. This increase in resistance, appearing in d^2V/dI^2 as N-shaped feature, can reach several percent of the PC resistance. The N-shaped feature is superposed on the PC EPI spectrum with very smoothed phonon maxima, which indicate a reduction of the electron mean free path. The latter conclusion is based on numerous experimental studies (Lysykh et al., 1980; Naidyuk et al., 1984), where a degradation of phonon maxima was observed upon adding impurities in the contact region, such as 3–10% of Ni in Cu (Lysykh et al., 1980) and 1–3% of Be in Ni (Naidyuk et al., 1984). Thus the probability of observing unipolar N-shaped features increases for PCs with degraded phonon maxima. That is a transition from the ballistic to diffusive and even thermal transport regime favors the appearance of these features.

The last is illustrated by Fig. 3, where d^2V/dI^2 are presented for PCs with nearly equal resistance, but their spectra have pronounced or smeared phonon maxima with respect to the background at high bias. This illustrates a very typical behavior we observe, namely that the phonon features are suppressed when the N-shaped features are present. A more detailed discussion

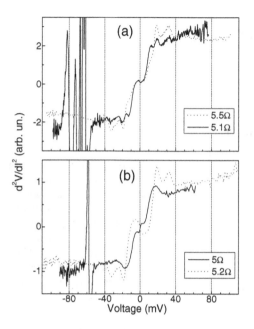

Figure 3. A comparison of the PC spectra for pairs of contacts of nearly equal resistance having different current flow regimes: (a) spectra between bulk Co and Ag; (b) spectra between a 100 nm Co film and Ag (solid line, $B=3T$) or Cu (dotted line, $B=4T$). The threshold current density for the magnetic peaks is about $5 \cdot 10^9$ A/cm^2.

of this behavior and the underlying physics is given elsewhere (Fisun et al., 2004; Yanson et al., 2005). We did not observe spectra containing simultaneously N-shaped features, illustrated in Fig. 2, and pronounced phonon maxima, shown in Fig. 1. One must note, that not all non-ballistic contacts showed unipolar N-shaped peculiarities. This means that the regime of current flow is a pre-condition for the observed phenomena, which we connect (Fisun et al., 2004; Yanson et al., 2005) with spin torque effects. However, the specific micromagnetic structure in the contact area, affected by possible dislocations, impurities, domain boundaries, etc., must play a role.

A study of Ni–Ag heterocontacts (Fig. 4) revealed, that the N-shaped features in the second derivative of the IVC appear with equal probability in both polarities. The magnetic field behavior is similar, i.e. the field shifts the peaks to higher bias. In Fig. 4(b) spectra with N-shaped features for positive bias polarity, corresponding to an electron flow from ferromagnetic metal into N are shown. Our observation is that such magnetic features appear at positive bias for heterocontacts of Co or Fe with Cu or Ag, however with much lower probability (not more that 10% of spectra having N-anomalies). Moreover, their behavior lacked systematics: they appeared suddenly and unexpectedly disappeared under small changes in contact resistance or in an external magnetic field. We also note, that the probability of occurrence of "magnetic" features increased after polishing the polycrystal Ni with a fine-

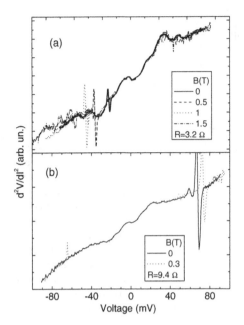

Figure 4. d^2V/dI^2 of Ni–Ag contacts in a magnetic field applied along needle (saturation field for Ni is 0.6 T). The magnetic N-peculiarities are seen both at negative (a) and positive (b) bias. Magnetic field shifts these features to higher voltage (current) in both cases.

grained emery paper. Nevertheless, such processing still allowed observation of Ni phonon maxima [see Fig. 4(a)]. We will not dwell further on the nature of these irregular effects, which are likely connected with the specific microscopic structure of the contact(s).

4.   Spectra of film samples

We first present data for 100 nm thick Co films and argue that PCs in this case should not deviate in behavior from those for a bulk Co sample, since the film is thicker than all characteristic length scales in Co – the exchange and spin diffusion lengths. This is because the exchange length in Co is of the order of several nanometers and the spin-lattice relaxation is several tens of nanometers.

Figure 5 shows d^2V/dI^2 for a contact between a 100-nm Co film and Cu. Well resolved Co phonon peaks indicate that this contact is close to the ballistic regime. The contribution of Cu in the spectrum, estimated from low intensity of 18-mV peak in comparison with a maximum at 30 mV, is relatively weak. A N-shaped anomaly at $V=0$ in d^2V/dI^2 is likely caused by the Kondo effect due to Co atoms penetrating into Cu (Naidyuk and Yanson, 2005). It is interesting to track the behavior of noise at high bias. In Fig. 5 magnetic field of 4 T leads to a reduction of chaotic noise at large negative bias. For more diffusive contacts application of a magnetic field reduces such

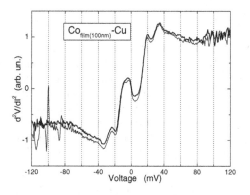

Figure 5. d^2V/dI^2 spectra of PC between a 100 nm Co film and Cu showing phonon maxima in Co and a Kondo-anomaly at $V=0$ in zero external magnetic field (thin line) and at 4 T (thick line). $R_0 = 2.2\,\Omega$.

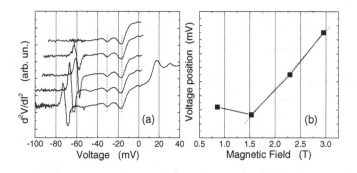

Figure 6. (a) Magnetic field dependence of PC spectrum with N-shaped feature for a Co(10 nm)–Cu heterocontact, $R_0=22\,\Omega$. Magnetic field is: 2.95, 2.29, 1.53, 0.86 and 0 T from bottom. The vertical dashed lines are drawn to mark the positions of the Cu phonon maxima. (b) The position of the N-shaped feature versus magnetic field. Linear approximation of three last points by a dashed line gives a slope $k_I=0.37$ mA/T.

magnetic noise at the same time often causing N-shaped features at negative bias voltages. In other words, a magnetic field higher than the saturation field for Co ($B_s \approx 1.7$ T) can freeze the spins in one direction, reducing the noise connected with spin reorientation in the PC region.

The position of the N-shaped features, found at negative bias for a PC with bulk Co as mentioned above, is shifted by an external magnetic field. Such behavior is clearly seen in the PC spectra of Co(10 nm)–Cu contact (Fig. 6). Here the N-shaped features in d^2V/dI^2 are observed at negative polarity in fields 0.86–3 T. Their position is shifted proportional to the magnitude of the field [Fig. 6(b)]. At low magnetic fields, the position of the N-shaped feature is unaffected and its intensity falls to nearly zero. Because of the small thickness of the Co film, phonon peaks of the the underlying Cu are seen at 18 and

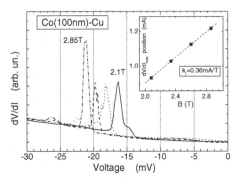

Figure 7. Behavior of a peak in dV/dI for a non-spectral PC Co(100 nm)–Cu, R_0=17.5 Ω, in magnetic field (B=2.85, 2.6, 2.35 and 2.1 T from left to right). Inset: dependence of the peak position versus field. The slope is k_I=0.36 mA/T.

30 mV. As mentioned above, the first phonon peak of Cu at about 18 mV has a larger intensity than the second peak at about 30 mV, whereas for Co the two phonon peaks are usually comparable in magnitude.

In film based contacts, where the PC spectra do not contain phonon peaks or where these are significantly smeared, similar to what is found for PCs to bulk Co, the magnetic peaks in $dV/dI(V)$ appear much more regularly than in contacts with pronounced EPI maxima. For magnetic fields larger than the saturation field, the position of the magnetic peaks on the bias axis is generally linear versus the field [see Fig. 7 and Fig. 6(b)]:

$$I_c \approx I_0 + k_I B \quad or \quad V_c \approx V_0 + k_V B \tag{1}$$

The transformation in (1) from variable I to V neglects a small (10%) nonlinearity of the IVC in the given interval of bias. Unipolar maxima in dV/dI have been observed for PCs in (Tsoi et al., 1998; Ji et al., 2003; Rippard et al., 2003) and explained as due to magnetization excitations in the ferromagnet at the interface with the normal metal, caused by a spin-polarized current of high density (about 10^8 A/cm^2). Features observed by us, predominatingly in the non-spectral regime, apparently have the same origin. We would like to point out, that the absence of clear phonon features in the spectra of contacts with magnetic anomalies indicates a degradation of the Fermi-step in the electron distribution function by at about the value of the excess energy eV, i. e. indicates a current regime close to thermal. The temperature of the lattice, however, is not necessarily in the thermal limit (Naidyuk and Yanson, 2005). Stronger interactions of electrons with magnons (or others nonphononic excitations) can lead to that other sub-system (for example, the spin sub-system) being more intensively warmed up, leading to a loss of the spectral regime of a current flow.

The theory of spin transfer for alternating non-magnetic–ferromagnetic metal structures (Slonczewski, 1996; Berger, 1996) predicts a linear dependence of the critical current I_c on magnetic field, Eq.(1). This critical current

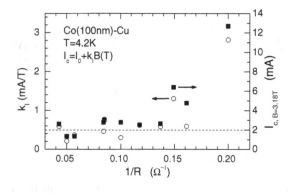

Figure 8. Dependence of k_I (open circles) and critical current (solid squares), calculated according to Eq.(1) at $B=3.18\,\text{T}$, versus inverse contact resistance $1/R$, proportional to contact diameter d^α, where $1<\alpha<2$.

produces a jump-like change in the PC resistance. Thus, the derivative of the critical current with respect to field, dI_c/dB, should weakly depend on the microstructure of the contact and is defined only by its area and attenuation of spin waves in the ferromagnet (Slonczewski, 1996; Berger, 1996). In the ballistic regime the conductivity of a contact $1/R$ is proportional to its area $S = \pi d^2/4$ (where, d is the PC diameter). In the diffusive regime, $1/R = d/\rho$. Even though the resistivity of a metal in a PC, ρ, can considerably differ from that of the respective bulk material, we have attempted to construct the dependence of $dI_c/dB = k_I$ on conductance in Fig. 8. Despite of the significant scatter of points an increase in k_I with increasing d is clearly seen. At small d k_I has a tendency to saturation. $k_I \approx 0.5\,\text{mA/T}$ for the intermediate d range approximately coincides with the values reported by Rippard et al. (Rippard et al., 2003), where the authors studied exchange-coupled Co–Cu multilayer structures and also observed a significant scatter in k_I. The I_c versus $1/R$ dependence is consistent with the above behavior at sufficiently high values of the magnetic field. The black squares in Fig. 8 denote I_c at $B=3.18\,\text{T}$. On the other hand, our attempts to construct a dependence of the critical bias V_c and the corresponding factor k_V on $1/R$ were unsuccessful. It means that the determining parameter in the phenomenon of the magnetic order excitation by a spin-polarized electron flow is the current rather than the voltage drop on the contact.

A step like change in the resistance of N–F PCs is observed not only under action of a high current density, but also under an application of an external magnetic field at a constant bias. In Fig. 9 $dV/dI(V)$ exhibits a magnetic peak at a negative bias of -22 mV. When the bias is fixed at -18 mV and the field is swept in sequence $3.3 \rightarrow 0.75 \rightarrow 3.3\,\text{T}$ (Fig. 9, inset) a peak is seen in $dV/dI(H)$ at $B=2.2\,\text{T}$ (Fig. 9, inset). The likely origin of this anomaly is that the spins in the contact area form a single domain at high fields which does

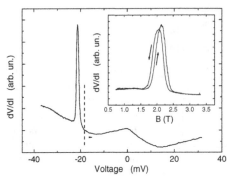

Figure 9. A peak in $dV/dI(V)$ at $B=2.96\,$T (main panel) transformed into a peak in $dV/dI(B)$ (inset) at bias $V=-18\,$mV ($I=2.3\,$mA, marked by a vertical dashed line) for a contact between a Co($10\,$nm) film and Cu, $R = 7.81\,\Omega$.

not contribute additional interface resistance. At low fields, the current from the non-magnetic metal causes a non-uniform spin distribution in the contact, which results in an additional domain wall-like resistance.

5. Conclusions

We have analyzed PC spectra of heterocontacts between non-magnetic (normal) metals and ferromagnets where, along with the maxima caused by the electron-phonon interaction, non-spectral features were observed. These features are not connected with the spectral regime of current flow through the contact, where the applied voltage determines the energy of the excitation processes. These non-spectral features have a threshold character. They are observed at some critical values of the current through the contacts and are connected with excitations of the magnetic subsystem by a current of high density. Such magnetic excitations are observed in diffusive or even more likely in thermal contacts, where phonon maxima are smeared or are absent due to the electrons experiencing strong impurity scattering. These results demonstrate the importance of the regime of electron flow through a PC for the observed phenomena, namely, that the impurity scattering is found to be at the origin of the new mechanism of single interface spin torque effects. Thus spin torque effects, earlier observed in more complex systems like F–N–F nanopillars and multilayered magnetic structures, are possible for single N–F interfaces provided the transport regime is non-ballistic.

Future research on magnetic point contacts will be directed towards further reduction of the contact size down to dimensions comparable with the ferromagnetic exchange length. PC spectroscopy can be expected to provide additional information on the nature of the spin torque effects in nanometer size magnetic structure, that are beyond the fabrication limits for lithographic techniques.

Acknowledgements

The work in Ukraine was partially supported by the National Academy of Sciences within the Program "Nanosystems, nanomaterials and nanotechnologies".

References

Berger, L. (1996) Emission of spin waves by a magnetic multilayer transversed by a current, *Phys. Rev. B* **54**, 9353–9358.

Fisun, V. V., Balkashin, O. P., Naidyuk, Y. G., Bashlakov, D. L., Yanson, I. K., Korenivski, V., and Shekhter, R. I. (2004) Peculiarities in nonlinear electroconductivity of nanocontacts based on ferromagnetic metals Co and Fe, *Metallofiz. Noveishie Technol.* **26**, 1439–1446.

Ji, Y., Chien, C. L., and Stiles, M. D. (2003) Current-Induced Spin-Wave Excitations in a Single Ferromagnetic Layer, *Phys. Rev. Lett.* **90**, 106601-1–106601-4.

Khotkevich, A. V. and Yanson, I. K. (1995) *Atlas of Point-Contact Spectra of Electron-Phonon Interactions in Metals*, Kluwer Academic Publishers.

Lysykh, A. A., Yanson, I. K., Shklyarevski, O. I., and Naydyuk, Y. G. (1980) Point-contact spectroscopy of electron-phonon interaction in alloys, *Solid State Commun.* **35**, 987–989.

Myers, E. B., Ralph, D. C., Katine, J. A., Louie, R. N., and Buhrman, R. A. (1999) Current-Induced Switching of Domains in Magnetic Multilayer Devices, *Science* **285**, 867–870.

Naidyuk, Y. G. and Yanson, I. K. (2005) *Point-Contact Spectroscopy*, Springer Science+Business Media Inc.

Naidyuk, Y. G., Yanson, I. K., Lysykh, A. A., and Shitikov, Y. L. (1984) Microcontact spectroscopy of Ni-Be alpha-solution, *Fizika Tverdogo Tela* **26**, 2734–2738.

Ralph, D. C. (1994) Microscopy with nanoconstrictions studies of electron scattering from defects and impurities within metals, Ph.D. thesis, Materials Science Center, Cornell University.

Rippard, W. H., Pufall, M. R., and Silva, T. J. (2003) Quantitative studies of spin-momentum-transfer-induced excitations in Co/Cu multilayer films using point-contact spectroscopy, *Appl. Phys. Lett.* **82**, 1260–1262.

Slonczewski, J. C. (1996) Current-driven excitation of magnetic multilayers, *J. Magn. Magn. Mater.* **159**, L1–L7.

Tsoi, M., Jansen, A. G. M., Bass, J., Chiang, W.-C., Seck, M., Tsoi, V., and Wyder, P. (1998) Excitation of a Magnetic Multilayer by an Electric Current, *Phys. Rev. Lett.* **80**, 4281–4284.

Tsoi, M., Jansen, A. G. M., Bass, J., Chiang, W.-C., Tsoi, V., and Wyder, P. (2000) Generation and detection of phase-coherent current-driven magnons in magnetic multilayers, *Nature* **406**, 46–49.

Yanson, I. K., Naidyuk, Y. G., Bashlakov, D. L., Fisun, V. V., Balkashin, O. P., Korenivski, V., Konovalenko, A., and Shekhter, R. I. (2005) Spectroscopy of Phonons and Spin Torques in Magnetic Point Contacts, *Phys. Rev. Lett.* **95**, 186602-1–186602-4.

PART II

Superconductivity

II.1 Magnesium diboride and the two-band scenario

SUPERCONDUCTIVITY IN MAGNESIUM DIBORIDE AND ITS RELATED MATERIALS

Satoshi Akutagawa, Takahiro Muranaka and Jun Akimitsu[*]
Department of Physics and Mathematics, Aoyama-Gakuin University, Fuchinobe 5-10-1, Sagamihara, Kanagawa 229-8558, Japan

Abstract. After the discovery of MgB_2, much attention has been paid to the boride and carbide systems. In this paper, we mainly review the following subjects. First, we review the present situation of the superconductivity in MgB_2, which can be interpreted in terms of BCS-type phonon-mediated pairing. In particular, we focus on the relationship between the superconductivity and the electronic structure in MgB_2 and two gap feature. Second, we review our present status for the new superconductors including the carbon systems. Recently, we have reported a relatively high-T_c superconductivity in Y_2C_3 at 18 K whose T_c could be changed by synthesis conditions from 10 to 18 K [J. Phys. Soc. Jpn. 73 (2004) 530.]. We synthesized a high-purity sample of the medium-T_c phase (T_c=13.9 K) in Y_2C_3 and examined its physical properties. From a specific heat measurement, $\square C(T_c)/\gamma T_c$ value is calculated to be 3.3 and the superconducting gap is estimated to be $2\square/k_B T_c = 4.5$, indicating that the superconductivity in Y_2C_3 can be described by s-wave strong coupling regime. From specific heat in various magnetic fields, the upper critical field $H_{c2}(0)$ is estimated to be 24.7 T.

Key words: Superconductivity; MgB_2; Y_2C_3

1. Introduction

MgB_2 has opened a new frontier into the physical properties of intermetallic superconductors. In the short period since the discovery of its high-T_c superconductivity[1], a large number of experimental and theoretical works has been performed[2,3,4], and the interpretation of this superconductivity focusses on the metallic nature of the 2D layer formed by B atoms. This discovery also stimulated the search for other high-T_c materials in similar systems. MgB_2 is a particularly fascinating material because of its multiple superconducting gaps which are caused by the characteristic electronic structure derived from 2D σ-bands and 3D π-bands. Although multiple-gap superconductivity had been proposed by Suhl et al. [5] and since then discussed in relation to other materials, for example, Nb-doped $SrTiO_3$ by Binnig et al. [6], MgB_2 is the first material containing intrinsic multiple gaps, and many characteristic features of

---

[*]Corresponding author: Jun Akimitsu; e-mail: jun@phys.aoyama.ac.jp

K. Scharnberg and S. Kruchinin (eds.),
Electron Correlation in New Materials and Nanosystems, 73–92.
© 2007 *Springer.*

multiple-gap superconductivity have been investigated from various experimental and theoretical aspects.

Because the limit of T_c in metallic superconductors had been believed to be about 30 K in the framework of the BCS theory, the discovery of unexpectedly-high-T_c superconductivity in this simple binary intermetallic compound has triggered an enormous interests in the world. The enthusiasm caused by the dis-covery of superconductivity in MgB_2 has led to a search for a new high-T_c material in a similar system containing light elements B and C. Thus, we concentrated our search for a new materials with higher-T_c's on transition metal borides and carbides, as well as lanthanide carbides. Recently, we discovered new superconductors in related compounds and discussed their superconducting properties, Re-B system (Re_3B T_c=4.8 K, Re_7B_3 T_c=3.3 K) [7], $W_7Re_{13}X$ (X=B and C) (T_c=7.1 K and 7.3 K, respectively) [8] and Y_2C_3 (T_c=18 K) [9].

In this paper, we mainly review the present situation of the superconductivity in MgB_2, which can be interpreted in terms of BCS-type phonon-mediated pairing. In particular, we focus on the relationship between the superconductivity and the electronic structure in MgB_2 and two gap feature. Moreover, we review our present status for the new superconductor including the carbon systems, Y_2C_3.

2. Superconductivity in MgB_2

Soon after the discovery of MgB_2, it was basically classified as a conventional phonon-mediated BCS superconductor [2,10-12]. However, deviations from the BCS framework have been inferred from several experiments, in particular specific heat, photoemission spectroscopy (PES) and tunneling spectroscopy, using polycrystalline samples or single crystals [13-21]. The striking feature is the direct observation of two kinds of superconducting gaps in MgB_2. Theoretically, the two-band model was proposed to be appropriate for MgB_2 because the system consists of two qualitatively different charge carriers derived from 3D π and 2D σ bands [22]. Holes in σ bands are strongly coupled to phonons confined within the boron honeycomb plane. The large gap $\Box_\sigma$ opens in σ bands, whereas a relatively small gap $\Box_\pi$ opens in π bands due to the difference in the strength of electron-phonon (EP) coupling in each band. This scenario for superconductivity in MgB_2 is supported by an increasing number of experimental results. The concept of two-band superconductivity was first suggested by Suhl *et al.* [5] soon after the advocation of the BCS theory, and has been discussed, since then, in relation to other superconductors, including Nb, Ta [23], doped $SrTiO_3$ [6] and borocarbide [24]. Several experimental reports presented clear signatures for two-gap superconductivity and offered the opportunity for comparison with theoretical predictions. Thus, MgB$_2$ is the first example of this model and opens up the possibility of interesting new phenomena.

2.1. ELECTRONIC STRUCTURE OF MgB$_2$

Despite its crystal structure being similar to that of a graphite intercalated compounds, MgB$_2$ has a qualitatively different and uncommon structure of the conducting states. The band structure has been calculated in several studies since the discovery of superconductivity [25-28] (see Fig. 1). The band structure of MgB$_2$ is similar to that of graphite, and is formed by three bonding σ bands (in-plane sp_xp_y hybridization) and two π bands (bonding and antibonding; p_z hybridization). MgB$_2$ has two imperfectly filled σ bands, and these σ bands correspond to the sp^2-hybrid bonding within the 2D honeycomb layer. The holes of σ bands along the ΓA line localized within the 2D boron layer manifest 2D properties. In contrast, the electrons and holes in 3D π bands are delocalized. The 2D σ bands and 3D π bands at E_F contribute equally to the total density of states, while 2D σ bands have a strong interaction with longitudinal vibrations within the 2D boron layer. Moreover, because k_z dispersion of σ bands is weak, two cylindrical sheets appear around the ΓA line (see Fig. 2). On the other hand, π bands form two tubular networks: an antibonding electron-type and a bonding hole-type sheet. These two sheets touch at one point on the KH line. Belashchenko *et al.* examined the relationship between the band structures of MgB$_2$ and graphite by comparing the following lattices; primitive graphite lattice with no displacement between layers, as in MgB$_2$,

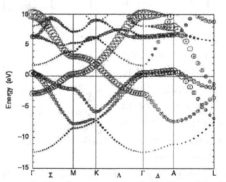

Figure 1. Band structure of MgB$_2$ with Bp character [24].

using graphite lattice parameters; boron in the primitive graphite lattice with a as in MgB$_2$, and c/a as in graphite (see Fig. 3) [28]. The band structure of the primitive graphite lattice with no dis-placement between layers is similar to that of graphite, and that of boron in the primitive graphite lattice shows a natural enhance-ment of the out-of-plane dispersion of the π bands when the interlayer distance is reduced. The distinct difference between MgB_2 and the primitive graphite lattice is the position of σ bands to E_F at the Γ point.

An investigation of the charge density distribution would give a better understanding of how the supercon-ductivity is related to the electronic and crystal structures of MgB_2. Precise x-ray structure analyses by Nishibori *et al.* (in a polycrystalline sample) [29], Lee *et al.* (in a single crystal) [30] and Mori *et al.* (in a single crystal; $Mg_{1-x}B_2$ x=0.045) [31] yielded accurate charge densities in MgB_2. The charge density obtained at room temperature revealed a strong B-B covalent bonding feature. On the other hand, there was no bond electron between Mg and B atoms, and Mg atoms were

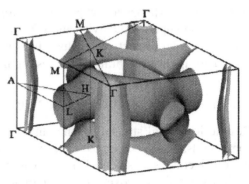

Figuer 2. Fermi surface of MgB_2 [24]. Two cylinders along ΓAΓ (hole-like) come from the bonding $p_{x,y}$ bands, the tubular network along MKM (hole-like) from the bonding p_z bands, and the tubular network along LHL (electron-like) from the antibonding p_z band.

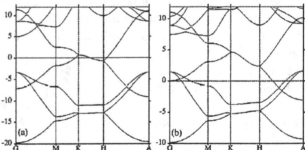

Figure 3. Band structures of (a) primitive (*AA* stacking) graphite (PG), a=2.456 Å, c/a=1.363; (b) PG boron, a=3.085 Å, c/a=1.142 as in MgB_2 [27].

found to be fully ionized and in the divalent state. Nishibori *et al.* also reported that these characteristic density features were preserved in the charge density obtained at 15 K and were consistent with the calculated band structures indicating a two-band model [26]. Moreover, Nishibori *et al.* examined the valence of the atom by accumulating the number of electrons around a certain atom in the MEM density (see Fig. 4). The numbers of electrons at room temperature and 15 K were estimated to be 10.0(1)e and 10.0(1)e around the Mg atom and 9.9(1)e and 10.9(1)e around the boron 2D sheets, respectively. The values for Mg atoms are very close to the number of electrons around Mg^{2+} ions, so Mg atoms are fully ionized in the MgB_2 crystal at whole temperatures. On the other hand, the total numbers of electrons around boron 2D sheets show significant difference, which can be attributed to the valence of the whole boron 2D sheet changing from neutral to monovalent at 15 K. These results suggest that the electrons transfer from π bands (p_z orbitals) to in-plane σ bands (p_{xy} orbitals) at 15 K. However, these results do not agree with the valence electron distribution at room temperature determined by Wu *et al.* using synchrotron x-ray and electron diffraction techniques [32]. They reported that two electrons from each Mg atom moved to the B plane. Therefore, the boron layer had the same number of valence electrons at whole temperatures, and these electrons were mainly located in the p_x p_y orbitals between neighboring boron atoms. This disagreement is not yet resolved.

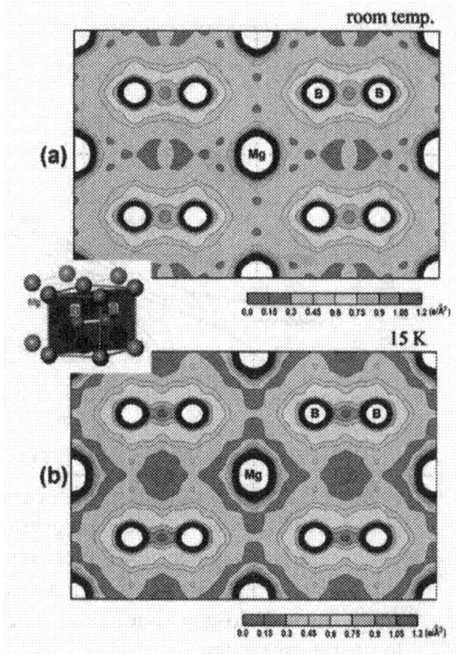

Figure 4. The (110) sections of the MEM charge density of MgB$_2$ at room temperature (a) and 15K (b) [28].

2.2. SPECIFIC HEAT OF MgB$_2$

In an early stage, specific heat measurements of a polycrystalline sample provided direct experimental evidence of the presence of a second gap in MgB$_2$ [13-15,33,34]. Several experimental results on higher quality polycrystalline samples and single crystals were subsequently reported [19,35,36]. Bouquet *et al.* reported the temperature dependence of specific heat, $C(T)$, under magnetic fields [13-15,36]. In a zero field, deviations from the BCS curve were observed in the amplitude and sharpness of the specific heat jump, and the large excess weight near $0.2T_c$ was a robust feature (see Fig. 5)

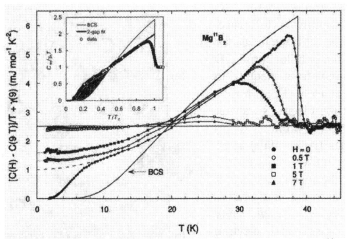

Figure 5. Temperature dependence of specific heat in $Mg^{11}B_2$ [C(H)-C(9T)]/T at various magnetic fields [12]. Inset shows temperature dependence of electronic part of specific heat in superconducting state C_{es}/T normalized to γ_n [35]. The thick line is fitted to a two-gap line, and hatched area marks the low-temperature excess.

Similar features were observed by other groups. Yang *et al.* and Bouquet *et al.* noted that the exponential decrease of $C(T)$ at low temperature is representative of a small gap, ~ 1-$1.5k_BT_c$, whereas Kremer *et al.* reported a larger gap of $\sim 4k_BT_c$ and strong coupling from specific heat jump [33-36]. These unusual behaviors were consequences of the anisotropic band structure of MgB_2, in which two main bands cross the Fermi level. One consists of electrons and holes in the 3D π band and represents $\sim 56\%$ of the total DOS. The other consists of holes in the 2D σ band and represents $\sim 44\%$ of the total DOS. These different types of carriers induce two gaps at $T = 0$, $\square_{0\pi}\sim 2$ meV and $\square_{0\sigma}\sim 7$ meV; both gaps disappear at the same T_c [22,25,37]. The specific heat data were analyzed using the two-gap model ("α model" [38]) with the gap values $\square_{0\pi}$ and $\square_{0\sigma}$, and relative weights of the Sommerfeld constant in each band, $\gamma_{\pi n}/\gamma_n$ and $\gamma_{\sigma n}/\gamma_n$ (see inset of Fig. 5). The superconducting state entropy is described by the following equation in the two-gap superconductor model:

$$S_{es} = -2k_B \sum \sum [f_k \ln f_k + (1 - f_k)\ln(1 - f_k)],$$
$$f_k = [1 + \exp(E_k/k_BT)]^{-1},$$

where the double sum is over the quasiparticle states in both bands, and E_k is the quasiparticle energy, which is different in the two bands.

$E_{k1}^2 = \varepsilon_{k1}^2 + \square_1^2$ and $E_{k2}^2 = \varepsilon_{k2}^2 + \square_2^2$, where ε_k is the normal state quasiparticle energy, and $\square_1(T)$ and $\square_2(T)$ are gap functions. The temperature dependence of each gap follows the weak-coupling BCS behavior, but the amplitude is scaled by the parameter $\alpha = \square(0)/k_B T_c$: $\square_1(T) = (\alpha_1/\alpha_{BCS})\square_{BCS}(T)$ and $\square_2(T) = (\alpha_2/\alpha_{BCS})\square_{BCS}(T)$, where $\alpha_{BCS} = 1.764$, α_1 and α_2 are free parameters, and the reduced gaps, $\square_1(T)/\square_1(0) = \square_2(T)/\square_2(0) = \delta(T)$, are assigned numerical values tabulated by Mühlshlegel [39]. Thus, $S_{es}(t)/\gamma_n T_c$ in Ref. 38 becomes

$$S_{es}/\gamma_n T_c = -(3/\pi^2)\alpha_1(\gamma_1/\gamma_n) \quad f_{x1}\ln f_{x1} + (1-f_{x1})\ln(1-f_{x1}) \quad dx$$
$$-(3/\pi^2)\alpha_2(\gamma_2/\gamma_n) \quad f_{x2}\ln f_{x2} + (1-f_{x2})\ln(1-f_{x2}) \quad dx,$$

where γ_1 and γ_2 represent the electron DOS in the two bands, and have the relationship

$$\gamma_1 + \gamma_2 = \gamma_n,$$
$$f_{x1} = \left\{1 + \exp\left[\alpha_1 t^{-1}(x^2 + \delta^2)^{1/2}\right]\right\}^{-1}$$
$$f_{x2} = \left\{1 + \exp\left[\alpha_2 t^{-1}(x^2 + \delta^2)^{1/2}\right]\right\}^{-1}.$$

Here $t \equiv T/T_c$, and the differential of $S_{es}(t)/\gamma_n T_c$ gives C_{es}. For a two-gap superconductor, C_{es} is the sum of two curves, one scaled by γ_1/γ_n and the other by γ_2/γ_n, and is obtained numerically. These parameters were estimated to be $\gamma_{\pi n}:\gamma_{\sigma n} \approx 45:55$-$50:50$, $2\square_{0\pi}/k_B T_c \approx 1.2$-$1.3$ and $2\square_{0\sigma}/k_B T_c \approx 3.8$-$4.3$ [14], which show in good agreement with band calculations [22] and other spectroscopic measurements [17,18,40]. Bouquet et al. also determined the Sommerfeld constant, γ_n, to be 2.6 mJ/mol·K^2, the Debye temperature, Θ, to be 1050 K, $[H_c(0)]^2/\gamma_n T_c^2$ to be 5.46 and $\square C(T_c)/\gamma_n T_c$ to be 1.32. From these results, they concluded that gap values $[H_c(0)]^2/\gamma_n T_c^2$ and $\square C(T_c)/\gamma_n T_c$ which are smaller than the BCS ones (5.95 and 1.43, respectively) can be ascribed to gap anisotropy or a multigap structure. In addition, they reported the single crystal results under magnetic fields up

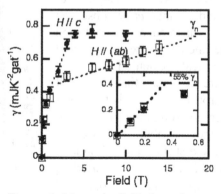

Figure 6. Magnetic field dependence of Sommerfeld constant $\gamma(H)$ for $H//c$ and $H//ab$ [18]. Inset shows expanded view of the low-field region.

to 14 T (H//ab and H//c). In the region of H<0.5 T, no anisotropy of the Sommerfeld constant Γ (γ_{ab}/γ_c) was distinguished for the two directions. On the other hand, in the region of H>0.5 T, Γ rapidly increased with increasing H (see Fig. 6). They reported that Γ increased toward ~5. Similarly, Angst et al. reported that Γ increased linearly from ~2 under zero field up to 3.7 under 10 T [41,42].

2.3. PHOTOEMISSION SPECTROSCOPY (PES) OF MgB$_2$

In an early stage after the discovery of MgB$_2$, the superconducting gap size □ of 2-5.9 meV ($2\square/k_BT_c$=1.2-3.5) was reported, as determined by spectroscopic measurements assuming an isotropic s-wave symmetry gap [11,12,43]. On the other hand, later experimental reports showed clear evidence for two-gap superconductivity. Theoretically, in MgB$_2$, superconducting gaps with significantly different magnitudes were expected to open in the σ and π bands, as a result of the strong k dependence of EP coupling [22,37]. Because angle-resolved photoemission spectroscopy (ARPES) can observe the k-dependence of the superconducting gap, this technique is a powerful probe for measuring the superconducting gap in both σ and π bands in MgB$_2$. ARPES of a high-density polycrystalline sample [17] and a single crystal [20,21,44] yielded experimental evidence of two-gap superconductivity and indicated the band structure of MgB$_2$ along ΓKM (AHL), by Uchiyama et al. and Souma et al. (see Fig. 7), and along ΓΣM by Tsuda et al. The observed band structure showed considerable agreement with the results of band calculations [25,26]. However, Uchiyama et al. and Souma et al. observed a difference between experimental results and theoretical calculations. The difference was that a small electron-like pocket was observed around the Γ(A) point, and they ascribed the observed state to a surface band. Tsuda et al. and Souma et al. reported □$_\pi$ and □$_\sigma$ to be 2.2±0.4 and 5.5±0.4 meV [20] and 1.5±0.5 and 6.5±0.5 meV [21] with s-wave symmetry. In the surface band observed by Souma et al.,

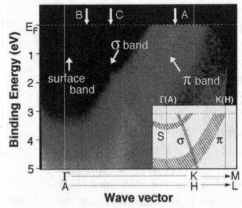

Figure 7. Band structure near E_F along ΓKM (AHL) at 45 K obtained by ARPES [20].

the gap size was close to □$_\sigma$ (6.0±0.5 meV). Tsuda et al. reported that both gaps Δ_π and Δ_σ closed at T_c (see Fig. 8). These behaviours indicate the existence of strong interband pairing interaction in MgB$_2$.

2.4. TUNNELING SPECTROSCOPY OF MgB$_2$

Tunnelling spectroscopy traditionally is one of the most sensitive probes for observing the superconducting gap structure because the tunnelling conductance (dI/dV) is equivalent to DOS near E_F [45]. Scanning tunnelling microscopy (STM), point contacts, planar junctions and break junctions are widely used in tunnelling spec-

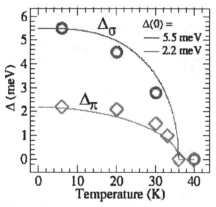

Figure 8. Temperature dependence of superconducting gap $\Delta(T)$ [19]. Open circles and diamonds show the gap values on the σ and π bands.

troscopy. At a finite temperature, the tunnelling conductance in the S-I-N junction can be obtained from

$$dI/dV = -\sigma_N \int N(E)\partial f(E + eV)/\partial eV \, dE ,$$

where $N(E)$ is the quasiparticle DOS and σ_N is the normal state conductance of the junction. To account for the experimentally observed broadening due to thermal smearing, the smearing parameter Γ is usually used in the BCS DOS [46].

$$N(E) = \mathrm{Re}\left| (|E| - i\Gamma) \Big/ \sqrt{(|E| - i\Gamma)^2 - \Delta^2} \right|$$

To analyze the tunneling conductance of MgB$_2$ under the assumption of two-band superconductivity, an analytical model requires the solutions of two simultaneous equations for two gaps:

$$\Delta_1(E) = \frac{\Delta_1^0 + \Gamma_1\Delta_2(E)\Big/\sqrt{\Delta_2^2(E) - (E - i\Gamma_2^*)^2}}{1 + \Gamma_1\Big/\sqrt{\Delta_2^2(E) - (E - i\Gamma_2^*)^2}}$$

$$\Delta_2(E) = \frac{\Delta_2^0 + \Gamma_2\Delta_1(E)\Big/\sqrt{\Delta_1^2(E) - (E - i\Gamma_1^*)^2}}{1 + \Gamma_2\Big/\sqrt{\Delta_1^2(E) - (E - i\Gamma_1^*)^2}} ,$$

where $\Delta_{1,2}^0$ are the intrinsic pairing amplitudes in the two bands, $\Gamma_{1,2}$ are scattering rates related inversely to the times spent in each band prior to

Table 1. Energy gaps of MgB_2 determined with different tunneling spectroscopy methods (from [46]).

	Energy gap (meV)			Reference
STM		$\Delta=5$-7		[47]
	$\Delta_\pi=2.0$			[48]
	$\Delta_\pi=2.3$		$\Delta_\sigma=7.1$	[10,49]
	$\Delta_\pi=3.5$		$\Delta_\sigma=7.5$	[50]
	$\Delta_\pi=2.2$		$\Delta_\sigma=6.9$	[51,52]
Point	$\Delta_\pi=2.6$			[53]
contact	$\Delta_\pi=1.7$		$\Delta_\sigma=7$	[54]
	$\Delta_\pi=2.8$		$\Delta_\sigma=7.1$	[17,55]
	$\Delta_\pi=2.8$		$\Delta_\sigma=9.8$	[56,57]
	$\Delta_\pi=2.8$		$\Delta_\sigma=7.0$	[15]
	$\Delta_\pi=2$-3		$\Delta_\sigma=6$-8	[58]
	$\Delta_\pi=2.3$		$\Delta_\sigma=6.2$	[59]
		$\Delta=3$-4		[60]
Break	$\Delta_\pi=2.5$		$\Delta_\sigma=7.6$	[46,61,63]
junction	$\Delta_\pi=1.7$-2			[62]
	$\Delta_\pi=2$-2.5		$\Delta_\sigma=8.5$-10	[64,65]

scattering to the other, and $\Gamma^*_{1,2}$ are smearing parameters in the two bands, which are added to account for lifetime effects.

Although many tunnelling results for MgB_2 have been reported, the gap values are scattered in the range of $\Delta = 1.7$-10, which are not consistent with each other (see Table 1). This is probably due to the sensitivity of the sample surface state. However, these observed values have been categorized into two groups, Δ_π and Δ_σ [47], which had already been predicted theoretically [22,37]. Many groups reported two distinct gap structures in the STM spectrum [11,48-53]. With increasing temperature, these two distinct gap structures disappeared at T_c of the bulk MgB_2. In particular, Iavarone et al. reported that the tempe-rature dependence of both gaps followed the BCS theory (see Fig. 9). This result confirmed the importance of Ferm-surface-sheet-dependent super-conductivity in MgB_2 that was proposed in the multigap model [22]. Results obtained with point contacts were also reported by many groups for MgB_2-Cu [16], -In [18,54],

-Pt [18,55,56], -Nb [57,58] and -Au [18,56,59-61] junctions. They also reported evidence for two distinct superconducting energy gaps, and the temperature dependences of both gaps were in good agreement with the BCS theory. Schmidt *et al.*, Gonnelli *et al.* and Ekino *et al.* reported the tunneling spectrum obtained by the break junction method [47,56,62-66], and their results were in basic agreement

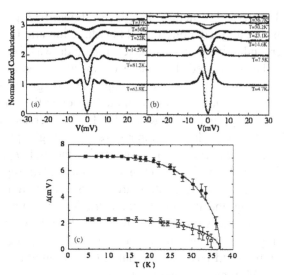

Figure 9. Tunneling spectrum with theoretical curves **(a)**, **(b)** and temperature dependence of the gap values **(c)** [49]. Solid lines show BCS curves.

with other results of STM, point contact and the break junction. However, the observed gaps revealed parallel contributions to the conductance from both bands because of the existence of the S-I-S junction. A subtle feature was observed near the $\square_\pi + \square_\sigma$ region, which was reminiscent of strong coupling effects. This feature provided important insight into the nature of two-gap superconductivity in MgB₂.

3. Superconductivity in Y₂C₃

The rare earth carbide system has been attracting renewed interests since the discovery of superconductivity in the layered yttrium carbide halides $Y_2C_2I_2$ (T_c = 9.97 K) and $Y_2C_2Br_2$ (T_c = 5.04 K) [67]. Binary and quasi-binary yttrium carbides (in particular Th-doped yttrium carbide ($Y_{1-x}Th_x)C_2$ (T_c = 17.0 K)) [68] have also attracted considerable attention since their T_c values are close to those of A15 compounds [69,70]. Recently, our group has reported the discovery of superconductivity at 18 K in Y_2C_3 [9]. In a previous study, Novokshonov showed that Y_2C_3 crystallizes in a body-centered cubic (bcc) Pu_2C_3-type structure and $I\bar{4}3d$ space group with eight formula units per unit cell by X-ray diffraction investigation [71].

The crystal structure of Y_2C_3 is shown in Fig. 10. In this structure, Y atoms are aligned along the <111> direction and C atoms form dimers. The superconductivity of yttrium sesquicarbide was found by the Krupka group [72]. They claimed that phase composition has a significant effect on T_c over the entire homogeneity range. The magnetic susceptibility indicated that the material was superconducting over the entire homogeneity range with T_c ranging from 6 K to 11.5 K. To increase T_c, we synthesized this material under high temperature and high pressure conditions, and consequently succeeded in synthesizing the high-T_c phase of Y_2C_3.

From a theoretical point of view, Shein *et al.* reported the band structure of Y_2C_3 [73]. The band structure shows that the hybridization of C-C dimer antibonding and the Y-4d characteristics are dominant at the Fermi level. The electronic structure of other metals with C dimers both at the tight binding [74] and density functional levels [75] suggests that the electronic structure could have substantial C-C antibonding character near the Fermi level. Moreover, Singh et al. reported the magnitude of the electron-phonon coupling in Y_2C_3 from density functional calcu-lations of the electronic structure using the general potential linea-rized augmented plane wave (LAPW) method in their recent report [76]. They suggest that low-frequency metal atom vibr-ations have the largest electron-phonon coupling whereas the contribution of the high-frequency C-C stretching vibrations is found to be comparatively small. It is interesting to have a deep insight into the mechanism of super-

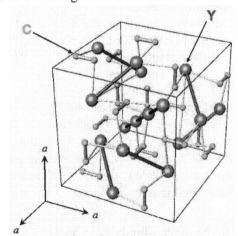

Figure 10. Crystal structure of Y_2C_3. The dark gray circles indicate Y atoms and the light gray circles indicate C atoms.

conductivity in Y_2C_3. From an experimental point of view, Nakane *et al.* examined the reproducibility of superconductivity in Y_2C_3 and succeeded in transport measurements after our report [77]. Because the upper critical field (H_{c2}) of this material was estimated to be approximately 30 T from their report and its T_c relatively high among simple binary intermetallic compounds, Y_2C_3 has attracted considerable attention. However, there still remains much to be understood on the mechanism of superconductivity because of the difficulty to synthesize pure samples and to stabilize the Y_2C_3 phase in air. We established a technique for synthesizing a high-purity sample. In this paper, we report on the physical properties of the medium-T_c phase (13.9 K) of Y$_2$C$_3$.

3.1. EXPERIMENTAL DETAILS

The samples were prepared by mixing appropriate amounts of Y (99.9%) and C (99.95%) powders in a dry box and synthesized by arc melting in high purity-Ar gas. Binary alloys were, in the first stage, prepared by arc melting, after which they were subjected to high-pressure and high-temperature treatment to produce the sesquicarbide phase. All subsequent sample manipulations, including preparations for high-pressure experiments, were performed in a dry He atmosphere. The melted samples were heated to 1673 – 1873 K for five min. and maintained for 10 - 30 min. under a pressure of 4 - 5.5 GPa using a cubic-anvil-type high-pressure apparatus and quenched to room temperature within a few seconds. The samples were examined by powder X-ray diffraction analysis, using a conventional X-ray spectrometer with a graphite monochromator (RINT-1100 RIGAKU). Intensity data were collected with CuKα radiation over a 2θ range from 5 to 80 degrees at 0.02 degree step widths. The magnetic susceptibility and magnetization measurements were performed with a SQUID magnetometer MPMSR2 (Quantum Design Co., Ltd.) and a PPMS system (Quantum Design Co., Ltd.). The specific heat measurement was performed using the PPMS system (Quantum Design Co., Ltd.) in the temperature range between 2 K and 30 K.

3.2. EXPERIMENTAL RESULTS OF Y$_2$C$_3$

Figure 11 shows the powder X-ray diffraction pattern at room temperature of Y$_2$C$_3$. The Y$_2$C$_3$ phase was obtained as the main phase and indexed as a cubic unit cell with the space group $I\bar{4}3d$. The lattice constant calculated from all indices is $a = 8.198$ Å. Weak additional reflections were observed, which are attributed to an impurity phase of Y$_2$O$_x$C$_y$, whose overall fraction is less than 6%. The superconductivity of Y$_2$C$_3$ was confirmed by DC susceptibility in a magnetic field of 10 Oe, as shown in Fig. 12. The magnetic susceptibility of Y$_2$C$_3$ significantly decreases at 13.9 K, suggesting the occurrence of superconductivity. The volume fraction of superconductivity at 1.8 K is estimated to be approximately 27% in the field cooling process. Fig. 13 shows magnetization vs. magnetic field (*M-H*) curves ($T_c = 15$ K), exhibiting the behavior of a typical type-II superconductor. From the M-H curves, we estimated the lower critical field, $H_{c1}(T)$ defined as a magnetic field where the initial slope meets the extrapolation curve of $(M_{up}+M_{down})/2$. The roughly estimated $H_{c1}(0)$ of Y_2C_3 is 3.2 mT. The penetration depth $\lambda(0)$ is calculated to be approximately 4400 Å from the relationship between $H_{c1}(0)$ and $\lambda(0)$, $\mu_0 H_{c1} \sim \Phi_0/\pi\lambda^2$, where μ_0 and Φ_0 are the magnetic permeability in vacuum and quantum flux, respectively.

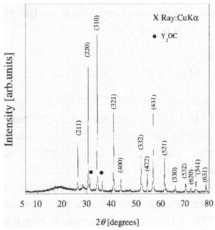

Figure 11. Powder X-ray diffraction pattern of Y_2C_3.

Figure 14 shows the specific heat of Y_2C_3 taken at zero field and 8 T. As shown in Fig. 14, the specific heat taken at zero field exhibits a sharp jump at T_c = 13.9 K, with a width of 0.7 K, exactly the same as the T_c determined from the susceptibility data. T_c is defined as the midpoint of the jump. We assume that the overall specific heat is composed of the electron and lattice parts, $C(T) = C_{el}(T) + C_{ph}(T)$. The lattice part is expressed by the $\beta T^3 + \delta T^5$ term at low temperature below the Debye temperature Θ_D, and the electron specific heat C_{el} is expressed by the γT term. Thus, the specific heat in the normal state can be expressed by the formula $C(T)/T = \gamma + \beta T^2 + \delta T^4$. From the $C(T)/T$-T^2 plot in Fig.14, the γ, β and δ values of Y_2C_3 was obtained to be 2.81 mJ/mol·K^2, 0.09

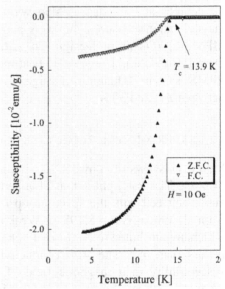

Figure 12. Temperature dependence of susceptibility in Y_2C_3.

mJ/mol·K^4 and 9.62×10^{-5} mJ/mol·K^6, respectively. The Debye temperature Θ_D is related to the coefficient of the T^3 term, which originates from the phonon contribution $\beta = N(12/5)\pi^4 R\Theta_D^{-3}$, where R = 8.314 J/mol·K and N = 40, because Y_2C_3 includes eight formula units per unit cell. The Θ_D of Y_2C_3

was calculated to be 920 K, which was unexpectedly high, being comparable to that of MgB$_2$ (Θ_D = 900 K) estimated from the specific heat measurement [78]. γ and Θ_D are larger than those of the previously reported low-T_c phase in Y$_2$C$_3$ (T_c = 10.0 K). Cort *et al.* reported that γ and Θ_D were 2.8 mJ/mol·K^2 and 557 K, respectively [79].

Figure 15 shows the temperature dependence of C_{el} in Y$_2$C$_3$. It is well known that $\Delta C(T_c)/\gamma T_c$ obtained from a heat capacity measurement shows the strength of electron-phonon coupling and its value is 1.43 in the BCS weak-coupling limit. We obtained a $\Delta C(T_c)/\gamma T_c$ of 3.3. Moreover, the fitting below T_c gives exp($-1/T$) dependence as predicted within the BCS theory rather than T^n dependence as in superconductors with an anisotropic symmetry of superconductivity. These results indicate that the

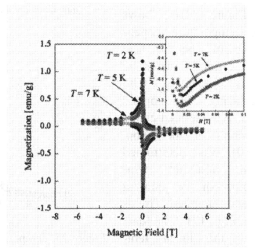

Figure 13. Magnetization vs. magnetic field (*M-H*) curves of Y$_2$C$_3$. The inset shows the *M-H* curves in the low-field region.

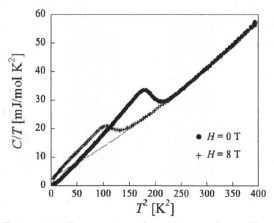

Figure 14. Temperature dependence of specific heat in Y$_2$C$_3$. The solid line shows a fitting result.

symmetry of the superconductivity of Y$_2$C$_3$ is that of an isotropic *s*-wave. From the fitting below T_c, we estimated $2\Delta/k_B T_c$ to be 4.5, which means that the $2\Delta/k_B T_c$ of Y$_2$C$_3$ is larger than that of MgB$_2$ ($2\Delta/k_B T_c$ = 3.8 ~ 4.3) [14]. From the $\Delta C(T_c)/\gamma T_c$ and $2\Delta/k_B T_c$ values determined from specific heat measurement, the superconductivity in Y$_2$C$_3$ indicates a strong coupling regime.

Figure 16 shows the temperature dependence of specific heat in various magnetic fields. T_c decreases with increasing an applied field. T_c is defined as the midpoint of the jump. $H_{c2}(0)$ was determined from midpoint temperature of the jump at several applied fields (see Fig. 17). The data show the linear temperature dependence, and the gradients dH_{c2}/dT of Y_2C_3 is found to be about -2.57 T/K. On the basis of relationship $H_{c2}(0)$ value ($H_{c2}(0) \approx 0.69 \times (-dH_{c2}/dT) \times T_c$) for a type-II superconductors in dirty limit, the dH_{c2}/dT values of Y_2C_3 is found to be about 24.7 T. The coherence length ξ is determined to be 36 Å using $\mu_0 H_{c2} \sim \Phi_0/\pi\xi^2$.

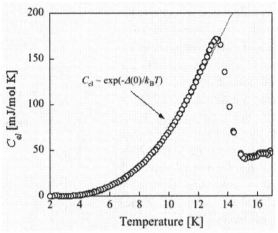

Figure 15. Temperature dependence of C_{el} in Y_2C_3. The dashed line shows a fitting result.

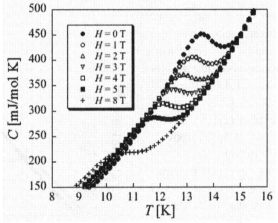

Figure 16. Temperature dependence of specific heat in Y_2C_3 at several magnetic fields.

4. Summary

The entire scenario of superconductivity in MgB_2 has essentially been clarified because a huge number of theoretical and experimental investigations has been reported in a short period because of its unexpectedly high-T_c. In the framework of BCS theory, the theoretical interpretation of this superconductivity focuses on the metallic nature of a 2D layer formed by boron atoms and also on strong electron-phonon coupling mainly caused by the E_{2g} phonon mode (B in-plane bond stretching), which is supported by experimental results. The results have become consistent, except for small inconsistencies which are probably due to some disorder of the boron atom and/or a slight defect at the Mg site. In particular, the two-gap (on the 2D σ-band and the 3D π-band) nature of the

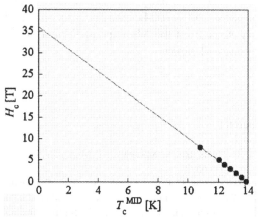

Figuer 17. Upper critical fields, $H_{c2}(T)$, of Y$_2$C$_3$ determined from midpoint of the jump of specific heat in magnetic fields. The solid line shows a fitting result.

superconducting state in MgB$_2$ has been confirmed by the results of several spectroscopic measurements and MgB$_2$ is now recognized as the first material showing intrinsic multiple supercon-ducting gaps.

We succeeded in synthesizing a high-quality sample of medium-T_c phase in Y$_2$C$_3$ (T_c = 13.9 K), and examined the physical properties in detail by magnetic susceptibility, magnetization and specific heat measurements. The magnetization (*M-H*) curves show typical type-II superconducting behavior, and $H_{c1}(0)$ and $\lambda(0)$ are estimated to be 3.2 mT and 4400 Å, respectively. From specific heat measurement, γ, $\Box C(T_c)/\gamma T_c$, Θ_D, $2\Box/k_B T_c$, $H_{c2}(0)$ and ξ are determined to be 2.81 mJ/mol·K^2, 3.3, 920 K, 4.5, 24.7 T and 36 Å respectively. We conclude from these results that the symmetry of superconductivity in Y$_2$C$_3$ can be described as that of an isotropic *s*-wave strong coupling regime.

MgB$_2$ has opened a new frontier in the investigation of the physical properties of intermetallic superconductors. Now, one of the most exciting questions is "whether MgB$_2$ is merely a special example in compounds including *p*-electron elements". Because the "BCS limit" has been broken in our recent quest for new superconductors, other exotic high-T_c superconductors may yet be awaiting their discovery in *p*-electron system.

Acknowledgements

This work was partly supported by the 21st COE program, "High-Tech Research Center" Project for Private Universities: matching fund subsidy from MEXT (Ministry of Education, Culture, Sports, Science and Technology; 2002-2004), and a Grant-in-Aid for Scientific Research on Priority Area from the Ministry of Education, Culture, Sports, Science and Technology of Japan.

References

[1]   J. Nagamatsu, N. Nakagawa, T. Muranaka, Y. Zenitani and J. Akimitsu, *Nature* **410** (2001) 63.

[2]   For review, C. Buzea and T. Yamashita, *Supercond. Sci. Technol.* **14** (2001) R115.

[3]   *Physica C*, Volume **385**, Issues 1-2, (2003).

[4]   T. Muranaka, Y. Zenitani, J. Shimoyama and J. Akimitsu, "Frontiers in Superconducting Materials" Springer edited by A. Narlikar. (2005) pp.937-981.

[5]   H. Suhl, B. T. Matthias and L. R. Walker, *Phys. Rev. Lett.* **3** (1959) 552.

[6]   G. Binnig, A. Baratoff, H. E. Hoenig and J. G. Bednorz, *Phys. Rev. Lett.* **45** (1980) 1352.

[7]   A. Kawano, Y. Mizuta, H. Takagiwa, T. Muranaka and J. Akimitsu, *J. Phys. Soc. Jpn.* **72** (2003) 1724.

[8]   K. Kawashima, A. Kawano, T. Muranaka and J. Akimitsu, *J. Phys. Soc. Jpn.* **74** (2005) 700.

[9]   G. Amano, S. Akutagawa, T. Muranaka, Y. Zenitani and J. Akimitsu, *J. Phys. Soc. Jpn.* **73** (2004) 530.

[10]   H. Kotegawa, K. Ishida, Y. Kitaoka, T. Muranaka and J. Akimitsu, *Phys. Rev. Lett.* **87** (2001) 127001.

[11]   G. Karapetrov, M. Iavarone, W.K. Kwok, G.W. Crabtree and D.G. Hinks, *Phys. Rev. Lett.* **86** (2001) 4374

[12]   T. Takahashi, T. Sato, S. Souma, T. Muranaka and J. Akimitsu, *Phys. Rev. Lett.* **86** (2001) 4915.

[13]   F. Bouquet, R.A. Fisher, N.E. Phillips, D.G. Hinks and J.D. Jorgensen, *Phys. Rev. Lett.* **87** (2001) 047001.

[14]   F. Bouquet, Y. Wang, R.A. Fisher, D.G. Hinks, J.D. Jorgensen, A. Junod and N.E. Philips, *Europhys. Lett.* **56** (2001) 856.

[15]   Y. Wang, T. Plackowski and A. Junod, *Physica C* **355** (2001) 179.

[16]   P. Szabó, P. Samuely, J. Kačmarčík, T. Klein, J. Marcus, D. Fruchart, S. Miraglia, C. Marcenat and A.G.M. Jansen, *Phys. Rev. Lett.* **87** (2001) 137005.

[17]   S. Tsuda, T. Yokoya, T. Kiss, Y. Takano, K. Togano, H. Kito, H. Ihara and S. Shin, *Phys. Rev. Lett.* **87** (2001) 177006.

[18]   R.S. Gonnelli, D. Daghero, G.A. Ummarino, V.A. Stepanov, J. Jun, S.M. Kazakov and J. Karpinski, *Phys. Rev. Lett.* **89** (2002) 247004.

[19]   F. Bouquet, Y. Wang, I. Sheikin, T. Plackowski, A. Junod, S. Lee and S. Tajima, *Phys. Rev. Lett.* **89** (2002) 257001.

[20]   S. Tsuda, T. Yokoya, Y. Takano, H. Kito, A. Matsushita, F. Yin, J. Itoh, H. Harima and S. Shin, *Phys. Rev. Lett.* **91** (2003) 127001.

[21]   S. Souma, Y. Machida, T. Sato, T. Takahashi, H. Matsui, S.-C. Wang, H. Ding, A. Kaminski, J.C. Campuzano, S. Sasaki and K. Kadowaki, *Nature* **423** (2003) 65.

[22]   A.Y. Liu, I.I. Mazin and J. Kortus, *Phys. Rev. Lett.* **87** (2001) 087005.

[23] L. Shen, N. Senozan and N.E. Philips, *Phys. Rev. Lett.* **14** (1965) 1025.
[24] S.V. Shulga, S.-L. Drechsler, G. Fuchs, K.-H. Müller, K. Winzer, M. Heinecke, and K. Krug, *Phys. Rev. Lett.* **80** (1998) 1730.
[25] J. Kortus, I.I. Mazin, K.D. Belashchenko, V.P. Antropov and L.L. Boyer, *Phys. Rev. Lett.* **86** (2001) 4656.
[26] J. M. An and W.E. Pickett, *Phys. Rev. Lett.* **86** (2001) 4366.
[27] G. Satta, G. Profeta, F. Bernardini, A. Continenza, and S. Massidda, *Phys. Rev. B* **64** (2001) 104507.
[28] K.D. Belashchenko, M. van Schilfgaarde and V.P. Antropov, *Phys. Rev. B* **64** (2001) 092503.
[29] E. Nishibori, M. Takata, M. Sakata, H. Tanaka, T. Muranaka and J. Akimitsu, *J. Phys. Soc. Jpn.* **70** (2001) 2252.
[30] S. Lee, H. Mori, T. Masui, Y. Eltsev, A. Yamamoto and S. Tajima, *J. Phys. Soc. Jpn.* **70** (2001) 2255
[31] H. Mori, S. Lee, A. Yamamoto, S. Tajima and S. Sato, *Phys. Rev. B* **65** (2002) 092507.
[32] L. Wu, Y. Zhu, T. Vogt, H. Su and J.W. Davenport, *Phys. Rev. B* **69** (2004) 064501.
[33] R.K. Kremer, B.J. Gibson and K. Ahn, cond-mat/0102432 (2001).
[34] H.D. Yang, J.-Y. Lin, H.H. Li, F.H. Hsu, C.J. Liu, S.-C. Li, R.-C. Yu and C.-Q. Jin, *Phys. Rev. Lett.* **87** (2001) 167003.
[35] R.A. Fisher, G. Li, J.C. Lashley, F. Bouquet, N.E. Philips, D.G. Hinks, J.D. Jorgensen and G.W. Crabtree, *Physica C* **385** (2003) 180.
[36] F. Bouquet, Y. Wang, I. Sheikin, P. Toulemonde, M. Eisterer, H.W. Weber, S. Lee. S. Tajima and A. Junod, *Physica C* **385** (2003) 192.
[37] H.J. Choi, D. Roundy, H. Sum, M.L. Cohen and S.G. Louie, *Nature* **418** (2002) 758.
[38] H. Padamasee, J.E. Neighbor and C.A. Schiffman, *J. Low Temp. Phys.* **12** (1973) 387.
[39] B. Mühlshlegel, *Z. Phys.* **155** (1959) 313.
[40] J.W. Quilty, S. Lee, A. Yamamoto and S. Tajima, *Phys. Rev. Lett.* **88** (2002) 087001.
[41] M. Angst, R. Puzniak, A. Wisniewski, J. Jun, S.M. Kazakov, J. Karpinski, J. Roos and H. Kreller, *Phys. Rev. Lett.* **88** (2002) 167004.
[42] M. Angst, R. Puzniak, A. Wisniewski, J. Roos, H. Kreller, P. Miranović, J. Jun, S.M. Kazakov and J. Karpinski, *Physica C* **385** (2003) 143.
[43] A.D. Caplin, Y. Bugoslavsky, L.F. Cohen, L. Cowey, J. Driscoll, J. Moore and G.K. Perkins, *Supercond. Sci. Technol.* **16** (2003) 176.
[44] H. Uchiyama, K.M. Shen, S. Lee, A. Damascelli, D.H. Lu, D.L. Feng, Z.-X. Shen and S. Tajima, *Phys. Rev. Lett.* **88** (2002) 157002.
[45] I. Gieaver, *Phys. Rev. Lett.* **5** (1960) 147.
[46] R.C. Dynes, V. Narayanamurti and J.P. Garno, *Phys. Rev. Lett.* **41** (1978) 150.
[47] For example, H. Schmidt, J.F. Zasadinski, K.E. Gray and D.G. Hinks, *Physica C* **385** (2003) 221.
[48] A. Sharoni, I. Felner and O. Millo, *Phys. Rev. B* **63** (2001) 220508(R).
[49] G. Rubio-Bollinger, H. Suderow and S. Vieira, *Phys. Rev. Lett.* **86** (2001) 5582.
[50] M. Iavarone, G. Karapetrov, A.E. Koshelev, W.K. Kwok, G.W. Crabtree and D.G. Hinks, W.N. Kang, E.-M. Choi, H.J. Kim, H.-J. Kim and S.I. Lee, *Phys. Rev. Lett.* **89** (2002) 187002.
[51] F. Guibileo, D. Roditchev, W. Sacks, R. Lamy, D.X. Thanh, J. Klein, S. Miraglia, D. Fruchart, J. Marcus and Ph. Monod, *Phys. Rev. Lett.* **87** (2001) 177008.

[52]   M.R. Eskildsen, M. Kugler, S. Tanaka, J. Jun, S.M. Kazakov, J. Karpinski, and Ø. Fischer, *Phys. Rev. Lett.* **89** (2002) 187003.

[53]   J. Karpinski, M. Angst, J. Jun, S.M. Kazakov, R. Puzniak, A. Winsniewski, J. Roos, H. Keller, A. Perucchi, L. Degiorgi, M. Eskildsen, P. Bordet, L. Vinnikov and A. Mironov, *Supercond. Sci. Technol.* **16** (2003) 221.

[54]   A. Plecenik, Š. Beňačka, P. Kúš and M. Grajcar, *Physica C* **368** (2001) 251.

[55]   F. Laube, G. Goll, J. Hagel, H.v. Löhneysen, D. Ernst and T. Wolf, *Europhys. Lett.* **56** (2001) 296.

[56]   R.S. Gonnelli, A. Calzolari, D. Daghero, G.A. Ummarino, V.A. Stepanov, P. Fino, G. Giunchi, S. Ceresara and G. Ripamonti, *J. Phys. Chem. Solids* **63** (2002) 2319.

[57]   Z.-Z. Li, Y. Xuan, H.-J. Tao, Z.-A. Ren, G.-C. Che, B.-R. Zhao and Z.-X. Zhao, *Supercond. Sci. Technol.* **14** (2001) 944.

[58]   Z.-Z. Li, H.-J. Tao, Y. Xuan, Z.-A. Ren, G.-C. Che and B.-R. Zhao, *Phys. Rev. B* **66** (2002) 064513.

[59]   S. Lee, Z.G. Khim, Y. Chong, S.H. Moon, H.N. Lee, H.G. Kim, B. Oh and E. J. Choi, *Physica C* **377** (2002) 202.

[60]   Y. Bugoslavsky, Y. Miyoshi, G.K. Perkins, A.V. Berenov, Z. Lockman, J.L. MacManus-Driscoll, L.F. Cohen, A.D. Caplin, H.Y. Zhai, M.P. Paranthaman, H.M. Christen and M. Blamire, *Supercond. Sci. Technol.* **15** (2002) 526.

[61]   A. Kohen and G. Deutscher, *Phys. Rev. B* **64** (2001) 060506(R).

[62]   For example; H. Schmidt, J.F. Zasadzinski, K.E. Gray and D.G. Hinks, *Phys. Rev. B* **63** (2001) 220504(R).

[63]   R.S. Gonnelli, A. Calzolari, D. Daghero, G.A. Ummarino, V.A. Stepanov, G. Giunchi, S. Ceresara and G. Ripamonti, *Phys. Rev. Lett.* **87** (2001) 097001.

[64]   H. Schmidt, J. F. Zasadzinski, K. E. Gray, and D.G. Hinks, *Phys. Rev. Lett.* **88** (2002) 127002 and *Physica C* **385** (2003) 221.

[65]   T. Takasaki, T. Ekino, T. Muranaka, H. Fujii and J. Akimitsu, *Physica C* **378-381** (2002) 229.

[66]   T. Ekino, T. Takasaki, T. Muranaka, J. Akimitsu and H. Fujii, *Phys. Rev. B* **67** (2003) 094504.

[67]   R. W. Henn, W. Schnelle, R. K. Kremer, and A. Simon, *Phys. Rev. Lett.* **77** (1996) 374.

[68]   M. C. Krupka *et al.*, *J. Less-Common Metals* **19** (1969) 113.

[69]   S.V. Vonsovsky *et al.*, Superconductivity in Transition Metals, their Alloys and Compounds., Springer-Verlag, Berlin, 1982.

[70]   C. P. Poople and H. A. Farach, *J. Supercond.* **13** (2000) 47.

[71]   V. I. Novokshonov, *Russian J. Inorganic Chem.* **25**(3) (1980) 375.

[72]   M. C. Krupka *et al.*, *J. Less-Common Metals* **17** (1969) 91.

[73]   I. R. Shein and A. L. Ivanovskii, cond-mat/0312391, *Solid State Commun.* **131** (2004) 223

[74]   S. Lee *et al.*, *Inorg. Chem.* **28** (1989) 4094.

[75]   Th. Gulden R, W. Henn, O. Jepsen, R. K. Kremer, W. Schnelle, A. Simon, and C. Felser, *Phys. Rev. B* **56** (1997) 9021.

[76]   D. J. Singh and I. I. Mazin, *Phys. Rev. B* **70** (2004) 052504.

[77]   T. Nakane, T. Mochiku, H. Kito, J. Itoh. M. Nagao. H. Kumakura, and Y. Takano, *Appl. Phys. Lett.* **84** (2004) 2859.

[78]   F. Bouquet *et al.*, Physica C 385, Issues 1-2 (2003) 192-204.

[79]   B. Cort, G. R. Stewart, and A. L. Giorgi, *J. Low Temp. Phys.* **54** (1980) 149.

SHAPE RESONANCES IN THE INTERBAND PAIRING IN NANOSCALE MODULATED MATERIALS

A. Bianconi[1], M. Filippi[1], M. Fratini[1], E. Liarokapis[2], V. Palmisano[1], N. L. Saini[1], L. Simonelli[1]

[1]Dipartimento di Fisica, Università di Roma "La Sapienza", P. Aldo Moro 2, 00185 Roma, Italy. [2]Department of Applied Mathematics and Physics, National Technical University of Athens, GR-157 80 Athens, Greece

Abstract. The Feshbach shape resonance in the interband pairing is shown to provide the mechanism for evading temperature de-coherence effects in macroscopic quantum condensates. This mechanism provides the T_c amplification in particular nanoscale material architectures like superlattices of metallic layers. The Feshbach resonance is reached by tuning the chemical potential to a Van Hove-Lifshitz singularity (vHs) in the electronic energy spectrum associated with the change of the Fermi surface dimensionality of one of the superlattice subbands. The case of light element diborides where Mg is substituted for Al in AlB_2 and its similarity with intercalated graphite compounds is discussed.

Key words: Feshbach resonance; Shape resonance; Diborides; Heterostructure at atomic limit

1. Introduction

The light element diborides AB_2 with (A=Al, Mg) are artificial intermetallics synthesized in the 20th century. The first compound to be synthesized was AlB_2 in 1935 [1]. It was followed by MgB_2 in 1953 [2,3], discovered as a byproduct during the search for new boron compounds for nuclear reactor bars in the fifties. The synthesis of $Al_{1-y}Mg_yB_2$ ternary compounds has first been reported in 1971 [4].

For fifty years, these light element diborides formed by low T_c, or non-superconducting elements have not been suspected to show supercon-ductivity according to the comprehensive review papers discussing super-conductivity in all the known elements, alloys, and compounds [11,12].

Transition element diborides AB_2 show superconducting (A=Nb,Ta) [5] or magnetic order (A=Mn,Cr) [6,7] at low temperatures. During the sixties and seventies the intermetallic transition element diborides attracted a great deal of interest being at the borderline between superconductivity and ferromagnetism. Therefore the research was extended to binary hexa- and

K. Scharnberg and S. Kruchinin (eds.),
Electron Correlation in New Materials and Nanosystems, 93–101.
© 2007 Springer.

dodecaborides as the ternary rare-earth rhodium borides where the very interesting reentrant superconductivity and magnetic ordering phenomena were found [8]. In fact, before 1986 the superconductors with the highest T_c have been found in alloys of transition metals and actinides at the borderline between superconducting and magnetic elements [9, 10].

The electronic structure of all diborides is characterized by π bands and σ bands like that of graphite. The π electrons provide the metallic bonding between the boron layers in the c axis direction and the σ electrons provide the covalent bonding within boron atoms in the a,b planes. The dispersion of the σ subbands in the c-axis direction is determined by the electron hopping between the graphene-like boron mono-layers that is controlled by their distance (i.e., the c-axis of the AlB_2 structure).

The two classes of diborides have a subtle but important difference in their electronic structure. In transition metal diborides the boron σ orbital is hybridized with the d orbital of the transition metal ions forming 3D bonding wavefunctions extending into the space between the boron layers.

On the other hand, in light element diborides there is no hybridization between σ and π orbitals. Therefore, the electronic structure of $Al_{1-y}Mg_yB_2$ [13] is similar to that of hole doped intercalated graphite [14,15]. In AlB_2 the Fermi level crosses only the π subband since it is in the partial gap between a filled and an empty σ subband like in graphite. In the ternary intermetallics $Al_{1-y}Mg_yB_2$ it is possible by changing the magnesium content y to dope holes into the band structure and thus move the Fermi level below the top of the σ subband.

Since the light element diborides were not expected to be supercon-ductors, their superconducting properties have escaped notice for 48 years, even though MgB_2 has been available in kilogram-size bottles for years from suppliers of inorganic chemicals. It was commonly used as a reagent in metathesis reactions (in which compounds change partners) and in some commercial preparations of elemental boron [16]. However, the attention should have been put on the particular architecture of these materials which makes them the simplest systems to get high T_c superconductivity via the shape resonance process described in the 1993 patent [17,18] and in references [19,20].

The process to increase the superconducting critical temperature via the Feshbach shape resonance [17-20] requires particular material architectures: i) a superlattice of metallic layers, as in doped graphite-like layered materials; or ii) superlattices of wires, as in doped crystals of nanotubes; or iii) superlattices of dots, as in doped fullerenes. These lattice architectures produce an electronic structure formed by several subbands with different symmetry and no hybridization.

The T_c amplification process is activated by changing the charge density in order to tune the chemical potential to a Van Hove-Lifshitz singularity (vHs) in the electronic energy spectrum of the superlattice associated with the change of the Fermi surface dimensionality of one of the subbands. Under these conditions a Feshbach shape resonance in the interband exchange-like pairing shows up. The interband exchange-like pairing has been proposed since 1959 [21-30] as a non BCS pairing that plays a key role for thin films [31], cuprates [32-40], doped fullerenes [41], borocarbides [42], ruthenates [43] and intercalated graphite [44-45] however only rare superconductors are in the requested clean limit [46].

2. Feshbach shape resonances in diborides

The "shape resonances" (first described by Feshbach in nuclear elastic scattering cross-section for the processes of neutron capture and nuclear fission) deal with the configuration interaction between different excitation channels including quantum superposition of states corresponding to *different spatial locations* [46]. These resonances in the interband superconducting pairing where a pair is transferred between two Fermi surfaces show up only under special conditions. No hybridization between the electronic states of the two bands is required. The wave functions should have different parity and/or spatial locations. The impurity scattering rate for single electrons between the different bands should be suppressed by the disparity and negligible overlap between electron wave-functions of the different bands. Therefore the Feshbach resonances occur only in multiband superconductors in the clean limit.

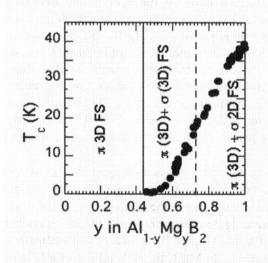

Figure 1. The superconducting critical temperature as function of the magnesium content y.

These Feshbach shape resonance (FSR) resonances in the interband pairing are similar to the Feshbach resonances in ultra-cold fermionic gases that are used to raise the ratio T_c/T_F of the superfluid critical temperature T_c on the Fermi temperature T_F [47].

In $Al_{1-y}Mg_yB_2$ the position of the Fermi level is tuned toward the filled σ subband by Mg for Al substitution. For $y \approx 0.44$ the Fermi level reaches the top of the σ subband and for $0.44 < y < 0.7$ the large π Fermi surface typical of graphite coexists with a second small hole-like 3D σ Fermi surface that for $0.7 < y < 1$ has a 2D topology. Therefore the electronic structure of $Al_{1-y}Mg_yB_2$ for $y > 0.44$ is like that of heavily hole doped graphite where the additional σ subband appears at the Fermi level so that the conduction elec-tron Fermi gas is made up of two components: the σ and π electrons having two different spatial locations (in the graphene-layers and in the interlayer space between them, respectively) characterized by the disparity between their wavefunction. Figure 1 shows the superconducting critical temperature as a function of y in $Al_{1-y}Mg_yB_2$.

The disparity and the negligible overlap between electron wave functions in the different σ and π subbands suppress both the hybridization between the two components and the single-electron interband impurity scattering rate that makes possible multiband superconductivity with a key role of the interband proximity effect. Since the Feshbach shape resonance (FSR) in the interband pairing occurs where there is a change of the Fermi surface topology, i.e., an electronic topological transition (ETT) in one of the subbands it has been proposed that the high T_c in magnesium diboride is determined by the proximity to the FSR in the interband pairing term centered at the 2D to 3D ETT in the σ Fermi surface and the superconducting properties of $Al_{1-y}Mg_yB_2$ can be assigned only to band filling effects [48-53] with minor effects of impurity scattering.

Recently, several experiments have proven that the aluminium [54,55], carbon [55,56] and scandium [57,58] substitution for magnesium tune the chemical potential, with minor effects on the impurity scattering, keeping the superconductor in the clean limit. At y=0.7 in $Al_{1-y}Mg_yB_2$ there is a cross-over in the hierarchy of superconducting gap energies in the σ and π bands [53].

Therefore for y>0.7 the σ gap is the largest but for $0.4 < y < 0.7$ the system is like a superconducting hole doped graphite with the main gap in the π(3D) band and a smaller gap in the small pocket of the σ(3D) band. In this work we have investigated the variation of the superconducting properties as a function of y in $Al_{1-y}Mg_yB_2$ in order to investigate this interesting superconducting phase. In MgB_2 the σ band is partially unoccupied due to the electron transfer from the boron layers to the magnesium layers. In fact the chemical potential E_F in MgB_2 is at about 750 meV below the energy E_A

of the top of the σ band. Moreover the chemical potential E_F in MgB_2 is also at about 350 meV below the energy of the Γ point in the band structure ($E_\Gamma > E_F$). Therefore the σ Fermi surface of MgB_2 where $E_F < E_\Gamma < E_A$ has the corrugated tubular shape with a two-dimensional topology for the case $E_F < E_\Gamma$. Going from below ($E_F < E_\Gamma < E_A$) to above the energy of the Γ point the σ Fermi surface becomes a closed Fermi surface with 3D topology like in $AlMgB_4$, y=0.5, that belongs to the Fermi surface type for the case $E_\Gamma < E_F < E_A$. Therefore by tuning the chemical potential E_F it is possible to reach the point where E_F is tuned at the 2D/3D ETT at y=0.7. This is a type (II) 2.5 Lifshitz electronic topological transition (ETT) with the disruption of a "neck" in the σ Fermi surface with the critical point at $E_F = E_\Gamma$.

The total and partial Density of States (DOS) are plotted in Fig. 2 as functions of the reduced Lifshitz parameter $\zeta = (E_\Gamma - E_F)/(E_A - E_\Gamma)$, where $D = E_A - E_\Gamma$ is the energy dispersion in the c-axis direction due to electron hopping between the boron layers (D=0.4 eV in MgB_2, but it changes with chemical substitutions). Figure 3 shows the ratio T_c/T_F of the critical temperature T_c and the Fermi temperature $T_F = E_F/K_B$ for the holes in the σ subband as function of the reduced Lifshitz parameter ζ.

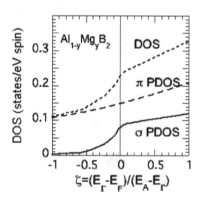

Figure 2. The total density of states (DOS) and the partial DOS (PDOS) of the σ and π band at the Fermi level E_F in $Al_{1-y}Mg_yB_2$ as a function of the reduced Lifshitz parameter ζ where E_A is the energy of the top of the σ band at the A point in the band structure, and E_Γ is where the σ Fermi surface changes from a 2D corrugated tube for $E_\Gamma > E_F$ to a closed 3D Fermi surface for $E_\Gamma < E_F$. The type (I) electronic topological transition (ETT) with the appearance of the closed 3D σ Fermi surface occurs by tuning the Fermi energy to the critical point $E_F = E_A$ ($\zeta = -1$). The type (II) ETT with the disruption of a "neck" in the σ Fermi surface occurs by tuning E_F to the critical point in the band structure $E_F = E_\Gamma$ ($\zeta = 0$) while the large π Fermi surface keeps its 3D topology.

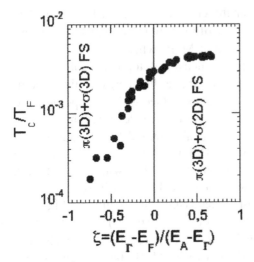

Figure 3. The ratio T_c/T_F of the critical temperature T_c and the Fermi temperature $T_F=E_F/K_B$ for the holes in the σ subband in magnesium for aluminum substituted diborides as a function of the reduced Lifshitz parameter ζ in $Al_{1-y}Mg_yB_2$.

In Fig. 4 we report preliminary results on the variation of the isotope coefficient of the critical temperature for the ^{11}B and ^{10}B replacement. The results clearly show that the isotope exponent decreases approaching the centre of the shape Feshbach resonance at z=0 as expected [23] if the interband pairing becomes dominant.

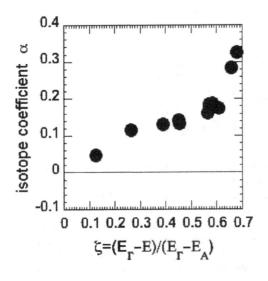

Figure 4. The isotope coefficient on the critical temperature as a function of the reduced Lifshitz parameter ζ in $Al_{1-v}Mg_vB_2$.

The influence of the proximity to a type (II) electronic topological transition on the anomalous electronic and lattice properties of MgB_2 is shown by the anomalous behaviour of the Raman spectra, which is under investigation

3. Conclusions

Evidence for the Feshbach shape resonance is provided by the measured isotope coefficient going to zero at the electronic topological transition. Therefore, we have shown here the relevance of the Feshbach shape resonances around the dimensional crossover of the topology of the σ Fermi surface for the particular case of doped MgB_2, which is a clear case of a multi-band superconductor in the clean limit, where T_c can be controlled by tuning the Fermi level in a particular nanoscale architecture.

Acknowledgements

This work is supported by European STREP project 517039 "Controlling Mesoscopic Phase Separation" (COMEPHS).

References

1. W. Hoffmann, and W. Jänicke, *Naturwiss* 23, 851 (1935).
2. V. Russel, R. Hirst, F. Kanda and A. King, *Acta Cryst.* 6, 870 (1953).
3. E. Jones and B. Marsh, *J Am. Chem. Soc.* 76, 1434 (1954).
4. N.V. Vekshina, L. Ya. Markovskii, Yu. D. Kondrashev, and T.K. Voevodskyaya, *Zh. Prikl. Khim.* (Leningrad) 44 958-963 (1971).
5. L. Leyarovska *et al. J. Less-common Metals* 67, 249 (1979).
6. L. Andersson *et al.*, *Solid State Communications* 4, 77 (1966).
7. J. Castaing *et al.*, *J. Phys.Chem.Solids* 33, 533 (1972).
8. B. T. Matthias D. C. Johnston, W. A. Fertig, and M. B. Maple, *Sol. St. Comm.* 26, 141 (1978).
9. B. T. Matthias T. H. Geballe and V. B. Compton, *Rev. Mod. Phys.* 35:1. Errata *Rev. Mod. Phys.* 35, 414 (1963).
10. J.L. Smith and E.A. Kmetko, *J. Less Common Metals* 90, 83 (1983); J. L. Smith and R.S. Riseborough *Journal of Magnetism and Magnetic Materials* 47&48 , 545 (1985).
11. B. T. Matthias, *Phys. Rev.* 97, 74 (1955).
12. B. T. Matthias, *Superconductivity in the periodic system* in *Progress in Low Temperature Physics*, ed. C. J. Gorter, p. 138. Amsterdam, North-Holland Publishing Co. (1957).
13. O. de la Pena, A. Aguayo, and R. de Coss, *Phys. Rev. B* 66, 012511 (2002).
14. M.S. Dresselhaus, G. Dresselhaus, *Adv. in Phys.* 30, 139 (1981).
15. M.S. Dresselhaus, G. Dresselhaus and P.C. Eklund, *Science of Fullerenes and Carbon Nanotubes,* Academic Press Inc. San Diego (1996).
16. E. G. Killian, and R. B. Kaner, *Chem. Mater.* 8, 333 (1996).
17. A. Bianconi *"Process of increasing the critical temperature T_c of a bulk superconductor by making metal heterostructures at the atomic limit"* United State Patent: No.:US 6, 265, 019 B1. (priority date: 07 12 1993).

18. A. Bianconi *"High T_c superconductors made by metal heterostuctures at the atomic limit"* European Patent N. 0733271 (priority date: 07 12 1993)
19. A. Bianconi, *Sol. State Commun.* 89, 933 (1994)
20. A. Bianconi, A. Valletta, A. Perali, and N. L. Saini, *Physica C* 296, 269 (1998).
21. H. Suhl, B.T. Matthias, and L.R. Walker, *Phys. Rev. Lett.* 3, 552 (1959).
22. V.A. Moskalenko, *Phys. Met. and Metallog.* 8, 25 (1959).
23. J. Kondo, *Prog. Theor. Phys.* 29, 1 (1963).
24. J.R. Schrieffer, *Theory of Superconductivity*, 1964, see p.p.300.
25. A. J. Leggett, *Prog. Theor. Phys.* 36, 901 (1966); 36, 931 (1966).
26. B.T. Geilikman, R.O. Zaitsev, and V. Z. Kresin, *Sov. Phys. Solid State* 9, 642 (1967).
27. V.A. Moskalenko and M.E. Palistrant, *Sov. Phys. JETP* 22, 536 (1966).
28. L.Z. Kon, *Phys. Met. Metallogr.* (USSR). 23, (1967).
29. P. Entel, and M. Peter, *Jour. of Low Temp. Phys.* 22, 613 (1976).
30. N. Schopohl and K. Scharnberg, *Solid State Commun.* 22, 371 (1977).
31. J.M. Blatt and C.J. Thompson, *Phys. Rev. Lett.* 10, 332 (1963); C.J. Thompson and J.M. Blatt, *Phys. Lett.* 5, 6 (1963); J. M. Blatt in *Theory of Superconductivity* Academic Press New York 1964, (pag. 362 and 215).
32. D.H. Lee, and J. Ihm, *Solid State Commun.* 62, 811 (1987).
33. K. Yamaji, and A. Abe, *J. Phys. Soc. Japan* 56, 4237 (1987).
34. V.Z. Kresin, and S.A. Wolf, *Phys. Rev. B* 46, 6458 (1992); *Phys. Rev. B* 51, 1229 (1995).
35. N. Bulut, D. J. Scalapino, and R.T. Scalettar, *Phys. Rev. B* 45, 5577 (1992).
36. B.K. Chakraverty, *Phys Rev B* 48, 4047 (1993).
37. R. Combescot, and X. Leyronas, *Phys. Rev. Lett.* 75, 3732 (1995).
38. N. Kristoffel, P. Konsin, and T. Ord, *Rivista Nuovo Cimento* 17, 1 (1994).
39. J.F. Annett, and S. Kruchinin, *"New Trends in Superconductivity"* Kluwer Academic Publishers, Dordrecht, Netherlands (2002).
40. A. Bianconi, *J. Phys. Chem. Sol.* 67, 566 (2006).
41. R. Friedberg, and T.D. Lee, *Phys. Rev.* B40, 6745 (1989).
42. S.V. Shulga, S.L. Drechsler, G. Fuchs, K.H. Muller, K. Winzer, M. Heinecke and K. Krug, *Phys. Rev. Lett.* 80, 1730 (1998).
43. M.E. Zhitomirsky, and T. M. Rice, *Phys. Rev. Lett.* 87, 057001 (2001).
44. G. Cs'anyi, P. B. Littlewood, and A. H. Nevidomskyy, C. J. Pickard and B. D. Simons *arXiv:cond-mat/0503569* (23 Mar 2005).
45. T. E. Weller, M. Ellerby, S. S. Saxena, R. P. Smith, and N. T. Skipper, *cond-mat/0503570* (23 Mar 2005).
46. A. Bianconi and M. Filippi, in *" Symmetry and Heterogeneity in high temperature superconductors"* A. Bianconi (ed.) Spinger, Dordrecht, The Netherlands. Nato Sciente Series II Mathematics, Physics and Chemistry vol. 214, pp. 21-53, (2006).
47. C.A. Regal, M. Greiner, and D.S. Jin, *Phys. Rev. Lett.* 92, 040403 (2004).
48. A. Bianconi et al., *cond-mat/0102410* (22 Feb 2001); A. Bianconi et al., *cond-mat/0103211* (9 March 2001).
49. A. Bianconi, D. Di Castro, S. Agrestini, G. Campi, N.L. Saini, A. Saccone, S. De Negri, and M. Giovannini, Presented at *APS March meeting*, March 12, 2001 Seattle, Washington, *"Session on MgB$_2$"* tak 79, http://www.aps.org/MAR01/mgb2/talks.html#talks79
50. A. Bianconi, D. Di Castro, S. Agrestini, G. Campi, N. L. Saini, A. Saccone. S. De Negri, M. Giovannini, *J. Phys.: Condens. Matter* 13, 7383 (2001).
51. A. Bianconi, S. Agrestini, D. Di Castro, G. Campi, G. Zangari, N.L. Saini, A. Saccone, S. De Negri, M. Giovannini, G. Profeta, A. Continenza, G. Satta, S. Massidda, A. Cassetta, A. Pifferi and M. Colapietro, *Phys. Rev. B* 65, 174515 (2002).
52. A. Bussmann-Holder and A. Bianconi, *Phys. Rev. B* 67, 132509 (2003).

53. G.A. Ummarino, R.S.Gonnelli, S.Massidda and A. Bianconi, *Physica C* 407, 121 (2004).
54. L. D. Cooley A. J. Zambano, A.R. Moodenbaugh, R. F. Klie, Jin-Cheng Zheng, and Yimei Zhu *Phys. Rev. Lett.* 95, 267002 (2005)
55. P. Samuely, P. Szabó, P.C. Canfield, and S.L. Bud'ko *Phys, Rev. Lett.* 95, 099702 (2005).
56. S. Tsuda, T. Yokoya, T. Kiss, T. Shimojima, S. Shin, T. Togashi, S. Watanabe, C. Zhang, C. T. Chen, S. Lee, H. Uchiyama, S. Tajima, N. Nakai, and K. Machida *Phys. Rev. B* 72, 064527 (2005). S. Tsuda et al. *cond-mat/0409219*.
57. S. Agrestini, C. Metallo, M. Filippi, L. Simonelli, G. Campi, C. Sanipoli, E. Liarokapis, S. De Negri, M. Giovannini, A. Saccone, A. Latini, and A. Bianconi, *Phys. Rev. B* 70, 134514 (2004).
58. M. Filippi, A. Bianconi and A. Bussmann-Holder *J. Phys. IV France* 131, 49 (2005).

MAGNETIC AND MICROWAVE PROPERTIES OF THE TWO GAP SUPERCONDUCTOR MgB$_2$

Thomas Dahm (dahm@uni-tuebingen.de)
Institut für Theoretische Physik, Universität Tübingen, Auf der Morgenstelle 14, D-72076 Tübingen, Germany

Key words: Magnesium diboride, Two gap superconductivity, Magnetic properties, Microwave properties

A large amount of experimental and theoretical work in the past has convincingly shown that the binary compound MgB$_2$ is a two gap superconductor (Liu et al., 2001; Golubov et al., 2002; Choi et al., 2002; Dahm, 2005). The two significantly different gaps exist on disconnected parts of the Fermi surface: a small gap (~2 meV) on the π bands and a large gap (~7 meV) residing on the σ bands. This leads to a number of peculiar consequences both in its magnetic properties and its microwave properties.

Experimentally, the upper critical field anisotropy shows a surprisingly strong temperature dependence (Angst et al., 2002; Lyard et al., 2002). We have used a simple model for the Fermi surface topology and have taken into account the presence of the two gaps (Dahm and Schopohl, 2003). The parameters were taken from bandstructure calculations (Liu et al., 2001). Using this model we calculated the upper critical field solving Eilenberger's equations. We were able to show that indeed a strong temperature dependence of the upper critical field anisotropy in quantitative agreement with the experimental results has to be expected for the type of Fermi surface and gap distribution present in MgB$_2$.

Recently, we have also studied the core structure of a single vortex in a two gap superconductor (Gumann et al., 2005). At low temperatures a Kramer-Pesch effect occurs, i.e. a shrinkage of the size of the vortex core. Interestingly, this core shrinkage even exists, if only the σ band is in the clean limit. This situation is believed to be realistic for high quality MgB$_2$ samples (Mazin et al., 2002; Quilty et al., 2003) and opens the possibility to observe the Kramer-Pesch effect in this compound.

The temperature dependence of the microwave conductivity in MgB$_2$ thin films at a fixed microwave frequency shows a peak at temperatures around $0.5T_c$ (Jin et al., 2003). This is in contrast to conventional superconductors

K. Scharnberg and S. Kruchinin (eds.),
Electron Correlation in New Materials and Nanosystems, 103–105.
© 2007 *Springer.*

where a peak close to T_c ("Hebel-Slichter peak") appears and also in contrast to the high-T_c cuprates where a huge conductivity peak at very low temperatures has been observed (Hosseini et al., 1999). We have calculated the microwave conductivity within a two gap generalization of Mattis-Bardeen theory. These calculations showed that the Hebel-Slichter peak is moved towards much lower temperatures due to the presence of the small gap (Jin et al., 2003). This behavior is expected to have significant implications for the nonlinear microwave response as well (Dahm and Scalapino, 2004; Nicol et al., 2006).

References

Angst, M., Puzniak, R., Wisniewski, A., Jun, J., Kazakov, S. M., Karpinski, J., Roos, J., and Keller, H. (2002) Temperature and Field Dependence of the Anisotropy of MgB$_2$, *Phys. Rev. Lett.* **88**, 167004.

Choi, H. J., Roundy, D., Sun, H., Cohen, M. L., and Louie, S. G. (2002) The origin of the anomalous superconducting properties of MgB$_2$, *Nature* **418**, 758.

Dahm, T. (2005) Superconductivity of magnesium diboride: theoretical aspects, In A. V. Narlikar (ed.), *Frontiers in Superconducting Materials*, Springer, pp. 983–1009.

Dahm, T. and Scalapino, D. J. (2004) Nonlinear Microwave Response of MgB$_2$, *Appl. Phys. Lett.* **85**, 4436.

Dahm, T. and Schopohl, N. (2003) Fermi surface topology and the upper critical field in two-band superconductors - application to MgB$_2$, *Phys. Rev. Lett.* **91**, 017001.

Golubov, A. A., Kortus, J., Dolgov, O. V., Jepsen, O., Kong, Y., Andersen, O. K., Gibson, B. J., Ahn, K., and Kremer, R. K. (2002) Specific heat of MgB$_2$ in a one- and a two-band model from first-principles calculations, *J. Phys.: Condens. Matter* **14**, 1353.

Gumann, A., Graser, S., Dahm, T., and Schopohl, N. (2005) Induced Kramer-Pesch Effect in a Two Gap Superconductor: Application to MgB$_2$, cond-mat/0511520, to appear in Phys. Rev. B.

Hosseini, A., Harris, R., Kamal, S., Dosanjh, P., Preston, J., Liang, R., Hardy, W. N., and Bonn, D. A. (1999) Microwave spectroscopy of thermally excited quasiparticles in YBa$_2$Cu$_3$O$_{6.99}$, *Phys. Rev. B* **60**, 1349.

Jin, B. B., Dahm, T., Gubin, A. I., Choi, E.-M., Kim, H.-J., Lee, S.-I., Kang, W. N., and Klein, N. (2003) Anomalous coherence peak in the microwave conductivity of c-axis oriented MgB$_2$ thin films, *Phys. Rev. Lett.* **91**, 127006.

Liu, A. Y., Mazin, I. I., and Kortus, J. (2001) Beyond Eliashberg Superconductivity in MgB$_2$: Anharmonicity, Two-Phonon Scattering, and Multiple Gaps, *Phys. Rev. Lett.* **87**, 087005.

Lyard, L., Samuely, P., Szabo, P., Klein, T., Marcenat, C., Paulius, L., Kim, K. H. P., Jung, C. U., Lee, H.-S., Kang, B., Choi, S., Lee, S.-I., Marcus, J., Blanchard, S., Jansen, A. G. M., Welp, U., Karapetrov, G., and Kwok, W. K. (2002) Anisotropy of the upper critical field and critical current in single crystal MgB$_2$, *Phys. Rev. B* **66**, 180502(R).

Mazin, I. I., Andersen, O. K., Jepsen, O., Dolgov, O. V., Kortus, J., Golubov, A. A., Kuz'menko, A. B., and van der Marel, D. (2002) Superconductivity in MgB$_2$: Clean or Dirty?, *Phys. Rev. Lett.* **89**, 107002.

Nicol, E. J., Carbotte, J. P., and Scalapino, D. J. (2006) Nonlinear current response of one- and two-band superconductors, *Phys. Rev. B* **73**, 014521.

Quilty, J. W., Lee, S., Tajima, S., and Yamanaka, A. (2003) c-Axis Raman Scattering Spectra of MgB$_2$: Observation of a Dirty-Limit Gap in the π Bands, *Phys. Rev. Lett.* **90**, 207006.

FREE ENERGY FUNCTIONAL AND CRITICAL MAGNETIC FIELDS ANISOTROPY IN MAGNESIUM DIBORIDE

T. Örd[1], N. Kristoffel[1,2] and K. Rägo[2]
[1]*Institute of Theoretical Physics, University of Tartu,*
Tähe 4, 51010 Tartu, Estonia
[2]*Institute of Physics, University of Tartu,*
Riia 142, 51014 Tartu, Estonia

Abstract. Several characteristics of pure and doped MgB_2 are calculated basing on the model (Kristoffel et al., 2003). The free energy expansion is derived and expressions for thermodynamic characteristics obtained. The dependence of superconducting gaps on Al doping are presented with the $\Delta_{\sigma,\pi}$ crossing at $x = 0.38$. The calculated anisotropy coefficients for the MgB_2 critical magnetic fields agree with the experiment.

Key words: Magnesium diboride; Critical fields, Anisotropy, Doping, Gaps

1. Introduction

In multiband superconductors the interplay of multiple order parameters enriches and influences the behaviour of various properties of the system in a characteristic manner. The contribution of the interband channel to the pairing interactions opens an effective way to reach high transition temperatures out of exotic conditions. The interest in multigap systems has been recently especially grown in connection with the nature (Akimitsu and Murakama, 2003; Mazin and Antropov, 2003) of the MgB_2 superconductivity. Wide experimental and theoretical investigations have illuminated magnesium diboride to be a two-gap superconductor with pairings in 2D σ- and 3D π-bands driven by both intraband and interband interactions. The basic properties of MgB_2 have been understood and quantitatively desribed. Experimental investigations affirm that the interband pairing channel must be included into the MgB_2 superconductivity description (Tsuda et al., 2003; Geerk et al., 2005; Zehetmayer et al., 2004). Here we mention some models from an extended list (Liu et al., 2001; Kristoffel et al., 2003b; Ummarino et al., 2004; Bussmann-Holder and Bianconi, 2003; Kruchinin and Nagao, 2005; Kristoffel et al., 2003a) expressing the participation of the $\sigma - \pi$ pairing coupling.

K. Scharnberg and S. Kruchinin (eds.),
Electron Correlation in New Materials and Nanosystems, 107–115.
© 2007 *Springer.*

Current work on magnesium diboride is concentrated on related mixed (doped) compounds. On the other hand, the deepening understanding of the basic properties of MgB$_2$ is connected to the thermodynamic characteristics such as the critical magnetic fields. Especially the pronounced anisotropy of these properties arising from the layered structure of MgB$_2$ is actively discussed. However, in this field there are experimental difficulties with the self-consistenty of the data, e.g. for the high and low magnetic field regions (Zehetmayer et al., 2004). Furthermore, the applicability of the anisotropic Ginzburg-Landau type approach to MgB$_2$ superconducting phase transition is estimated to be valid only in a narrow region near T$_c$ (Koshelev and Golubov, 2004).

The authors have developed a MgB$_2$ superconductivity model with the allowance of intraband electron-phonon (attractive) and Coulomb interaction besides the repulsive coupling between effective $\sigma - \pi$ bands (Kristoffel et al., 2003b; Kristoffel et al., 2003a). Various observed dependencies of MgB$_2$ basic superconducting characteristics have been reproduced without fittings. In the present communication we represent the gaps dependence on electron doping (Al) for the Mg$_{1-x}$Al$_x$B$_2$ type system and anisotropy coefficients for the first and second critical magnetic fields of MgB$_2$ near T$_c$ according to this model.

2. The model

The Hamiltonian of a superconductor with two ($l = \sigma, \pi$) bands intersecting the Fermi level and pairs formed from the particles of the same band is taken as

$$H = \sum_{l,\vec{k}} \tilde{\epsilon}_l(\vec{k})(a^+_{l\vec{k}\uparrow}a_{l\vec{k}\uparrow} + a^+_{l-\vec{k}\downarrow}a_{l-\vec{k}\downarrow}) +$$

$$+ 2 \sum_{l,l'} \sum_{\vec{k},\vec{k}'} \sum_{\vec{q}} W_{ll'}(\vec{k}, -\vec{k} + \vec{q}, -\vec{k}' + \vec{q}, \vec{k}') a^+_{l\vec{k}\uparrow} a^+_{l-\vec{k}+\vec{q}\downarrow} a_{l'-\vec{k}'+\vec{q}\downarrow} a_{l'\vec{k}'\uparrow} . \quad (1)$$

Here the band energies read as $\epsilon_l = \tilde{\epsilon}_l + \mu$, where μ is the chemical potential and $\vec{q}$ the pair momentum. The pairing channels characterized by $W_{ll'}(\vec{k}, \vec{k}')$ are supposed to be active in different regions of the momentum space. This means that such superconductivity characteristics as band gaps Δ_l will be energy-dependent. The intraband coupling constant $W_{ll} = V_l + U_l$ includes the Coulomb part $U_l \geq 0$ in the band besides the electron-phonon attraction $V_l \leq 0$ in the layer of the width $2\hbar\omega_D$ around the Fermi energy. The interband coupling is characterized by a constant $W = W_{\sigma\pi} \geq 0$ including repulsive Coulomb and electron - phonon contributions. Interactions U_l and W cover

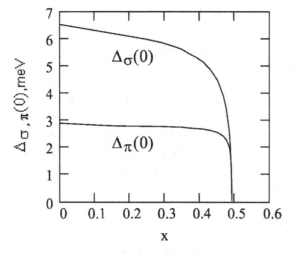

Figure 1. The σ- and π- band superconducting zero temperature gaps *vs* Al content in $Al_xMg_{1-x}B_2$.

the bands overlap region from $E_c < 0$ to zero (see (Mazin and Antropov, 2003)).

The parameter set of (Kristoffel et al., 2003b) will be used where $V_\sigma = -1.01$; $U_\sigma = 1$; $W = 0.53$ (eV), $V_\pi = U_\pi = 0$; the constant densities of states per one cell and spin $\rho_\sigma = 0.25$, $\rho_\pi = 0.11$ (eV), $\hbar\omega_D = 0.06$, $E_c = -2$ and $\mu = -0.6$ (eV) for the pure MgB_2. The anisotropic average Fermi velocities are taken from (Brinkman et al., 2002) as $v_{ab}^\sigma = 4.4$, $v_{ab}^\pi = 5.35$, $v_c^\sigma = 0.72$, $v_c^\pi = 6.23$ (10^5 ms^{-1}).

3. Superconducting gaps in $Al_xMg_{1-x}B_2$

Various theoretical and experimental problems on the family of MgB_2 relative compounds are widely investigated. The behaviour of the superconducting gaps with doping offers recently a special interest. The experimental situation can be followed by (Putti et al., 2003; Karpinski et al., 2005; Kortus et al., 2005; Putti et al., 2005) and there are also theoretical approaches (Ummarino et al., 2004; Bussmann-Holder and Bianconi, 2003; Floris et al., 2005).

The well-known quenching of MgB_2 superconductivity by doping in the mixed compound $Al_xMg_{1-x}B_2$ has been reproduced also by our model (Kristoffel et al., 2003b). At this the change in the band densities of states taking a linear μ *vs* x dependence with $\mu = -0.6$ eV for $x = 0$; $\mu = 0$ for $x = 0.5$ and using the data of (Kortus et al., 2001) has been taken into account. Correspondingly also the integration limits change. The $\Delta_{\sigma,\pi}$ dependences on x have not been presented in (Kristoffel et al., 2003b) and are given here in Fig.1 for $T = 0$. The observed suppression of the σ-band gap is reproduced, whereas the 3D π-band gap shows no essential change until $x = 0.4$. The $\sigma - \pi$ bands crossing appears at $x_c = 0.485$ where $\Delta_{\sigma,\pi}(0) = 1.09$ meV.

Further the gaps decrease rapidly and vanish at the same T_c in accordance with the presence of interband pairing channel. The measured value for the merging point of the gaps is $x_c = 0.34$ (Bianconi et al., 2002). The leading σ-gap depression reflects the deviation from the optimal "self-doping" of pure MgB$_2$ which weakens the σ-intraband pairing. Near x_c the dimensionality of the σ-band changes as also the chemical potential relation to the bands. This effect is an example to the conclusion that in multigap systems the band components spectral overlap dynamics (with the account of μ position) can act as a source of critical points on the phase diagram (Kristoffel and Rubin, 2002). The ratio Δ_σ/Δ_π calculated by us seems to fall off too slowly as compared with the measurement (Putti et al., 2005).

4. The free energy expansion

The free energy expansion corresponding to the Hamiltonian (1) in powers of the nonequilibrium gap order parameters $\delta_{l\vec{k}\vec{q}}$ has been derived in the form $F = F_2 + F_4$, where ($\Theta = k_B T$)

$$F_2 = \frac{1}{2} \sum_{l,\vec{k},\vec{q}} \left[\zeta_{l\vec{k}\vec{q}} |\delta_{l\vec{k}\vec{q}}|^2 + \sum_{l'\vec{k}'} W_{ll'}(\vec{k},\vec{k}') \zeta_{l\vec{k}\vec{q}} \zeta_{l'\vec{k}'\vec{q}} \delta_{l\vec{k}\vec{q}} \delta^*_{l'\vec{k}'\vec{q}} \right] \qquad (2)$$

with

$$\zeta_{l\vec{k}\vec{q}} = [\tilde{\epsilon}_l(\vec{k}) + \tilde{\epsilon}_l(\vec{k} - \vec{q})]^{-1} \left[\tanh\left(\frac{\tilde{\epsilon}_l(\vec{k})}{2\Theta} \right) + \tanh\left(\frac{\tilde{\epsilon}_l(\vec{k} - \vec{q})}{2\Theta} \right) \right], \qquad (3)$$

and

$$F_4 = \sum_{l\vec{k}} \frac{3}{4} \nu_{l\vec{k}} |\delta_{l\vec{k}}|^4 + \frac{1}{2} \sum_{l\vec{k}} \sum_{l'\vec{k}'} W_{ll'}(\vec{k},\vec{k}') \delta^*_{l\vec{k}} \delta_{l'\vec{k}'} \times$$

$$\times \left[\zeta_{l\vec{k}} \nu_{l'\vec{k}'} |\delta_{l'\vec{k}'}|^2 + \zeta_{l'\vec{k}'} \nu_{l\vec{k}} |\delta_{l\vec{k}}|^2 \right], \qquad (4)$$

with

$$\nu_{l\vec{k}} = \frac{1}{2} \frac{1}{[\tilde{\epsilon}_l(\vec{k})]^3} \left[\tanh\frac{\tilde{\epsilon}_l(\vec{k})}{2\Theta} - \frac{\tilde{\epsilon}_l(\vec{k})}{2\Theta} \operatorname{ch}\frac{\tilde{\epsilon}_l(\vec{k})}{2\Theta} \right], \qquad (5)$$

and $\delta_{l\vec{k}} \equiv \delta_{l\vec{k}0}, \zeta_{l\vec{k}} \equiv \zeta_{l\vec{k}0}$. A further expansion of F in powers of $\tau = (T - T_c)/T_c$ and $\vec{q}$ with the following minimizing of the functional leaves one with the system of equations for superconductivity order parameters. The linearized equations for the gaps on the Fermi level read ($i = x, y, z$)

$$[q_\sigma - a_\sigma(2\tau - \sum_i \beta_{\sigma i} \nabla_i^2)]\Delta_\sigma + [Q_\pi - A_\pi(2\tau - \sum_i \beta_{\pi i} \nabla_i^2)]\Delta_\pi = 0 \qquad (6)$$

$$[q_\sigma - A_\sigma(2\tau - \sum_i \beta_{\sigma i} \nabla_i^2)]\Delta_\sigma + [q_\pi - a_\pi(2\tau - \sum_i \beta_{\pi i} \nabla_i^2)]\Delta_\pi = 0 .$$

Here

$$q_l = a_l g + b_l , \quad Q_l = A_l g + B_l , \tag{7}$$

$$a_l = \rho_l(V_l + U_l + \rho_l V_l U_l G) , \quad b_l = 1 + \rho_l U_l G , \tag{8}$$

$$A_l = \rho_l W(1 + \rho_l V_l G) , \quad B_l = \rho_l W G , \tag{9}$$

$$g = 2 \ln\left(\frac{1.13\hbar\omega_D}{\Theta_c}\right) , \tag{10}$$

$$G = \ln\left[\frac{-\mu(\mu - E_c)}{(\hbar\omega_D)^2}\right] , \tag{11}$$

$$\beta_{li} = \frac{14\zeta(3)\hbar^2 v_{li}^2}{(4\pi\Theta_c)^2} . \tag{12}$$

The superconductivity transition temperature is determined from the condition

$$Q_\sigma Q_\pi = q_\sigma q_\pi , \tag{13}$$

which guarantees the existence of nontrivial solutions of the system (6) in the homogeneous case. The terms from the gradient expansions of both bands enter both components of (6), c.f. (Zhitomirski and Dao, 2004).

5. Coherence lengths and critical fields anisotropy

An orthogonal transformation to new variables Δ_- (soft) and Δ_+ (rigid) splits (6) into independent equations

$$\Delta_\mp = \sum_i \xi_{\mp i}^2 \nabla_i^2 \Delta_\mp . \tag{14}$$

The role of the leading order parameter for the superconducting phase transition is played by Δ_-. Its nonzero value at $T < T_c$ is determined by the combination of both gaps $\Delta_{\sigma,\pi}$ vanishing simultaneously at T_c. The corresponding coherence lengths ξ_{-i} reveal the usual critical behaviour with divergency at T_c.

The scaling factors ξ_{+i} depend weakly on temperature and remain finite at T_c. These quantities turn out to be imaginary and characterize spatially periodic fluctuations of Δ_+ which relax rapidly (Örd and Kristoffel, 2000) to the homogeneous value equal to zero. The critical coherence length is expressed as

$$\xi_i^2 = \frac{(Q_\pi A_\sigma - q_\pi a_\sigma)\beta_{\sigma i} + (Q_\sigma A_\pi - q_\sigma a_\pi)\beta_{\pi i}}{Q_\sigma A_\pi + Q_\pi A_\sigma - q_\sigma a_\pi - q_\pi a_\sigma} |2\tau|^{-1} . \tag{15}$$

The thermodynamic critical field determined by the difference of the equilibrium free energies of the normal and superconducting state reads (Kristoffel et al., 2003b)

$$H_c = 8\pi\Theta_c \left[\frac{\pi}{14\zeta(3)} (\rho_\sigma \Gamma_\sigma^4 + \rho_\pi \Gamma_\pi^4) \right]^{1/2} |\tau|;, \tag{16}$$

with

$$\Gamma_\sigma = \left[\frac{Q_\pi q_\pi (Q_\sigma A_\pi + Q_\pi A_\sigma) - q_\sigma a_\pi - q_\pi a_\sigma}{Q_\sigma^2 q_\sigma A_\pi + Q_\pi^2 q_\pi A_\sigma - Q_\sigma q_\sigma^2 a_\pi - Q_\pi q_\pi^2 a_\sigma} \right]^{1/2}, \tag{17}$$

and an analogous expression for Γ_π, with $\sigma \leftrightarrow \pi$.

The critical magnetic fields are expressed as $(i \neq j \neq j')$

$$H_{c2}^i = \frac{\Phi_0}{2\pi\xi_j\xi_{j'}}, \qquad H_{c1}^i = \frac{H_c \ln \kappa_i}{\sqrt{2}\kappa_i}. \tag{18}$$

The Maki parameters

$$\kappa_i = \frac{H_{c2}^i}{\sqrt{2}H_c} \tag{19}$$

are independent on temperature near T_c. The gradient expansions in the anisotropic Ginzburg-Landau type scheme developed here are justified if the scaling factors

$$\xi_{li} = \sqrt{\frac{\beta_{li}}{2}} \tag{20}$$

are smaller than the coherence lengths: $\xi_{li} < \xi_i(T)$. (Koshelev and Golubov, 2004) For MgB$_2$ in connection with ξ_c and $\xi_{\pi c}$ the applicability of the Ginzburg-Landau type approach can be expected to be valid in a very narrow region near T_c (Koshelev and Golubov, 2004). Our model of MgB$_2$ superconductivity gives according to (12), (15), (20) for the c-axis coherence characteristics $\xi_c = 8|\tau|^{-1/2}$ and $\xi_{\pi c} = 28$ (nm). Correspondingly the appropriate temperature region is limited by the condition $|\tau| < 0.08$, which is somewhat weaker as found in (Koshelev and Golubov, 2004). The main physical reason for this limitation is the anisotropy of the Fermi surface (small v_c^σ) of the 2D σ-band which reduces $\xi_{\sigma c}$ and leads by that to the decrease of ξ_c. We mention also the following calculated results for $|\tau| = 1 : H_c(0) = 0.27T, \xi_{ab}(0) = 20 \ nm, \kappa_{ab}(0) = 5.1, \kappa_c(0) = 2.1$.

With the aim to stay in the narrow region near T_c where the theory can be expected to work we limit ourselves to the discussion of MgB$_2$ critical fields anisotropy calculated according to (15). In this temperature interval the coefficients of anisotropy are expected to be constant and are given by

$$\gamma_{H_{c2}} = \frac{H_{c2}^{ab}}{H_{c2}^c} = \left[\frac{(Q_\pi A_\sigma - q_\pi a_\sigma)\beta_{\sigma ab} + (Q_\sigma A_\pi - q_\sigma q_\pi)\beta_{\pi ab}}{(Q_\pi A_\sigma - q_\pi a_\sigma)\beta_{\sigma c} + (Q_\sigma A_\pi - q_\sigma q_\pi)\beta_{\pi c}} \right]^{1/2}, \tag{21}$$

$$\gamma_{H_{c1}} = \frac{H_{c1}^c}{H_{c1}^{ab}} = \gamma_{H_{c2}} \frac{\ln \kappa_c}{\ln \kappa_{ab}} . \tag{22}$$

Experimentally one has found (Lyard et al., 2002; Zehetmayer et al., 2002) that γ_{c2} diminishes with temperature, however, for γ_{c1} the trend is opposite. Such a behaviour of $\gamma_{c2}(T)$ has been obtained also theoretically (Dahm and Schopohl, 2003). Our calculation (see also (Floris et al., 2005)) has given for the higher critical field $\gamma_{H_{c2}} = 2.46$ which is comparable with the experimental data $\gamma_{H_{c2}}(T_c) \approx 2$ (Eltsev et al., 2002; Welp et al., 2003), 2.2 (Shi et al., 2003), 2.3÷2.7 (Bando et al., 2004; Angst et al., 2002), and 1.5±0.5 (Zehetmayer et al., 2004). The calculated $\gamma_{H_{c1}} = 1.94$ is close to the measured values $\gamma_{H_{c1}}(T_c) \approx 2$ (Lyard et al., 2004; Perkins et al., 2002), 2.16 (Bando et al., 2004). The present two-band superconductivity model with intra- and interband pairing channels reproduces the observed MgB_2 critical magnetic fields anisotropies near T_c.

Acknowledgements

This work was supported by Estonian Science Foundation grant No 6540.

References

Akimitsu, J. and Murakama, T. (2003) Superconductivity in MgB_2, *Physica C* **388–389**, 98–102.

Angst, M., Puzniak, R., Wisniewski, A., Jun, J., Kazakov, S. M., Karpinski, J., Roos, J., and Keller, H. (2002) Temperature and field dependence of the anisotropy of MgB_2, *Phys. Rev. Lett.* **88**, 167004-4.

Bando, H., Yamaguchi, Y., Shirakawa, N., and Yanagisawa, T. (2004) Anisotropy in the upper and lower critical fields of MgB_2 single crystals, *Physica C* **412–414**, 258–261.

Bianconi, A., Agrestini, S., Castro, D. D., Campi, G., Zangari, G., Saini, N. L., Saccone, A., Negri, S. D., Giovannini, H., Profeta, G., Continenca, A., Satta, G., Massidda, S., Cassetta, A., Pifferi, A., and Calapietro, H. (2002) Scaling of the critical temperature with the Fermi temperature in diborides, *Phys. Rev.* **65**, 174515-5.

Brinkman, A., Golubov, A. A., Rogalla, H., Dolgov, O. V., Kortus, J., Kong, Y., Jepsen, O., and Andersen, O. K. (2002) Multiband model for tunneling in MgB_2 junctions, *Phys. Rev. B* **65**, 180517-4.

Bussmann-Holder, A. and Bianconi, A. (2003) Raising the diboride superconductor transition temperature using quantum interference effects, *Phys. Rev. B* **67**, 132509-4.

Dahm, T. and Schopohl, N. (2003) Fermi surface topology and the upper critical field in two-band superconductors: application to MgB_2, *Phys. Rev. Lett.* **91**, 017001-4.

Eltsev, Y., Lee, S., Nakao, K., Chikumoto, N., Tajima, S., Koshizuka, N., and Murakami, M. (2002) Anisotropic superconducting properties of MgB_2 single crystals, *Physica C* **378–381**, 61–64.

Floris, A., Profeta, G., Zalhiotakis, N. N., Lüders, M., Marques, M. A. L., Franchini, C., Gross, E. K. U., Continenca, A., and Massidda, S. (2005) Superconducting properties of MgB_2 from first principles, *Phys. Rev. Lett.* **94**, 037004-4.

Geerk, J., Schneider, L., Linken, G., Zaitsev, A. G., Heid, R., Bohnen, K. P., and von Löhneysen, H. (2005) Observation of interband pairing interaction in a two-band superconductor MgB_2, *Phys. Rev. Lett.* **94**, 227005-4.

Karpinski, A., Zhigadlo, N. D., Schuck, G., Kazakov, S. M., Batlogg, B., Rogacki, K., Pozniak, R., Jun, J., Müller, E., Wägli, P., Gonelli, R., Daghero, D., Ummarino, G. A., and Stepanov, V. A. (2005) Al substitution in MgB_2 crystals: Influence on superconducting and structural properties, *Phys. Rev.* **71**, 174506-15.

Kortus, J., Dolgov, O. V., and Kremer, R. K. (2005) Band filling and interband scattering effect in MgB_2, *Phys. Rev. Lett.* **94**, 027002-4.

Kortus, J., Mazin, I. I., Belashchenko, K. D., Antropov, V. P., and Boyer, L. L. (2001) Superconductivity of metallic boron in MgB_2, *Phys. Rev. Lett.* **86**, 4656-4659.

Koshelev, A. E. and Golubov, A. A. (2004) Why Magnesium Diboride is not described by anisotropic Ginzburg-Landau theory, *Phys. Rev. Lett.* **22**, 107008-4.

Kristoffel, N., Örd, T., and Rägo, K. (2003)a A description of MgB_2 superconductivity including $\sigma - \pi$ bands coupling, *J. Supercond.* **16**, 517-519.

Kristoffel, N., Örd, T., and Rägo, K. (2003)b MgB_2 two-gap superconductivity with intra- and interband couplings, *Europhys. Lett.* **61**, 109-115.

Kristoffel, N. and Rubin, P. (2002) Pseudogap and superconductivity gaps in a two-band model with the doping determined components, *Solid State Commun.* **122**, 265-268.

Kruchinin, S. P. and Nagao, H. (2005) Two-gap superconductivity in MgB_2, *Phys. Particles and Nuclei* **36**, 127-130, Suppl.1.

Liu, A. Y., Mazin, I. I., and Kortus, J. (2001) Beyond Eliashberg superconductivity in MgB_2: Anharmonicity, two-phonon scattering and multiple gaps, *Phys. Rev. Lett.* **87**, 087005-4.

Lyard, L., Samuely, P., Szabo, P., Klein, T., Marcenat, C., Paulius, L., Kim, K. H. P., Yung, C. U., Lee, H.-S., Kang, B., Choi, S., Lee, S.-I., Marcus, J., Blanchard, S., Jansen, A. G. M., Welp, U., Karapetrov, G., and Kwok, W. K. (2002) Anisotropy of the upper critical field and critical current in single crystal MgB_2, *Phys. Rev. B* **66**, 180503-4 (R).

Lyard, L., Szabo, P., Klein, T., Marcus, J., Marcenat, C., Kim, K. H., Kang, B. W., Lee, H. S., and Lee, S. I. (2004) Anisotropy of the lower and upper critical fields in MgB_2 single crystals, *Phys. Rev. Lett.* **92**, 057001-4.

Mazin, I. I. and Antropov, V. P. (2003) Electronic structure, electron-phonon coupling, and multiband effects in MgB_2, *Physica C* **385**, 49-65.

Örd, T. and Kristoffel, N. (2000) Two relaxation times and the high-temperature superconductivity two-component scenario, *Physica C* **331**, 13-17.

Perkins, G. K., Moore, J., Bugoslavsky, Y., Cohen, L. F., Jun, J., Kazakov, S. M., Karpinski, J., and Caplin, A. D. (2002) Superconducting critical fields and and anisotropy of MgB_2 single crystal, *Supercond. Sci. Technol.* **15**, 1156-1159.

Putti, M., Affonte, M., Manfrinetti, P., and Palenzona, A. (2003) Effects of Al doping on the normal and superconducting properties of MgB_2: A specific heat study, *Phys. Rev. B* **68**, 094514-6.

Putti, M., Ferdeghini, C., Monni, M., Pallecchi, I., Tarantini, C., Manfrinetti, P., Palenzona, A., Daghero, D., Gonelli, R. S., and Stepanov, V. A. (2005) Critical field of Al-doped MgB_2 samples: correlation with the suppression of the σ-band gap, *Phys. Rev. B* **71**, 144505-6.

Shi, Z. X., Tokunaga, M., Tamegai, T., Takano, Y., Togano, K., Kito, H., and Ihara, H. (2003) Out-of-plane and in-plane anisotropy of upper critical field in MgB_2, *Phys. Rev. B* **68**, 104513-7.

Tsuda, S., Tsuda, S., Yokoya, T., Takano, Y., Kito, H., Matsushita, A., Yin, E., Itoh, J., Harima, H., and Shin, S. (2003) Definitive experimental evidence for two-band superrconductivity in MgB_2, *Phys. Rev. Lett.* **91**, 127001-4.

Ummarino, G. A., Gonelli, R. S., Massidda, S., and Bianconi, A. (2004) Two-band Eliashberg equations and the experimental T_c of the diboride $Mg_{1-x}Al_xB_2$, *Physica C* **407**, 121–127.

Welp, U., Rydh, A., Karapetrov, G., Kwok, W. K., Crabtree, G. W., Marcenat, C., Paulius, L., Klein, T., Marcus, J., Kim, K. H. P., Jung, C. U., Lee, H.-S., Kang, B., and Lee, S.-I. (2003) Superconducting transition and phase diagram of single-crystal MgB_2, *Phys. Rev. B* **67**, 012505–4(R).

Zehetmayer, M., Eisterer, M., Jun, J., Kazakov, S. M., Karpinski, J., and Weber, H. W. (2004) Magnetic field dependence of the reversible mixed state properties of superconducting MgB_2 single crystals, *Phys. Rev. B* **70**, 214516–12.

Zehetmayer, M., Eisterer, M., Jun, J., Kazakov, S. M., Karpinski, J., Wisniewski, A., and Weber, H. W. (2002) Mixed state properties of superconductor MgB_2 single crystals, *Phys. Rev. B* **66**, 052505–4.

Zhitomirski, M. E. and Dao, V. H. (2004) Ginzburg-Landau theory of vortices in a multigap superconductor, *Phys. Rev. B* **69**, 054508–11.

NANOSIZE TWO-GAP SUPERCONDUCTIVITY

Hidemi Nagao[a] (nagao@wriron1.s.kanazawa-u.ac.jp),
Hiroyuki Kawabe[b] (kawabe@kinjo.ac.jp),
and Sergei P. Kruchinin[c] (skruchin@i.com.ua)

[a]Division of Mathematical and Physical Science, Graduate School of Natural Science and Technology, Kanazawa University, Kakuma, Kanazawa 920-1192, Japan

[b]Department of Social Work, Faculty of Social Work, Kinjo University, 1200 Kasama, Hakusan Ishikawa 924-8511, Japan

[c]Bogolyubov Institute for Theoretical Physics, The Ukrainian National Academy of Science, Kiev 252143, Ukraine

Abstract. We investigate properties of nanosize two-gap superconductivity by using a two-sublevel model in the framework of a mean field approximation. A model corresponding to a nanosize two-gap superconductivity is presented, and the partition function of the nanosize system is analytically derived by using a path integral approach. A definition of the critical level spacing of the two-gap superconductivity is also presented, and we discuss condensation energy and parity gap of the two-gap superconductivity in relation to the size dependence of those properties with two bulk gaps and effective pair scattering process between two sublevels.

Key words: Two-gap superconductivity, Ultrasmall grain

1. Introduction

The recent discovery of superconductivity of MgB_2(Nagamatsu et. al., 2001) has been much attracted great interest in the properties and for elucidation of its mechanism from both experimental and theoretical view points. A crucial role of the electron-phonon (e-p) interaction has been strongly suggested in the superconductivity of MgB_2(Liu et.al., 2001).Recent band calculations of MgB_2(Kortus et. al., 2001; An and Picket, 2001; Kato et. al., 2004) with the McMillan formula (McMillan, 1968) of transition temperature have supported the e-p interaction mechanism for the superconductivity. Since the discovery, the possibility of two-band superconductivity arising from other

117

K. Scharnberg and S. Kruchinin (eds.),
Electron Correlation in New Materials and Nanosystems, 117–127.

mechanisms has also been discussed in relation to two gap functions theoretically.

Recently, two-band or multi-band superconductivity has been theoretically investigated in relation to superconductivity arising from coulomb repulsive interactions (Konsin et. al., 1988; Yamaji and Shimoi, 1994; Combescot and Leyronas, 1995; Konsin and Sorkin, 1998; Nagao et. al., 2000; Kondo, 2001; konbo, 2002). The concept of the two-band model has been discussed in 1958 (Suhl et. al., 1959; Moskalenko, 1959; konbo, 1963). Recently, we have pointed out importance of many-band effects in superconductivity (Nagao et. al., 1999; Nagao and Yaremko et. al., 2002; Nagao and Kruchinin et. al., 2002; Nagao et. al., 2003; Kruchinin and Nagao, 2005). We have also investigated anomalous phases in two-band model by using the Green function techniques (Nagao et.al., 2000; Nagao, Kawabe and Kruchinin, to appear.). The expressions of the transition temperature for several phases have been derived, and the approach has been applied to superconductivity in several crystals by charge injection (Nagao et. al., 1997; Nagao et. al., 1998; Nagao et. al., 2000).

Recent experiments (Black et. al., 1996; Ralph et. al., 1995) by Black et al. have also generated much interest in the size dependence of the superconductivity. Properties of ultrasmall superconducting grains have been theoretically investigated by many groups (Jankó et. al., 1994; von Delft et. al., 1996; Smith and Ambegaokar, 1996; Matveev and Larkin, 1997; Braun and von Delft, 1998; Gladilin et. al., 2002; Braun and von Delft, 1999). In such ultrasmall grains, the fundamental theoretical question for the size dependency of the superconductivity was noticed by Anderson (Anderson, 1959). The standard BCS theory gives a good description of the phenomenon of superconductivity in large sample. However, as the size of a superconductor becomes small, the BCS theory fails. In ultrasmall Al grains, the bulk gap has been discussed in relation to physical properties in ultrasmall grain such as the parity gap (Matveev and Larkin, 1997), condensation energy (Gladilin et. al., 2002), electron correlation (von Delft, 1996) etc. with the dependence of level spacing (Smith and Ambegaokar, 1996) of samples.

In this paper, we investigate properties of nanosize two-gap superconductivity by using a two-sublevel model in the framework of a mean field approximation. A model corresponding to a nanosize two-gap superconductivity is presented, and the partition function of the nanosize system is analytically derived by using a path integral approach. A definition of the critical level spacing of the two-gap superconductivity is also presented, and we discuss condensation energy and parity gap of the two-gap superconductivity in relation to the size dependence of those properties with two bulk gaps and effective pair scattering process between two sublevels.

2. Nanosize two-gap superconductivity

In nanosize grain of a superconductor, the quantum level spacing approaches the superconducting gap. In the case of two-gap superconductor, we can consider a model with two sublevels corresponding to two independent bands. In this section, we present a model for nanosize two-gap superconductivity and an expression of the partition function of the system.

2.1. HAMILTONIAN FOR NANOSIZE GRAINS

We consider a pairing Hamiltonian with two sublevels corresponding to two bands 1 and 2 written as

$$H = H_0 + H_{int} \ , \tag{1}$$

where

$$H_0 = \sum_{j,\sigma} \left[\varepsilon_{1j} - \mu \right] a_{j\sigma}^\dagger a_{j\sigma} + \sum_{k,\sigma} \left[\varepsilon_{2k} - \mu \right] b_{k\sigma}^\dagger b_{k\sigma} \ , \tag{2}$$

$$H_{int} = -g_1 \sum_{j,j' \in I} a_{j\uparrow}^\dagger a_{j\downarrow}^\dagger a_{j'\downarrow} a_{j'\uparrow} - g_2 \sum_{k,k' \in J} b_{k\uparrow}^\dagger b_{k\downarrow}^\dagger b_{k'\downarrow} b_{k'\uparrow}$$

$$+ g_{12} \sum_{j \in I, k \in J} a_{j\uparrow}^\dagger a_{j\downarrow}^\dagger b_{k\downarrow} b_{k\uparrow} + g_{12} \sum_{j \in I, k \in J} b_{k\uparrow}^\dagger b_{k\downarrow}^\dagger a_{j\downarrow} a_{j\uparrow} \ . \tag{3}$$

$a_{j\sigma}^\dagger (a_{j\sigma})$ and $b_{j\sigma}^\dagger (b_{j\sigma})$ are the creation (annihilation) operators in sublevels 1 and 2 with spin σ and energies ε_{1j} and ε_{2j}, respectively. The operators for each sublevel satisfy the anticommutation relations, and the operators between sublevels are independent. μ is the chemical potential. The second term in Eq.(1) is the interaction Hamiltonian. g_1 and g_2 are the effective interaction constant for sublevels 1 and 2. g_{12} is an effective interaction constant, which corresponds to the pair scattering process between two bands. The sums with respect to j and k in Eq.(3) are over the set I of N_{1I} states corresponding to a half-filled band 1 with fixed width $2\omega_{1D}$ and the set J of N_{2J} states for band 2, respectively.

In this study, we assume that the Deby energies for the two sublevel systems are the same $\omega_{1D} = \omega_{2D} = \omega_D$. With this assumption, the ratio of N_{1I} and N_{2J} can be related to the (constant) densities of state (DOS) ρ_1 and ρ_2 for two bands as follows: $N_{1I}/N_{2J} = \rho_1/\rho_2$. We write the interaction constants g_1 and g_2 as $d_1\lambda_1$ and $d_2\lambda_2$, respectively. $d_1 = 2\omega_D/N_{1I}$ and $d_2 = 2\omega_D/N_{2J}$ are the energy level spacings, and λ_1 and λ_2 are dimensionless parameters. In(3) we have introduced a pairing interaction $g_{12} = \sqrt{d_1 d_2}\lambda_{12}$. between the two sublevel systems. In summary, we have a relation of $\rho_1/\rho_2 = N_{1I}/N_{2J} = d_2/d_1$.

2.2. PATH INTEGRAL APPROACH

It is convenient to introduce a path integral approach for treatment of the fluctuations of the order parameters. This approach gives an exact expression for the grand partition function of a superconductor. $Z(\mu, T) = Tr \exp [-(H - \mu N)/T]$, where T is the temperature, and N is the number operator in the grain. The idea of the path integral approach is to replace the formulation for system of the problem in terms of electronic operators by equivalent formulation in terms of the superconducting order parameter.

By the path integral approach, we obtain an expression of the grand partition function for the Hamiltonian of Eq. (1).

$$Z(\mu, T) = \int D\Delta_1 D\Delta_1^* D\Delta_2 D\Delta_2^* e^{-S[\Delta_1, \Delta_2]} , \qquad (4)$$

where the action $S[\Delta_1, \Delta_2]$ is defined as

$$S[\Delta_1, \Delta_2] = -\sum_j \left[Tr \ln G_{1j}^{-1} - \frac{\xi_{1j}}{T} \right] - \sum_k \left[Tr \ln G_{2k}^{-1} - \frac{\xi_{2k}}{T} \right]$$

$$+ \int_0^{1/T} d\tau \frac{1}{g_1 g_2 - g_{12}^2} \left[g_2 |\Delta_1(\tau)|^2 + g_1 |\Delta_2(\tau)|^2 \right.$$

$$+ g_{12} \left(\Delta_1(\tau) \Delta_2(\tau)^* + \Delta_1(\tau)^* \Delta_2(\tau) \right) \right] . \qquad (5)$$

Δ_1 and Δ_2 are bulk gaps for sublevels 1 and 2, respectively. Here, $\xi_{1j} = \varepsilon_{1j} - \mu$ and $\xi_{2k} = \varepsilon_{2k} - \mu$, and the inverse Green functions

$$G_{aj}^{-1}(\tau, \tau') = \left[-\frac{d}{d\tau} - \xi_{aj} \sigma^z - \Delta_a(\tau) \sigma^+ - \Delta_a^*(\tau) \sigma^- \right] \delta(\tau - \tau') , \qquad (6)$$

where $a = 1, 2$ for band label, $\sigma^\pm = \sigma^x \pm i\sigma^y$, and $\sigma^{x,y,z}$ are the Pauli matrices. G_1^{-1} and G_2^{-1} satisfy antiperiodic boundary conditions.

In the case of stronger interaction, $\Delta_1 \gg d_1$ and $\Delta_2 \gg d_2$, we consider a mean field approximation for the order parameters in the path integral approach. Substituting a time-independent order parameters into the action of Eq. (5), we have

$$\Omega(\mu) = \sum_j \left(\xi_{1j} - \epsilon_{1j} \right) + \sum_k \left(\xi_{2k} - \epsilon_{2k} \right)$$

$$+ \frac{1}{g_1 g_2 - g_{12}^2} \left[g_2 \Delta_1^2 + g_1 \Delta_2^2 + g_{12} \left(\Delta_1^* \Delta_2 + \Delta_1 \Delta_2^* \right) \right] , \qquad (7)$$

where $\epsilon_{1j} = (\xi_{1j}^2 + \Delta_1^2)^{1/2}$, and $\epsilon_{2k} = (\xi_{2k}^2 + \Delta_2^2)^{1/2}$. In Eq.(7), the values of Δ_1 and Δ_2 must be chosen in a way which minimizes Ω. From the minimization of Ω, we obtain a coupled gap equation at zero temperature for the two-gap system:

$$\begin{pmatrix} \Delta_1 \\ \Delta_2 \end{pmatrix} = \begin{pmatrix} g_1 \sum_j \frac{1}{2\epsilon_{1j}} & -g_{12} \sum_k \frac{1}{2\epsilon_{2k}} \\ -g_{12} \sum_j \frac{1}{2\epsilon_{1j}} & g_2 \sum_k \frac{1}{2\epsilon_{2k}} \end{pmatrix} \begin{pmatrix} \Delta_1 \\ \Delta_2 \end{pmatrix} . \tag{8}$$

From the coupled gap equation of Eq.(8), we formally obtain an expression of bulk gap for two-gap superconductivity at zero temperature:

$$|\Delta_a| = \omega \sinh^{-1}\left(\frac{1}{\eta_a}\right) , \quad (a = 1, 2) , \tag{9}$$

where

$$\frac{1}{\eta_1} = \frac{\lambda_2 + \alpha_\pm [\eta_1, \eta_2] \lambda_{12}}{\lambda_1 \lambda_2 - \lambda_{12}^2} , \quad \frac{1}{\eta_2} = \frac{\lambda_1 + \alpha_\pm^{-1} [\eta_1, \eta_2] \lambda_{12}}{\lambda_1 \lambda_2 - \lambda_{12}^2} , \tag{10}$$

and $\alpha_\pm [\eta_1, \eta_2] = \pm \sinh\left(\frac{1}{\eta_1}\right) / \sinh\left(\frac{1}{\eta_2}\right)$ In two-band superconductivity, we can consider two cases for phase of the gaps: $\text{sgn}(\Delta_1) = \text{sgn}(\Delta_2)$, and $\text{sgn}(\Delta_1) = -\text{sgn}(\Delta_2)$. For the same phase, α_+ is used in Eqs.(10), and we use α_- for the opposite phase. Note that $\Delta_1 = -\Delta_2$ in the limit of strongly positive intersublevel coupling λ_{12}, that is, opposite phase. On the other hand, we find the same phase for negative λ_{12}. The transition temperature is defined from the absolute value of the gap functions. In $\lambda_{12} = 0$, we find the same results of two bulk gaps derived from conventional BCS theory for two independent sublevels.

3. Discussion

In this section, we discuss properties such as condensation energy, critical level spacing, and parity gap of nanosize two-gap superconductivity by using the partition function derived in previous section.

3.1. CONDENSATION ENERGY

In nanosize superconductivity, the condensation energy can be defined as $E_{N,b}^C(\lambda) = E_{N,b}^G(0) - E_{N,b}^G(\lambda) - n\lambda d$, where $E_{N,b}^G$ is the ground state energy of N-electron system in the interaction band. b is the number of electron on single occupied levels, and λ and n are the dimensionless coupling parameter and the number of pair occupied level, respectively. In the case of nanosize two-band system, the condensation energy can be written as $E_{N_1,b_1;N_2,b_2}^C(\lambda_1, \lambda_2, \lambda_{12}) =$

$E^G_{N_1,b_1;N_2,b_2}(0,0,0) - E^G_{N_1,b_1;N_2,b_2}(\lambda_1,\lambda_2,\lambda_{12}) - n_1\lambda_1 d_1 - n_2\lambda_2 d_2$, where $E^G_{N_1,b_1;N_2,b_2}(\lambda_1,\lambda_2,\lambda_{12})$ means the ground state energy of $(N_1 + N_2)$-electron system. From Eq. (7), the condensation energy of two-sublevel system can be expressed by the condensation energy of independent single level systems:

$$E^C_{N_1,b_1;N_2,b_2}(\lambda_1,\lambda_2,\lambda_{12}) = E^C_{N_1,b_1}(\lambda_1) + E^C_{N_2,b_2}(\lambda_2)$$
$$- \frac{\lambda_{12}^2}{\lambda_1\lambda_2 - \lambda_{12}^2}\left(\frac{\Delta_1^2}{d_1\lambda_1} + \frac{\Delta_2^2}{d_2\lambda_2} + \frac{2\left(\Delta_1^*\Delta_2 + \Delta_1\Delta_2^*\right)}{\sqrt{d_1 d_2}\lambda_{12}}\right),$$

$$(11)$$

where $E^C_{N_1,b_1}(\lambda_1)$ and $E^C_{N_2,b_2}(\lambda_2)$ correspond to the condensation energy for single band case. In the same phases of Δ_1 and Δ_2, the condensation energy of Eq. (11) decreases, that is, appearing the instabilization by coupling constant λ_{12}. On the other hand, in the opposite phases, the condensation energy becomes larger, because $\Delta_1^*\Delta_2 + \Delta_1\Delta_2^* < 0$. We can expect that the condensation energy of two-gap superconductivity becomes more stable than that of two independent systems due to the intersublevel coupling λ_{12} and the opposite phases.

3.2.   CRITICAL LEVEL SPACING

To discuss the critical level spacing for two-gap system, which means both gap functions vanish at a level spacing, $\Delta_1 = \Delta_2 = 0$, we start from the coupled gap equation of Eq. (8). For the case of the critical level spacing of two-gap system, we have

$$1 = \lambda_1 \sum_j \frac{1}{2|\tilde{\xi}_{1j}|} + \lambda_2 \sum_k \frac{1}{2|\tilde{\xi}_{2k}|} - (\lambda_1\lambda_2 - \lambda_{12}^2) \sum_j \frac{1}{2|\tilde{\xi}_{1j}|} \sum_k \frac{1}{2|\tilde{\xi}_{2k}|}, \quad (12)$$

where $\tilde{\xi}_i = \xi_i/d_i$ for sublevel $i = 1,2$. For the odd or even electron number parity in the grain, Eq. (12) can be approximately solved by using the digamma function: For odd case, the critical level spacing becomes

$$d^o_{1c} = \omega_D e^\gamma \exp\left[-\frac{1}{\lambda}\right], \quad d^o_{2c} = \frac{d_2}{d_1}d^o_{1c}, \quad (13)$$

and for even case,

$$d^e_{1c} = 4\omega_D e^\gamma \exp\left[-\frac{1}{\lambda}\right], \quad d^e_{2c} = \frac{d_2}{d_1}d^e_{1c}. \quad (14)$$

Here, we use

$$\frac{1}{\lambda} = \frac{1}{2x}\left[\lambda_1 + \lambda_2 - ax + \sqrt{(\lambda_1 - \lambda_2 - ax)^2 + 4\lambda_{12}^2}\right] \quad (15)$$

with

$$x = \lambda_1\lambda_2 - \lambda_{12}^2 \ , \quad a = \log\frac{d_1}{d_2} \ . \tag{16}$$

From these expressions, we find some relations:

$$d_{1c}^e = 4d_{1c}^o \ , \quad d_{2c}^e = 4d_{2c}^o \ . \tag{17}$$

and

$$d_{1/2c}^o \approx \frac{e^\gamma}{2}\exp\left[\frac{1}{\eta_{1/2}} - \frac{1}{\lambda}\right]\tilde{\Delta}_{1/2} \ . \tag{18}$$

In the case of $|\lambda_1 - \lambda_2| \gg \lambda_{12}$, Eq. (18) can be approximately rewritten as

$$d_{1/2c}^o \approx \frac{e^\gamma}{2}\exp\left[\frac{\lambda_2 - \lambda_1 + 2\alpha\lambda_{12}}{\lambda_1\lambda_2 - \lambda_{12}^2}\right]\tilde{\Delta}_{1/2} \ . \tag{19}$$

On the other hand, in the limit of $|\lambda_1 - \lambda_2| \ll \lambda_{12}$, we have

$$d_{1/2c}^o \approx \frac{e^\gamma}{2}\exp\left[\frac{(1 + \alpha)\lambda_{12}}{\lambda_1\lambda_2 - \lambda_{12}^2}\right]\tilde{\Delta}_{1/2} \ . \tag{20}$$

For case of $\lambda_{12} = 0$, Eq. (18) can be rewritten as $d_{1/2c}^o \approx \exp[\gamma]/2\exp[1/\lambda_1 - 1/\lambda_2]\tilde{\Delta}_{1/2}$. Therefore, when the coupling constants λ_1 and λ_2 become same value, we have a similar relation to that for single level system; $d_{1/2c}^o \approx 0.89\tilde{\Delta}_{1/2}$. These results suggest the critical level spacing strongly depend upon λ_{12} and the difference between the effective interaction constants for sublevels. The relation in Eq. (17) is the same relation in the conventional nanosize BCS theory.

3.3. PARITY GAP

In this subsection, we consider the parity gap in the case of two-gap super-conductivity in ultrasmall grains. The parity gap in a single band system is the difference between the ground state energy of a grain containing $2n + 1$ electrons (odd parity) and the average ground state energy of grains containing $2n$ and $2n + 2$ electrons (even parity). It is a measure for the cost in energy of having one unpaired electron. In the case of two sublevel spacings, the chemical potential lies halfway between the highest occupied and the lowest unoccupied levels of smaller level spacing in the half-filled case as shown in Fig. 1(a). We assume that $d_1 < d_2$ and that the numbers of occupied levels corresponding to each sublevel are n_1 and n_2, respectively. Then, the total number of electron becomes $N = 2n_1 + 2n_2$. When we consider $N = 2n_1 + 2n_2 + 1$, the chemical potential lies on the level ε_{1n_1+1} as shown in

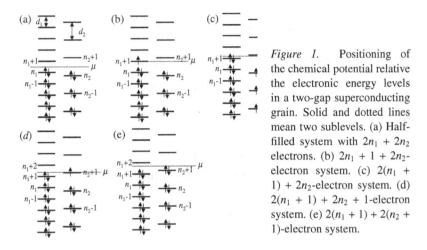

Figure 1. Positioning of the chemical potential relative the electronic energy levels in a two-gap superconducting grain. Solid and dotted lines mean two sublevels. (a) Half-filled system with $2n_1 + 2n_2$ electrons. (b) $2n_1 + 1 + 2n_2$-electron system. (c) $2(n_1 + 1) + 2n_2$-electron system. (d) $2(n_1 + 1) + 2n_2 + 1$-electron system. (e) $2(n_1 + 1) + 2(n_2 + 1)$-electron system.

Fig. 1(b). Figure 1(c) shows the position of the chemical potential in the case of $N = 2n_1 + 2n_2 + 2$. The parity gap of nanosize two-gap superconductivity is written as

$$\Delta_p^1 = E_{2n_1+1+2n_2,1}^G - \frac{1}{2}\left(E_{2n_1+2n_2,0}^G + E_{2(n_1+1)+2n_2,0}^G\right) . \tag{21}$$

From Eq. (7) and the ground state energy as $E_{N,b}^G = \Omega_{\mu N} + \mu_N N$, we obtain

$$\Delta_p^1 = \Delta_1 - \frac{d_1}{4}\left(\frac{\rho_1}{\rho_2} - 1\right) . \tag{22}$$

From Figs. 1(c), (d), and (e), we can define another parity gap:

$$\Delta_p^2 = E_{2(n_1+1)+2n_2+1,1}^G - \frac{1}{2}\left(E_{2(n_1+1)+2n_2,0}^G + E_{2(n_1+1)+2(n_2+1),0}^G\right) . \tag{23}$$

From the latter definition of Eq. (23), we have

$$\Delta_p^2 = \Delta_2 - \frac{d_2}{4}\left(\frac{3\rho_2}{\rho_1} - 1\right) . \tag{24}$$

The present results suggest two kinds of the dependence of the parity gap on the level spacing. The parity gap does not depend upon the effective interaction λ_{12}. The structure around Fermi level plays an important role of the contribution to the size dependence on the parity gap.

3.4. CONCLUDING REMARKS

We have investigated properties of nanosize two-gap superconductivity by using a two-sublevel model in the framework of a mean field approximation. From the discussion for the condensation energy in nanosize two-gap superconductivity, the phases of the gaps is very important to stabilize the superconductivity. In the same phases, the two-gap superconductivity instabilizes by coupling constant λ_{12}. On the other hand, in the opposite phases, the superconductivity becomes stable. We can expect that the condensation energy of two-gap superconductivity becomes more stable than that of two independent systems due to the intersublevel coupling λ_{12} and the opposite phases.

We have also discussed the critical level spacing for two-gap superconductivity in ultrasmall grain. These results suggest the critical level spacing strongly depend upon λ_{12} and the difference between the effective interaction constants for sublevels. These results suggest that the relation between the critical level spacing and the bulk gaps is modified to compare with the result obtained in the ultrasmall grain of Al superconductivity.

In the parity gap in two-gap superconductivity, the present results suggest two kinds of the dependence of the parity gap on the level spacing and that the structure around Fermi level plays an important role of the contribution to the size dependence on the parity gap. The prity gap does not depend upon the effective interaction λ_{12}.

In the case of cluster system, we have to treat a more accurate approach for investigating these physical properties beyond the mean field approximation presented in this study, and we have also to consider the contribution of the surface of samples to the level structure around Fermi level. We will present these problems elsewhere (Kawabe, Nagao and Kruchinin, to appear.). From the present results, we might expect the possibility of a new multi-gap superconductivity arising in nanosize region with higher critical transition temperature.

In summary, a model corresponding to a nanosize two-gap superconductivity has been presented, and an expression of the partition function of the nanosize system has been analytically derived by using a path integral approach. A definition of the critical level spacing of the two-gap superconductivity has been also presented, and we discuss condensation energy and parity gap of the two-gap superconductivity in relation to the size dependence of those properties with two bulk gaps and effective pair scattering process between two sublevels.

Acknowledgments

H.N is grateful for a financial support of the Ministry of Education, Science and Culture of Japan (Research No. 17064013, No.16032204). The authors thank Profs. S. Aono, M. Kimura, K. Nishikawa, K. Yamaguchi for their continued encouragement helpful discussion.

References

An, J. M. and Picket, W. E. (2001) *Phys. Rev. Lett.* **86**, 4366.

Anderson, P. W. (1959) *J. Phys. Chem. Solids* **11**, 28.

Black, C. T., Ralph, D. C. and Tinkham, M. (1996) *Phys. Rev. Lett.* **76**, 688.

Braun, F. and von Delft, J. (1998) *Phys. Rev. Lett.* **81**, 4712.

Braun, F. and von Delft, J. (1999) *Adv. Sol. State Phys.*, **39**, 341.

Combescot, R. and Leyronas, X. (1995) *Phys. Rev. Lett.* **75**, 3732.

Gladilin, V. N., Fomin, V. M. and Devreese, J. T. (2002) *Solid Sate Comm.* **121**, 519.

Jankó, B., Smith, A. and Ambegaokar, V. (1994) *Phys. Rev.* B **50**, 1152.

Kato, N., Nagao, H., Nishikawa, K., Nishidate, K., Endo, K. (2004) *Int. J. Quantum Chem.* **96**, 457.

Kawabe, H., Nagao, H. and Kruchinin, S. P., to appear.

Kondo, J. (1963) *Prog. Theor. Phys.* **29**, 1.

Kondo, J. (2001) *J. Phys. Soc. Jpn.* **70**, 808.

Kondo, J. (2002) *J. Phys. Soc. Jpn.* **71**, 1353.

Konsin, P., Kristoffel, N. and Örd, T. (1988) *Phys. Lett.* A **129**, 339.

Konsin, P. and Sorkin, B. (1998) *Phys, Rev. B* **58**, 5795.

Kortus, J., Mazin, I. I., Belashenko, K. D., Antropov, V. P. and Boyer, I. L. (2001) *Phys. Rev. Lett.* **86**, 4656.

Kruchinin, S. P. and Nagao, H. (2005) *Phys. Particle Nuclei*, **36** *Suppl.*, S127.

Liu, A. Y., Mazin, I. I., and Kortus, J. (2001) *Phys. Rev. Lett.* **87**, 087005.

Matveev, K. A. and Larkin, A. I. (1997) *Phys. Rev. Lett.* **78**, 3749.

McMillan, W. L. (1968) *Phys. Rev.* **167**, 331.

Moskalenko, V. A. (1959) *Fiz. Met. Metalloved* **8**, 503.

Nagamatsu, J., Nakamura, N., Muranaka, T., Zentani, Y., and Akimitsu, J. (2001) *Nature* **410**, 63.

Nagao, H., Nishino, M., Mitani, M., Yoshioka, Y. and Yamaguchi, K. (1997) *Int. J. Quantum Chem.* **65**, 947.

Nagao, H., Mitani, M., Nishino, M., Shigeta, Y., Yoshioka, Y. and Yamaguchi, K. (1998) *Int. J. Quantum Chem.* **70**, 1075.

Nagao, H., Mitani, M., Nishino, M., Shigeta, Y., Yoshioka, Y. and Yamaguchi, K. (1999) *Int. J. Quantum Chem.* **75**, 549.

Nagao, H., Nishino, M., Shigeta, Y., Yoshioka, Y. and Yamaguchi, K. (2000) *Int. J. Quantum. Chem.* **80**, 721.

Nagao, H., Nishino, M., Shigeta, Y., Yoshioka, Y. and Yamaguchi, K. (2000) *J. Chem. Phys.* **113**, 11237.

Nagao, H., Yaremko, A. M., Kruchinin, S. P. and Yamaguchi, K. (2002) *New Trends in Superconductivity*, P155-165, Kluwer Academic Publishers.

Nagao, H., Kruchinin, S. P., Yaremko, A. M. and Yamaguchi, K. (2002) *Int. J. Mod. Phys. B*, **16**, 3419.

Nagao, H., Kawabe, H., Kruchinin, S. P., Manske, D. and Yamaguchi, K. (2003) *Mod. Phys. Lett.* B **17**, 423.

Nagao, H., Kawabe, H., Kruchinin, S. P., to appear.

Ralph, D. C., Black, C. T. and Tinkham, M. (1995) *Phys. Rev. Lett.* **74**, 3241.

Smith, R. A. and Ambegaokar (1996) *Phys. Rev. Lett.* **77**, 4962.

Suhl, H., Matthias, B. T. and Walker, R. (1959) *Phys. Rev. Lett.* **3**, 552.

von Delft, J., Zaikin, A. D., Golubev, D. S. and Tichy, W. (1996) *Phys. Rev. Lett.* **77**, 3189.

Yamaji, K. and Shimoi, Y. (1994) *Physica C* **222**, 349.

EXACT SOLUTION OF TWO-BAND SUPERCONDUCTIVITY IN ULTRASMALL GRAINS

Hiroyuki Kawabe[a] (kawabe@kinjo.ac.jp),
Hidemi Nagao[b] (nagao@wriron1.s.kanazawa-u.ac.jp),
and Sergei P. Kruchinin[c] (skruchin@i.com.ua)

[a]*Department of Social Work, Faculty of Social Work, Kinjo University, 1200 Kasama, Hakusan, Ishikawa 924-8511, Japan*

[b]*Division of Mathematical and Physical Science, Graduate School of Natural Science and Technology, Kanazawa University, Kanazawa 920-1192, Japan*

[c]*Bogolyubov Institute for Theoretical Physics, The Ukrainian National Academy of Science, Kiev 252143, Ukraine*

Abstract. We investigate two-band superconductivity in ultrasmall grain. The Richardson's exact solution is extended to two-band systems, and a new coupled equation is derived according to the procedure of Richardson's works. Parity gap and condensation energy of ultrasmall two-band superconducting grain are numerically obtained by solving the coupled equations. We discuss these properties in ultrasmall grain in relation to the correlation, interband interaction, and size dependence.

Key words: Two-band superconductivity, Ultrasmall grain, Richardson's exact solution

1. Introduction

Black et. al. have revealed the presence of a parity dependent spectroscopic gap in tunnelling spectra of nanosize Al grains (Black, 1996; Ralph, 1995). Many groups have theoretically investigated the physical properties, such as critical level spacing, condensation energy, parity gap, etc. of ultrasmall grains of conventional superconductors (Jankó, 1994; Delft, 1996; Smith, 1996; Matveev, 1997; Braun, 1998; Delft, 1999; Gladilin, 2002). The questions arising from the nanosize of superconducting grains have first been discussed by Anderson (Anderson, 1959). The standard BCS theory fails when the level spacing approaches the superconducting gap. To investigate

K. Scharnberg and S. Kruchinin (eds.),
Electron Correlation in New Materials and Nanosystems, 129–139.
© 2007 *Springer.*

the properties in such nanosize systems, it is necessary to take more accurate treatment. Braun and von Delft (Braun, 1998; Delft, 1999; Braun, 1999; Sierra, 2000) have reintroduced the exact solution to the reduced BCS Hamiltonian developed by Richardson (Richardson, 1963; Richardson, 1964; Richardson, 1965; Richardson, 1966). It is noteworthy that the Richardson's solution is applicable at distributions of single-electron energy level. V. N. Gladioli et. al. (Delft, 1999) have investigated the pairing characteristics such as condensation energy, spectroscopic gap, parity gap, etc. by using the Richardson's exact solution for the reduced BCS Hamiltonian.

Recent discovery of superconductivity of MgB_2 (Nagamatsu, 2001) with $T_c = 39$ K has also been much attracted great interest for elucidation of its mechanism from both experimental and theoretical view points. Since this discovery, the possibility of two-band superconductivity has also been discussed in relation to two gap functions experimentally and theoretically. Two-band model has been introduced by several groups (Suhl, 1959; Moskalenko, 1959; Kondo, 1963). Recently, two-band or multi-band superconductivity has been theoretically investigated in relation to superconductivity arising from coulomb repulsive interactions and the possibility of new superconductivity arising from new mechanisms with higher transition temperature (Konsin, 1988; Yamaji, 1994; Combescot, 1995; Konsin, 1998; Nagao, 2000.2; Kondo, 2001; Kondo, 2002; Nagao, 1999; Nagao, 2002.1; Nagao, 2002.2; Nagao, 1997; Nagao, 1998; Nagao, 2000.1).

In this paper, we investigate two-band superconductivity in ultrasmall grain. The Richardson's exact solution is extended to two-band systems, and new coupled equation is derived according to the procedure of Richardson's works. Parity gap and condensation energy of ultrasmall two-band superconducting grain are numerically given by solving the coupled equation. We discuss these properties in ultrasmall grain in relation to the correlation, interband interaction, and size dependence.

2.   Exact solution for two-band superconductivity

In this section, we derive an exact solution of two-band superconductivity for reduced BCS Hamiltonian.

2.1.   HAMILTONIAN

We consider a Hamiltonian for two bands 1 and 2 written as

$$H = \sum_{j\sigma} \varepsilon_{1j} a_{j\sigma}^\dagger a_{j\sigma} - g_1 \sum_{jk} a_{j\uparrow}^\dagger a_{j\downarrow}^\dagger a_{k\downarrow} a_{k\uparrow} + \sum_{j\sigma} \varepsilon_{2j} b_{j\sigma}^\dagger b_{j\sigma} - g_2 \sum_{jk} b_{j\uparrow}^\dagger b_{j\downarrow}^\dagger b_{k\downarrow} b_{k\uparrow}$$

$$+ g_{12} \sum_{jk} a_{j\uparrow}^\dagger a_{j\downarrow}^\dagger b_{k\downarrow} b_{k\uparrow} + g_{12} \sum_{jk} b_{j\uparrow}^\dagger b_{j\downarrow}^\dagger a_{k\downarrow} a_{k\uparrow} . \tag{1}$$

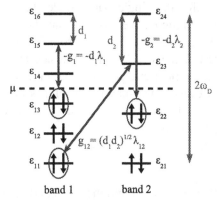

Figure 1. Two band system and parameters in Hamiltonian.

The first line of Eq. (1) corresponds to the reduced BCS Hamiltonian for bands 1 and 2. The third line means a coupling between them and corresponds to the pair scattering process between these two bands (see Fig.1). $a_{j\sigma}^{\dagger}$ ($a_{j\sigma}$) and $b_{j\sigma}^{\dagger}$ ($b_{j\sigma}$) are the creation (annihilation) operator in band 1 and 2 with spin σ and the single-particle levels ε_{1j} and ε_{2j}, respectively. The sums of j and k are over a set of N_1 states for band 1 with fixed width $2\hbar\omega_{1D}$ and a set of N_2 states for band 2 with fixed width $2\hbar\omega_{2D}$, respectively.

In this study, we assume that the Debye energies for two bands coincide with each other as

$$\omega_{1D} = \omega_{2D} = \omega_D . \tag{2}$$

The interaction constants g_1 and g_2 can be written as

$$g_1 = d_1\lambda_1 , \qquad g_2 = d_2\lambda_2 , \tag{3}$$

where d_1 and d_2 mean the mean single-particle level spacing,

$$d_1 = 2\hbar\omega_D/N_1 , \qquad d_2 = 2\hbar\omega_D/N_2 , \tag{4}$$

and λ_1 and λ_2 are dimensionality interaction parameters for two bands. We define the inter-band interaction constant as

$$g_{12} = \sqrt{d_1 d_2}\, \lambda_{12} . \tag{5}$$

The system we are considering consist of two half-filled bands, each of which has equally spaced N_n single-particle levels and M_n ($= N_n/2$) doubly occupied pair levels ($n = 1,2$). We take as our unit of energy the single-particle level spacing. Thus, the single-particle spectrum is given by

$$\varepsilon_{nj} = d_n j - \omega_D , \quad j = 1,2,\cdots,N_n \quad (n = 1,2) . \tag{6}$$

Richardson had solved for the single band model for arbitrary set of single-particle levels. For simplicity, we assume that there are not singly occupied single-particle levels. As can be seen from Eq. (1), these levels are decoupled from the rest of the system. They are said to be blocked and contribute with their single-particle energies to the total energy. The above simplification implies that every single-particle level j is either empty (i.e. $|vac\rangle$), or occupied by a pair of electrons (i.e. $a_{j\uparrow}^{\dagger}a_{j\downarrow}^{\dagger}\||vac\rangle$ and $b_{j\uparrow}^{\dagger}b_{j\downarrow}^{\dagger}|vac\rangle$). These are called as unblocked level.

2.2. EXACT SOLUTION

In order to extend Richardson's solution into the two-band system, we define two kind of hard-core boson operators as

$$c_j = a_{j\downarrow}a_{j\uparrow}, \qquad c_j^{\dagger} = a_{j\uparrow}^{\dagger}a_{j\downarrow}^{\dagger}, \qquad d_j = b_{j\downarrow}b_{j\uparrow}, \qquad d_j^{\dagger} = b_{j\uparrow}^{\dagger}b_{j\downarrow}^{\dagger}. \qquad (7)$$

The Hamiltonian Eq. (1) for the unblocked levels can be then written as

$$
\begin{aligned}
H_U &= 2\sum_{j}^{N_1} \varepsilon_{1j}c_j^{\dagger}c_j - g_1\sum_{jk}^{N_1} c_j^{\dagger}c_k + 2\sum_{j}^{N_2} \varepsilon_{2j}d_j^{\dagger}d_j - g_2\sum_{jk}^{N_2} d_j^{\dagger}d_k \qquad (8) \\
&+ g_{12}\sum_{j}^{N_1}\sum_{k}^{N_2} c_j^{\dagger}d_k + g_{12}\sum_{j}^{N_2}\sum_{k}^{N_1} d_j^{\dagger}c_k .
\end{aligned}
$$

We shall find the eigenstates $|M_1; M_2\rangle$ of this Hamiltonian with $M_1 + M_2$ pairs in the following form as

$$H_U|M_1; M_2\rangle_U = E(M_1; M_2)|M_1; M_2\rangle_U = \left(\sum_{J=1}^{M_1} E_{1J} + \sum_{K=1}^{M_2} E_{2K}\right)|M_1; M_2\rangle_U ,$$
$$(9)$$

where $E(M_1; M_2)$ is the eigenvalue and

$$|M_1; M_2\rangle_U = \prod_{J=1}^{M_1} C_J^{\dagger} \prod_{K=1}^{M_2} D_K^{\dagger}|vac\rangle , \qquad (10)$$

and

$$C_J^{\dagger} = \sum_{j}^{N_1} \frac{c_j^{\dagger}}{2\varepsilon_{1j} - E_{1J}} , \qquad D_J^{\dagger} = \sum_{j}^{N_2} \frac{d_j^{\dagger}}{2\varepsilon_{2j} - E_{2J}} . \qquad (11)$$

Now, we define $C_0^{\dagger}$ and $D_0^{\dagger}$ as

$$C_0^{\dagger} = \sum_{j}^{N_1} c_j^{\dagger} , \qquad D_0^{\dagger} = \sum_{j}^{N_2} d_j^{\dagger} , \qquad (12)$$

then, we can rewrite Eq. (8) as

$$H_U = 2 \sum_j^{N_1} \varepsilon_{1j} c_j^\dagger c_j - g_1 C_0^\dagger C_0 + 2 \sum_j^{N_2} \varepsilon_{2j} d_j^\dagger d_j - g_2 D_0^\dagger D_0 \qquad (13)$$

$$+ g_{12} C_0^\dagger D_0 + g_{12} D_0^\dagger C_0 .$$

Using these new operators, we find

$$H_U |M_1; M_2\rangle_U = \left(\sum_{J=1}^{M_1} E_{1J} + \sum_{K=1}^{M_2} E_{2K} \right) |M_1; M_2\rangle_U$$

$$+ C_0^\dagger \sum_{J=1}^{M_1} \left[1 - \sum_j^{N_1} \frac{g_1}{2\varepsilon_{1j} - E_{1J}} + \sum_{J' \neq J}^{M_1} \frac{2g_1}{E_{1J'} - E_{1J}} \right] |M_1(J); M_2\rangle_U$$

$$+ D_0^\dagger \sum_{J=1}^{M_1} \left(\sum_j^{N_1} \frac{g_{12}}{2\varepsilon_{1j} - E_{1J}} - \sum_{J' \neq J}^{M_1} \frac{2g_{12}}{E_{1J'} - E_{1J}} \right) |M_1(J); M_2\rangle_U \qquad (14)$$

$$+ C_0^\dagger \sum_{K=1}^{M_2} \left(\sum_j^{N_2} \frac{g_{12}}{2\varepsilon_{2j} - E_{2K}} - \sum_{K' \neq K}^{M_2} \frac{2g_{12}}{E_{2K'} - E_{2K}} \right) |M_1; M_2(K)\rangle_U$$

$$+ D_0^\dagger \sum_{K=1}^{M_2} \left[1 - \sum_j^{N_2} \frac{g_2}{2\varepsilon_{2j} - E_{2K}} + \sum_{K' \neq K}^{M_2} \frac{2g_2}{E_{2K'} - E_{2K}} \right] |M_1; M_2(K)\rangle_U ,$$

where

$$|M_1(L); M_2\rangle_U = \prod_{J=1}^{L-1} C_J^\dagger \prod_{J'=L+1}^{M_1} C_{J'}^\dagger \prod_{K=1}^{M_2} D_K^\dagger |vac\rangle , \qquad (15)$$

$$|M_1; M_2(L)\rangle_U = \prod_{J=1}^{M_1} C_J^\dagger \prod_{K=1}^{L-1} D_K^\dagger \prod_{K'=L+1}^{M_2} D_{K'}^\dagger |vac\rangle . \qquad (16)$$

Comparing Eq. (14) with Eq. (9), for arbitrary J and K we obtain

$$\begin{pmatrix} C_0^\dagger & D_0^\dagger \end{pmatrix} \begin{pmatrix} 1 + g_1 A_{1J} & -g_{12} A_{2K} \\ -g_{12} A_{1J} & 1 + g_2 A_{2K} \end{pmatrix} \begin{pmatrix} |M_1(J); M_2\rangle_U \\ |M_1; M_2(K)\rangle_U \end{pmatrix} = 0 , \qquad (17)$$

where

$$A_{nL} = - \sum_j^{N_n} \frac{1}{2\varepsilon_{nj} - E_{nL}} + \sum_{L' \neq L}^{M_n} \frac{2}{E_{nL'} - E_{nL}} . \qquad (18)$$

Non-trivial solution of Eq. (17) is derived from a determinantal equation;

$$F_{JK} = (1 + g_1 A_{1J})(1 + g_2 A_{2K}) - g_{12}^2 A_{1J} A_{2K} = 0 . \qquad (19)$$

This constitutes a set of $M_1 + M_2$ couples equations for $M_1 + M_2$ parameters E_{1J} and E_{2K} ($J = 1, 2, \cdots, M_1; K = 1, 2, \cdots, M_2$), which may be thought of as self-consistently determined pair energies. Equation (19) is exact eigenvalue equation for two-band superconducting system, and can be regarded as a generalization of the Richardson's original eigenvalue equation.

2.3. PREPROCESSING FOR NUMERICAL CALCULATION

To remove the divergences from the second term of A_{nL} in Eq. (18), we make changes of energy variables.

$$E_{n2\lambda} = \xi_{n\lambda} + i\eta_{n\lambda} , \qquad E_{n2\lambda-1} = \xi_{n\lambda} - i\eta_{n\lambda} , \qquad \lambda = 1, 2, \cdots, M_n/2 , \quad (20)$$

where we assume that the number of pairs is even. Since complex pair energies appear in complex conjugate pairs, the total energy is kept in real.

A further transformation is necessary in order to remove the divergences from the first term of A_{nL}. We define new variables $x_{n\lambda}$ and $y_{n\lambda}$ as

$$\xi_{n\lambda} = \varepsilon_{n2\lambda} + \varepsilon_{n2\lambda-1} + d_n x_{n\lambda} \qquad (x_{n\lambda} \leq 0) , \qquad (21)$$

$$\eta_{n\lambda}^2 = -\left\{ (\varepsilon_{n2\lambda} - \varepsilon_{n2\lambda-1})^2 - d_n^2 x_{n\lambda}^2 \right\} y_{n\lambda} \qquad (y_{n\lambda} \geq 0) . \qquad (22)$$

Then, we can rewrite F_{JK} by using new variables and extract the real and the imaginary part as

$$F_{\lambda\lambda'}^{+} = \frac{1}{2} \left(F_{\lambda\lambda'} + F_{\lambda\lambda'}^{*} \right) , \qquad F_{\lambda\lambda'}^{-} = \frac{1}{2i} \left(F_{\lambda\lambda'} - F_{\lambda\lambda'}^{*} \right) . \qquad (23)$$

Therefore, for arbitrary combination of λ and λ' we solve the following equations;

$$F_{\lambda\lambda'}^{+} = 0 , \qquad F_{\lambda\lambda'}^{-} = 0 \qquad (\lambda = 1, 2, \cdots, M_1/2; \lambda' = 1, 2, \cdots, M_2/2) . \quad (24)$$

3. Results and discussion

We now apply the exact solution for two-band system to discuss properties of the two-band superconducting in ultrasmall grain. The single-particle level patterns of $(2M_1 + m) + 2M_2$ electron system ($m = 0, 1$, and 2) that we are considering are represented in Figs. 2(a), (b), and (c), respectively. As seen these figures, the additional electrons first occupy the band 1, then the band 2. And numerical calculations given below are carried out under the condition that $N_1 : N_2 = 3 : 2$.

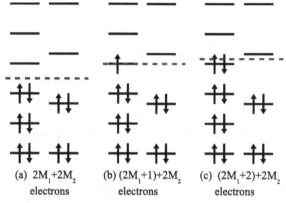

Figure 2. Single-particle levels near the Fermi level in a two band superconductivity. The dotted lines mean the chemical potential. The left and right bands are band 1 and 2, respectively. d_1 and d_2 are the mean level spacings. (a) $2M_1 + 2M_2$ electron system, where M_n is a number of pair levels. (b) $(2M_1 + 1) + 2M_2$ electron system. (c) $(2M_1 + 2) + 2M_2$ electron system.

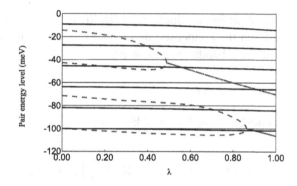

Figure 3. Typical behavior of pair energy levels of two bands for the ground state. Parameters used in calculation are $N_1 = 12$, $N_2 = 8$, $M_1 = 6$, $M_2 = 4$, $\hbar\omega_D = 50$, $0 < \lambda = \lambda_1 = \lambda_2 \leq 1.0$, and $\lambda_{12}/\lambda = 0.2$. Solid and broken lines are corrspond to the pair energy levels of the band 1 and the band 2, respectively.

3.1. PAIR ENERGY LEVEL

By minimizing the sum of squares of Eq. (24) for various interaction parameters, we obtain a behavior of pair energy levels E_{nJ} of two bands as shown in Fig. 3.

As seen in the figures, the band 2 condenses into degenerate levels, but the band 1 does not. In general, we can expect that the single-particle levels in band of which mean level spacing d is larger than the other band degenerate faster. The behavior of the condensing band is qualitatively the same as that

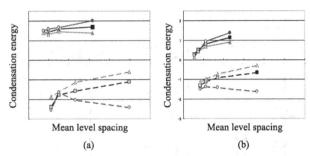

Figure 4. Condensation energy. Parameters used in the calculations are $\hbar\omega_D = 50$, and $\lambda = \lambda_1 = \lambda_2 = 0.5$. Values are normalized by the bulk gap, $\Delta = \omega_D \sinh^{-1}\left(\frac{\lambda}{\lambda^2 - \lambda_{12}^2}\right)$. The solid and broken lines correspond to the condensation energy for the band 1 and 2, respectively. Lines plotted by squares, by triangles and by circles are for $2M_1 + 2M_2$ electron system, for $(2M_1+1)+2M_2$ electron system, and for $(2M_1+2)+2M_2$ electron system, respectively. (a) The condensation energy for the interband coupling parameter $\lambda_{12} = 0.01$. (b) The condensation energy for $\lambda_{12} = 0.1$.

for the case of calculation for the single band (Delft, 1999). The co-existence of the normal band and the condensed one may be reflected in the opposite phase of the gaps of these bands (Nagao, in press).

3.2.   CONDENSATION ENERGY

The condensation energy of band n for $(2M_1 + m) + 2M_2$ electron system can be defined as

$$E_n^C(2M_1+m, 2M_2) = E_n(2M_1+m, 2M_2)+\left(M_n + \frac{m}{2}\right)g_n - E_n^0(2M_1+m, 2M_2),$$
(25)

where $E_n(2M_1 + m, 2M_2)$ and $E_n^0(2M_1 + m, 2M_2)$ are the ground state energy and the sum of the single-particle energy, respectively.

We calculate the condensation energies and show in Figs. 4(a) and (b). As seen in the figures, we can understand that the band 2 condenses, but the band 1 does not because of the sign of values. This difference of sign may also be reflected in the opposite phase of the gaps of these bands. The behaviors of the results for the condensed band (band 2) are qualitatively the same result as the case of the single band calculation. That of the condensation energy of band 2 for $(2M_1 + 2) + 2M_2$ electron system is, however, different from the others. We can also see that the condensation energy is affected by the interband interaction λ_{12} (Nagao, in press).

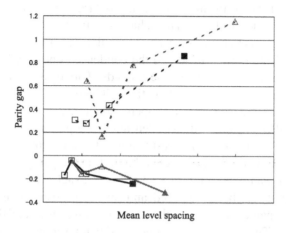

Figure 5. Parity gap. Parameters used in calculation are $\hbar\omega_D = 50$, and $\lambda_1 = \lambda_2 = 0.5$. The solid and broken lines correspond to the parity gap for the band 1 and 2, respectively. Lines plotted by triangles and by squares are for the interband coupling parameter $\lambda_{12} = 0.01$ and for $\lambda_{12} = 0.1$, respectively. Values are normalized by the bulk gap.

3.3. PARITY GAP

The parity gap of band n is defined as

$$\Delta_n^p = E_n(2M_1 + 1, 2M_2) - \frac{1}{2}\{E_n(2M_1, 2M_2) + E_n(2M_1 + 2, 2M_2)\}\ ,\quad (26)$$

which has been introduced by Matveev and Larkin and characterizes the even-odd ground state energy difference (Matveev, 1997).

The parity gaps we have calculated are shown in Fig. 5. For the condensed band, we obtain qualitatively the same result as the case of the single band calculation, i.e. there is a minimal point and a tendency toward 1 for $d \rightarrow 0$. The mean level spacing giving the minimal point is, however, much less than that for the single band. The parity gap is almost independent upon the interband interaction λ_{12}. This is also mentioned in our previous work (Nagao, in press).

4. Conclusions

We have extended the the Richardson's exact solution to the two-band system, and have derived a new coupled equation. To investigate the properties of the two-band superconductivity, we have solved the equation numerically, and have given the behavior of pair energy levels, the condensation energy, and the parity gap.

The band of which mean level spacing is larger than the other band degenerates and condenses faster. The behavior of the condensing band is qualitatively the same as that for the case of calculation for the single band. The co-existence of the normal band and the condensed one may be reflected in the opposite phase of the gaps of these bands. This phase character appears in every results of numerical calculations. Therefore, the phase of gap is important to stabilize the two-band superconductivity.

We have also calculated the condensation energy and the parity gap for two-band superconductivity. The result suggest that the interband interaction λ_{12} affects on the condensation energy, but not on the parity gap.

In summary, an expression of Richardson's exact solution for two-band superconductivity has been presented, and has been solved numerically. Then, the behavior of pair energy levels, the condensation energy, and the parity gap have presented. The results for the condensed band is almost qualitatively the same as those for the calculation of single band, and the co-existence of the normal band and the condensed one may be originated from the opposite phase of the gaps of these bands.

Acknowledgments

H.N is grateful for a financial support of the Ministry of Education, Science and Culture of Japan (Research No.17064013 and No.16032204). The authors thank Profs. S. Aono, M. Kimura, K. Nishikawa and K. Yamaguchi for their continued encouragement and helpful discussion.

References

Anderson, P. W., (1959) *J. Phys. Chem. Solids* **11**, 28.

Black, C. T., Ralph, D. C., and Tinkham, M., (1996) *Phys. Rev. Lett.* **76**, 688.

Braun, F. and von Delft, J., (1998) *Phys. Rev. Lett.* **81**, 4712.

Braun, F. and von Delft, J., (1999) *Adv. Sol. State Phys.* **39**, 341.

Combescot, R. and Leyronas, X., (1995) *Phys. Rev. Lett.* **75**, 3732.

von Delft, J., Zaikin, A. D., Golubev, D. S., and Tichy, W., (1996) *Phys. Rev. Lett.* **77**, 3189.

von Delft, J. and Braun, F, (1999) *Phys. Rev. B* **59**, 9527.

Gladilin, V. N., Fomin, V. M., and Devreese, J. T., (2002) *Solid Sate Comm.* **121**, 519.

Jankó, B., Smith, A., and Ambegaokar, V., (1994) *Phys. Rev. B* **50**, 1152.

Kondo, J., (1963) *Prog. Theor. Phys.* **29**, 1.

Kondo, J., (2001) *J. Phys. Soc. Jpn.* **70**, 808.

Kondo, J., (2002) *J. Phys. Soc. Jpn.* **71**, 1353.

Konsin, P., Kristoffel, N., and Örd, T., (1988) *Phys. Lett. A* **129**, 339.

Konsin, P., Sorkin, B., (1998) *Phys, Rev. B*, **58**, 5795.

Matveev, K. A. and Larkin, A. I., (1997) *Phys. Rev. Lett.* **78**, 3749.

Moskalenko, V. A., (1959) *Fiz. Met. Metalloved* **8**, 503.

Nagamatsu, J., Nakamura, N., Muranaka, T., Zentani, Y., and Akimitsu, J., (2001) *Nature* **410**, 63.

Nagao, H., Nishino, M., Mitani, M., Yoshioka, Y., and Yamaguchi, K., (1997) *Int. J. Quantum Chem.* **65**, 947.

Nagao, H., Mitani, M., Nishino, M., Shigeta, Y., Yoshioka, Y., and Yamaguchi, K., (1998) *Int. J. Quantum Chem.* **70**, 1075.

Nagao, H., Mitani, M., Nishino, M., Shigeta, Y., Yoshioka, Y. and Yamaguchi, K., (1999) *Int. J. Quantum Chem.* **75**, 549.

Nagao, H., Nishino, M., Shigeta, Y., Yoshioka, Y., and Yamaguchi, K., (2000) *Int. J. Quantum. Chem.* **80**, 721.

Nagao, H., Nishino, M., Shigeta, Y., Yoshioka, Y. Yamaguchi, K., (2000) *J. Chem. Phys.* **113**, 11237.

Nagao, H., Yaremko, A. M., Kruchinin, S. P., and Yamaguchi, K., (2002) *New Trends in Superconductivity* P155-165, Kluwer Academic Publishers.

Nagao, H., Kruchinin, S. P., Yaremko, A. M., and Yamaguchi, K., (2002) *Int. J. Mod. Phys. B* **16**, 3419.

Nagao, H., Kawabe, H., Kruchinin, S. P., to be submitted to *Phys. Rev. Lett.*.

Ralph, D. C., Black, C. T., and Tinkham, M., (1995) *Phys. Rev. Lett.* **74**, 3241.

Richardson, R. W., (1963) *Phys. Lett.* **3**, 277.

Richardson, R. W. and Sherman, N., (1964) *Nucl. Phys.* **523**, 221.

Richardson, R. W., (1965) *J. Math. Phys.* **6**, 1034.

Richardson, R. W., (1966) *Phys. Rev.* **141**, 949.

Sierra, G., Dukelsky, J., Dussel, G. G., von Delft, J. and Braun, F., (2000) *Phys. Rev. B* **61**, 11890.

Smith, R. A. and Ambegaokar, V., (1996) *Phys. Rev. Lett.* **77**, 4962.

Suhl, H., Matthias, B. T., and Walker, R., (1959) *Phys. Rev. Lett.* **3**, 552.

Yamaji, K. and Shimoi, Y., (1994) *Physica C* **222**, 349.

II.2 Cuprate and other unconventional superconductors

EXPERIMENTAL EVIDENCE FOR A TRANSITION TO
BCS SUPERCONDUCTIVITY IN OVERDOPED CUPRATES

Guy Deutscher
School of Physics and Astronomy, Tel Aviv University
Ramat Aviv, Tel Aviv 68978, Israel

Abstract. Recent measurements of the change of the electronic kinetic energy in the cuprates and more particularly its change upon condensation in the superfluid state are reviewed and discussed in the context of various theoretical frameworks. The increase observed in over-doped samples is compatible with a BCS condensation, while the decrease seen in underdoped samples is not. The significance of the sign reversal at or near optimum doping is given special emphasis.

Key words: Superconductivity, High-T_c cuprates, Overdoping, BCS condensation, Bose-Einstein condensation, Kinetic energy change

1. Introduction

Amongst the many theories that have been proposed to explain the high T_c of the superconducting cuprates, a cross-over between a BCS and a Bose-Einstein (BE) condensation may be the most natural extension of the BCS theory, that applies so well to the low T_c superconductors. Eagles was the first to show that the BCS wave function also applies to a Bose Einstein condensate (Eagles, 1969). Leggett showed that there is a continuous transition between a BCS condensate where there is a large overlap between Cooper pairs, and a BE condensate with no overlap (Leggett, 1980). For a given carrier concentration, the cross-over occurs as a function of the coupling strength. Nozieres and Schmitt-Rink extended the study of this cross-over to finite temperatures (Nozieres and Schmitt-Rink, 1985). In the BE limit, pairs form at high temperatures as a result of strong coupling, and condense at some lower temperature, while in the BCS limit pairs form and condense at the same temperature. BCS-BE models have been studied by several authors (see for instance (Alexandrov and Mott, 1994)). In these models, the well known phase diagram is interpreted as a manifestation of this cross-over, the overdoped range being in the weak coupling (BCS) regime and the underdoped one in the strong coupling (BE) regime.

K. Scharnberg and S. Kruchinin (eds.),
Electron Correlation in New Materials and Nanosystems, 141–148.
© 2007 *Springer.*

The existence of pairs above T_c is the characteristic feature of a BE condensation, but their direct detection is so far an unsolved experimental problem. Other, less direct signatures of this mode of condensation that have been quoted in the literature include the power law dependence of T_c on the penetration depth, $T_c \propto \lambda^{-2}$, interpreted as being due to the way the superfluid density n_s varies with doping, $T_c \propto (n_s/m^*)$, assuming the effective mass m^* to be constant (Uemura, 2003); the isotope effect on the pseudo-gap temperature T^* (Müller and Keller, 1989), and the existence of the pseudo-gap Δ_P in the underdoped strong coupling regime. However, these and other distinctive features of the cuprates have also received different interpretations.

In this contribution we focus on the sign and size of the change in the electronic kinetic energy upon condensation, ΔE_K, determined recently from reflectivity measurements (Deutscher et al., 2005). The central conclusion of these measurements is that in the Bi2212 compound, the sign of ΔE_K undergoes a reversal as a function of doping, going from positive (increase in the kinetic energy) in the overdoped regime to negative in the underdoped one. This behavior is consistent with a BCS to BE cross-over occuring around optimum doping.

2. Kinetic energy change in BCS and BE condensates

One of the characteristic features of a BCS condensate is that electron states with wave vectors k slightly larger than the Fermi wave vector k_F are occupied, as well as hole states below k_F. This is due to the mixed electron-hole character in the vicinity of k_F, in a range $\delta k = (\Delta/E_F)k_F$. Here Δ is the superconducting gap and E_F is the Fermi energy. Accordingly, the overall kinetic energy of the electron gas *increases* by a relative amount of the order of $(\Delta/E_F)^2$. The theoretical result is (de Gennes, 1966)

$$\frac{\Delta E_K}{E_F} = \frac{1}{NV}\left(\frac{\Delta}{E_F}\right)^2$$

and some additional terms that can be neglected to first order.

For a typical low temperature superconductor, $(\Delta/E_F)^2$ is of the order of 10^{-8}, much too small to be determined experimentally. And indeed, it never was. In fact this is one of the few fundamental predictions of the BCS theory that have remained unverified. But for the 100K class HTS, taking $\Delta = 20 meV$ and $E_F = 500 meV$, the relative change of kinetic energy could be of the order of 1%. On the other hand, in a BE condensate composed of pre-formed pairs, the kinetic energy drops to zero in the limit of zero temperature, while it is of order $k_B T_c$ just above the transition, per boson (here pre-formed pairs). Thus we expect the kinetic energy to *decrease* by that amount upon condensation.

The general expectation of a sign reversal of the kinetic energy change upon condensation is borne out by an explicit calculation in the attractive Hubbard model. Denoting by $8t$ the bandwidth and by U the potential, Kyung et al. have calculated the kinetic and potential energy in the normal and superconducting states as a function of U/t (Kyung et al., 2005a). They find that as long as $U < 6t$, the kinetic energy is higher in the superconducting state, like in the BCS limit, but that in the opposite case the order is reversed. In that same model, the critical temperature reaches a maximum at $U = 8t$. This is also the coupling value for which the condensation energy reaches a maximum. The decrease of the condensation energy and thus that of T_c, in the very strong coupling regime, is due to the diminishing difference between the density of states in the normal and superconducting states. A pseudo-gapped DOS (i.e. a DOS with states within the gap) already exists in the normal state when the coupling is sufficiently strong. All that happens in the superconducting state is that the remaining states within the gap are removed and pile up at the gap edge. For instance, when $U = 8t$, Kyung et al. find a pseudo-gap of $2t$ in the normal state, and this is also the value of the gap in the superconducting state. When the coupling gets stronger, the gap gets larger but the condensation energy diminishes. One gets in fact a weaker superconductor. While in the BCS case one may say that the transition to the condensed state is driven by the gain in potential energy proportional to $N(0)\Delta^2$, this gain is small in the BE case because of the existence of a pseudo-gap, and the transition becomes driven by the gain in kinetic energy. However, the majority view is that the physics of the cuprates is rather that of the repulsive Hubbard model. We return to the attractive versus repulsive alternative after a brief description of the recent results on the doping dependence of the kinetic energy change.

3.   Determination of the kinetic energy from reflectivity measurements

In a one band, nearest neighbor tight binding model, it has been shown (see for instance (Deutscher et al., 2005) and references therein) that the total kinetic energy of the electrons is proportional to the integral of the real part of the conductivity $\int_0^\infty \sigma_1(\omega)d\omega$. $\sigma_1(\omega)$ can be obtained from a measurement of the reflectivity over a broad frequency range by using the Kramers-Kronig relation. In practice, one must set the upper bound of the integral at a value that will eliminate interband transitions. The lower bound is itself dictated by the availability of low frequency reflectivity data. In what follows, it will be

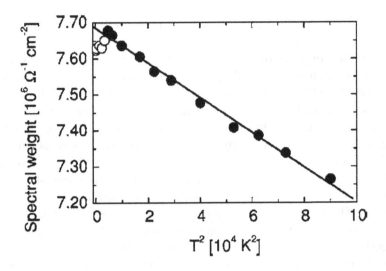

Figure 1. Temperature dependence of the spectral weight of an overdoped Bi2212 sample integrated up to 1 eV. Data in the superconducting state (open circles include the weight of the superfluid). After G. Deutscher, A. Santander-Syro and N. Bontemps (2005).

pointed out that the *temperature dependence* of the integral is of particular importance. It has been shown that this dependence is not very sensitive to different cut-off values in the integral, in a range of approximately 5,000 cm^{-1} to 10,000 cm^{-1} (Santander-Syro et al., 2004).

The reason for performing detailed measurements of the temperature dependence of the reflectivity is that in the normal state the value of the integral of the conductivity varies with temperature. Since one wishes to compare the values of the kinetic energy of the normal and condensed states *in the limit of zero temperature,* one must extrapolate the normal state value of the integral down to $T = 0$, and compare it to that in the superconducting state, itself measured well below the critical temperature. This is the correct procedure to be used, rather than taking for the normal state the value measured just above T_c. Additionally, the extrapolation procedure also increases the accuracy, since it makes use of a larger number of measurements. By comparing the extrapolated value of the normal state to that in the superconducting one, corrected for the spectral weight of the condensate (Santander-Syro et al., 2004), one obtains directly the *relative* change of the kinetic energy.

Figure 1 shows how this procedure works for an overdoped Bi2212 sample. In the normal state, the value of the integral of the conductivity is found to be a linearly decreasing function of T^2, which makes the extrapolation down to zero temperature straightforward. Corrected for the spectral weight

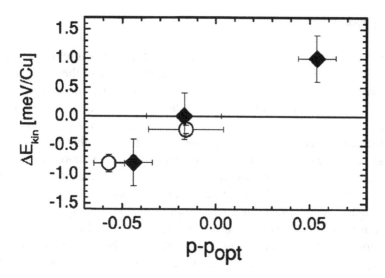

Figure 2. Change in the kinetic energy upon condensation in Bi 2212 samples as a function of doping, measured in hole/Cu around optimum doping. Open symbols are from Molegraaf et al. (2002), closed symbols are from Santander-Syro et al. (2004) for underdoped samples and Deutscher et al. (2005) for the overdoped sample.

of the superfluid, the value of the integral at low temperature is clearly smaller than that in the extrapolated normal state value. This corresponds to a kinetic energy which is higher in the condensed state than in the normal state (Deutscher et al., 2005). The relative variation is about 1%, which according to the BCS expression corresponds to $NV = 0.2$, not an unreasonable value. Santander et al. have found that in optimally doped and underdoped samples the temperature variation of the integral in the normal state seems to saturate below a temperature that is substantially higher than T_c (Santander-Syro et al., 2004). They have used this saturated value as the $T = 0$ normal state value. It is smaller than that in the superconducting state, which corresponds to a *decrease* of the kinetic energy in the condensed state. Others (Molegraaf et al., 2002) have found in the normal state of underdoped as well as optimally doped samples a T^2 dependence of the spectral weight, but nevertheless their results for the kinetic energy change are fully consistent with those of Santander-Syro et al. (2004).

Figure 2 gives a summary of the change in kinetic energy upon condensation as a function of doping in Bi 2212 samples. The sign reversal around optimum doping is quite clear.

4. Discussion

As shown by Nozieres and Schmitt-Rink (1985), the structure of the superconducting wave function in the low temperature limit is the same in the
(attractive) weak-coupling (BCS) and strong-coupling (BE) limits. Therefore,
the observed sign reversal of the kinetic energy change tells us more about
the normal state - here the state above the condensation temperature - than
about the condensed state. The experimental result is in that sense in line
with the general expectation that the overdoped regime is more similar to a
Fermi liquid than the underdoped one. The surprise, if there is one, is that that
the sign reversal occurs right at optimum doping, with the overdoped sample
still having a critical temperature close to optimum. One would not expect the
anomalous underdoped pseudo-gap normal state, whether due to the presence
of pre-formed pairs or to strong correlation effects, to become at once a usual
Fermi liquid as soon as samples are overdoped, even slightly (in terms of the
change of T_c).

The advantage of the attractive Hubbard model calculation of Kyung
et al. is indeed that it does give a sign reversal close to optimum doping
(Kyung et al., 2005a). The problem with this model is that, besides predicting
s-wave symmetry of the pairing state in contradiction to most experiments, it
also predicts a value of the pseudo-gap that is too large compared to experimental results. The maximum T_c is reached when $U = 8t$, the gap is then of
about $2t$, which for a bandwidth of 1 eV gives 250 meV, far above measured
values of the pseudo-gap as well as of the superconducting one.

On the other hand, as shown by Kyung et al., the repulsive Hubbard
model gives a pseudo-gap of the right order (Kyung et al., 2005b), but so far
there are no theoretical predictions for the DOS, the change in kinetic energy
and so on in the condensed state, that allow a comparison with experiment.
Finally, if one wishes to attribute the anomalous sign of the change in kinetic
energy to the strong electron-electron scattering rate in the normal state (a
strong correlation effect) and its known decrease upon condensation, one is
then faced with the difficulty that this strong reduction is supposed to be a
generic feature of the cuprates, and it is not at all clear why it should vanish
as soon as one enters into the overdoped regime. A further and possibly more
fundamental difficulty with the large negative U model is that it may not be
applicable at all to overdoped samples. A fingerprint of strong correlation
effects is the very broad energy range over which the optical spectral weight
is recovered in the superconducting state, as seen in underdoped samples
(Molegraaf et al., 2002; Santander-Syro et al., 2004). But in optimally doped
and even more clearly in overdoped samples of several cuprates the spectral
weight is recovered over a range of a few Δ, as in a weak-coupling BCS

superconductor (Lobo et al., 2001; Basov et al., 2001; Santander-Syro et al., 2004). This feature, together with the absence of a pseudo-gap in the normal state and the increase of the kinetic energy in the condensed state, suggests that strong correlation effects are absent in overdoped samples. In fact, the repulsive Hubbard model does not appear to predict the rather large positive change in the kinetic energy observed experimentally (Tremblay et al., 2005).

There remains the possibility that optimum doping is a very special point in the phase diagram, not just a cross-over concentration. The sign reversal of the change in kinetic energy at that concentration is one more property that signals that it is of special significance. Other such properties include the linear temperature dependence of the resistivity, the minimum temperature dependence of the Hall constant (Castro and Deutscher, 2005), and the appearance of an imaginary component of the order parameter (Dagan and Deutscher, 2001) although this last point remains controversial (Tsuei and Kirtley, 2000). The existence of a quantum critical point at or near optimum doping has been widely discussed in the literature (Sachdev, 2000), but again no specific theoretical predictions concerning a possible sign reversal of the kinetic energy change in these models are so far available.

Measurements of the kinetic energy change upon condensation are therefore of special interest because they challenge high T_c theories on a very specific and non trivial point. Additional data on different cuprates are of course desirable to check whether the sign reversal observed so far only on Bi-2212 samples is a general property.

Acknowledgements

I am indebted to Nicole Bontemps and Andres Santander for many discussions, and in particular on the significance of the data of (Santander-Syro et al., 2004). This work was supported in part by the Heinrich Hertz Minerva Center for High Temperature Superconductivity and by the Oren Family Chair for Experimental Solid State Physics.

References

Alexandrov, A. and Mott, N. (1994) Bipolarons, *Rep. Prog. Phys.* **57**, 1197.

Basov, D., Homes, C., Singley, E., Strongin, M., Blumberg, T. T. G., and van der Marel, D. (2001) Unconventional energetics of the pseudogap state and superconducting state in high-T_c cuprates, *Phys. Rev. B* **63**, 134514.

Castro, H. and Deutscher, G. (2005) Anomalous Fermi liquid behavior of overdoped high-Tc superconductors, *Phys. Rev. B* **70**, 174511.

Dagan, Y. and Deutscher, G. (2001) Doping and Magnetic Field Dependence of In-Plane Tunneling into $YBa_2Cu_3O_{7-x}$: Possible Evidence for the Existence of a Quantum Critical Point, *Phys. Rev. Lett.* **87**, 177004.

de Gennes, P. (1966) *Superconductivity of Metals and Alloys*, W.A. Benjamin Inc. New York.

Deutscher, G., Santander-Syro, A., and Bontemps, N. (2005) Kinetic energy change with doping upon superfluid condensation in high-temperature superconductors, *Phys. Rev. B* **72**, 092504.

Eagles, D. (1969) Possible Pairing without Superconductivity at Low Carrier Concentrations in Bulk and Thin-Film Superconducting Semiconductors, *Phys. Rev.* **186**, 456.

Kyung, B., Georges, A., and Tremblay, A.-M. (2005)a Potential-energy (BCS) to kinetic-energy (BEC)-driven pairing in the attractive Hubbard model, *cond-mat/0508645*.

Kyung, B., Kancharia, S., Sénéchal, D., Tremblay, A.-M., Civelli, M., and Kotliar, G. (2005)b Short-Range Spin Correlation Induced Pseudogap in Doped Mott Insulators, *cond-mat/0502565*.

Leggett, A. (1980), In A. Bekalsky and J. Przyspawa (eds.), *Modern trends in the Theory of Condensed Matter*, Berlin, , Springer, p. 13.

Lobo, R., Bontemps, N., Racah, D., Dagan, Y., and Deutscher, G. (2001) Pseudo-gap and supercondensate energies in the infrared spectra of Pr-doped YBa$_2$Cu$_3$O$_7$, *Europhys. Lett* **55**, 854.

Molegraaf, H., Presura, C., van der Marel, D., Kes, P., and Li, M. (2002) Superconductivity-induced transfer of in-plane spectral weight in $Bi_2Sr_2Ca_2O_{8+\delta}$, *Science* **295**, 2239.

Müller, K. and Keller, H. (1989), In E. L. E. Kaldis and K. Müller (eds.), *High Temperature Superconductivity 10 years after the discovery*, Series E, Kluwer Academic Publishers, Applied Science Volume 343:7.

Nozieres, P. and Schmitt-Rink, S. (1985) Bose condensation in an attractive fermion gas - from weak to strong coupling superconductivity, *J. Low Temp. Phys.* **59**, 195.

Sachdev, S. (2000) Quantum criticality: Competing ground states in low dimensions, *Science* **288**, 475.

Santander-Syro, A., Lobo, R., Bontemps, N., Lopera, W., Girat, D., Konstantinovic, Z., Li, Z., and Raffy, H. (2004) In-plane electrodynamics of the superconductivity in $Bi_2Sr_2CaCu_2O_{8+\delta}$: Energy scales and spectral weight distribution, *Phys.Rev. B* **70**, 134504.

Tremblay, A.-M., Kyung, B., and Sénéchal, D. (2005) Pseudogap and high-temperature superconductivity from weak to strong coupling: Towards quantitative theory, *cond-mat/0511334*.

Tsuei, C. and Kirtley, J. (2000) Pairing symmetry in cuprate superconductors, *Rev. Mod. Phys.* **72**, 969.

Uemura, Y. (2003) Superfluid density of high-T-c cuprate systems: implication on condensation mechanisms, heterogeneity and phase diagram, *Solid State Commun.* **126**, 23.

EXPERIMENTS USING HIGH-T_C VERSUS LOW-T_C JOSEPHSON CONTACTS

Ariando,[1] H. J. H. Smilde,[1] C. J. M. Verwijs,[1] G. Rijnders,[1] D. H. A. Blank,[1] H. Rogalla,[1] J. R. Kirtley,[2] C. C. Tsuei,[2] and H. Hilgenkamp[1]
(h.hilgenkamp@utwente.nl)

[1]*Faculty of Science and Technology and MESA+ Institute for Nanotechnology, University of Twente, P.O. Box 217, 7500 AE Enschede, The Netherlands*
[2]*IBM T. J. Watson Research Center, Yorktown Heights, New York 10598, USA*

Abstract. Remarkably rich physics is involved in the behavior of hybrid Josephson junctions, connecting high-T_c and low-T_c superconductors. This relates in particular to the different order parameter symmetries underlying the formation of the superconducting states in these materials. Experiments on high-T_c/low-T_c contacts have also played a crucial role in settling the decade-long d-wave versus s-wave debate in cuprate superconductors. Recently, such hybrid junctions have enabled more detailed pairing symmetry tests. Furthermore, with these junctions, complex arrays of π-rings have been realized, enabling studies on spontaneously generated fractional flux quanta and their mutual interactions. Steps toward novel superconducting electronic devices are taken, utilizing the phase-shifts inherent to the d-wave superconducting order parameter. This paper is intended to reflect the current status of experiments using high-T_c and low-T_c Josephson contacts.

Key words: Josephson junctions, Hybrid junctions, Pairing symmetry, High-temperature superconductors, Half-flux quanta

1. Introduction

The introduction of additional phase shifting elements in superconducting loops leads to remarkable effects (van Harlingen, 1995; Tsuei and Kirtley, 2000a). This can be achieved e.g. by using superconductors with unconventional pairing symmetry (Geshkenbein and Larkin, 1986; Geshkenbein et al., 1987; Sigrist and Rice, 1992; Tsuei and Kirtley, 2000a), by incorporating π-Josephson junctions (Bulaevskii et al., 1977), or by employing trapped flux quanta (Majer et al., 2002) or current injection (Goldobin et al., 2004). In phase-sensitive tests of the pairing symmetry, superconductors with unconventional pairing symmetry were used to create a π phase-shift in the ring

K. Scharnberg and S. Kruchinin (eds.),
Electron Correlation in New Materials and Nanosystems, 149–174.
© 2007 *Springer.*

(π-ring), leading to a complementary magnetic field dependence of the critical current of the ring (Wollman et al., 1993; Wollman et al., 1995) and to the half-flux quantum effect (Tsuei et al., 1994).

These interesting observations are due to the physics associated with d-wave symmetry taking place at the interface of the high-T_c cuprate superconductors, as reviewed for example in (van Harlingen, 1995; Tsuei and Kirtley, 2000a; Kashiwaya and Tanaka, 2000; Hilgenkamp and Mannhart, 2002; Tafuri and Kirtley, 2005). Other interesting phenomena primarily associated with d-wave symmetry include e.g. Andreev bound states (Hu, 1994; Kashiwaya and Tanaka, 2000; Chesca et al., 2005; Chesca et al., 2006), the presence of the second harmonic in the critical current versus phase relation (Golubov et al., 2004), or vortex splintering (Mints et al., 2002).

In the phase-sensitive experiments, superconducting loops were fabricated using a high-T_c cuprate connected to a low-T_c superconductor (Wollman et al., 1993; Wollman et al., 1995), or using tricrystal or tetracrystal grain boundary junctions (Tsuei et al., 1994; Tsuei and Kirtley, 2000a). In this article we will concentrate only on the first route, using a high-T_c material connected to a low-T_c superconductor.

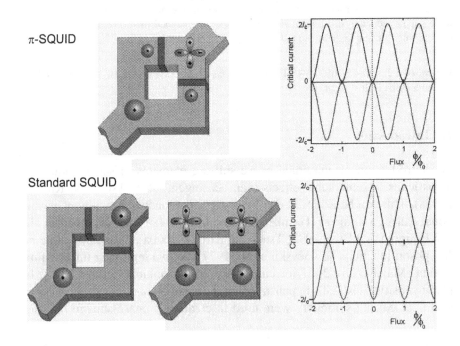

Figure 1. (Color) A schematic representation of a π-ring (top) and a 0-ring (bottom), and the expected magnetic field dependencies of their critical currents as shown at the right-hand side.

A schematic of a π-ring (π-SQUID) and 0-ring (standard SQUID) employing high-T_c and low-T_c Josephson contacts is depicted in Fig. 1. In the limit of a low inductance $L \ll \Phi_0/I_c$, with $\Phi_0 = h/2e = 2.07 \times 10^{-15}$ Wb the flux quantum and I_c the critical current of the junctions in the ring, the π-SQUIDs have $I_c(B)$ characteristics complementary to those of standard SQUIDs. For $L \gg \Phi_0/I_c$, the energetic ground state involves a spontaneously generated flux of $\frac{1}{2}\Phi_0$ in the SQUID loop. To compensate for the built-in π-phase shift a spontaneous circulating current flows in the ring, in either the clock- or counter-clockwise direction. The magnetic flux associated with this persistent circulating current is a fraction of a flux quantum, growing asymptotically to a half-flux quantum in the large inductance limit (Kirtley et al., 1997). In the limiting case when the inner diameter of the π-ring in Fig. 3 is reduced to zero, the structure represents the corner junction (Wollman et al., 1995). A corner junction behaves similarly to a π-ring. In this structure, the d-wave order parameter of the high-T_c cuprate induces a difference of π in the Josephson phase shift $\Delta\phi$ between the two junctions (facets). For facet lengths a in the small limit, that is, $a \ll \lambda_J$, the corner junction has $I_c(B)$ characteristics with zero critical current in the absence of magnetic fields, in stark contrast to the Fraunhofer pattern for a uniform junction. For facet lengths a in the wide limit, that is, $a \gg \lambda_J$, the lowest-energy ground state of the system is expected to be characterized by the spontaneous generation of a half-integer flux quantum at the corner. This half-fluxon provides a further π-phase change between neighbouring facets, either adding or subtracting to the d-wave induced π-phase shift, depending on the half-flux-quantum polarity. In both cases, this leads to a lowering of the Josephson coupling energy across the barrier, as this energy is proportional to $(1 - \cos \Delta\phi)$.

2. Preparation of high-T_c and low-T_c ramp-type Josephson contacts

Various early attempts have been made to prepare Josephson contacts between high-T_c and low-T_c superconductors, such as YBa$_2$Cu$_3$O$_7$ (YBCO) and Nb (Akoh et al., 1988; Akoh et al., 1989; Akoh et al., 1990; Fujimaki et al., 1990; Fujimaki et al., 1991; Hunt et al., 1990; Hunt et al., 1991; Foote et al., 1991; Terai et al., 1993; Wollman et al., 1993; Brawner and Ott, 1994a; Brawner and Ott, 1994b; Mathai et al., 1995; Gim et al., 1996; Wollman et al., 1995; Terai et al., 1995; Brawner and Ott, 1996; Usagawa et al., 1998). Oxygen migration due to the chemical reactivity of Nb with oxygen, combined with the sensitivity of YBCO to oxygen loss, presents a difficulty for the preparation of a good electrical connection between both superconductors. Another crucial step, especially using the ramp type configuration, is the structuring of the superconducting base electrode. Unfortunately, this procedure can severely degrade the quality of the base electrode near the interface. Transmission electron microscopy (TEM) studies of YBCO/Au ramp-type interfaces (Wen et al., 1999) clearly show an amorphous layer with a thickness up to 2 nm at the ramp edge between the high-T_c base electrode and the Au layer deposited at the freshly milled ramp edge.

Using the ramp-type interface configuration (Gao et al., 1990; Gao et al., 1991; Verhoeven et al., 1996), we have been able to establish a fabrication procedure (Smilde et al., 2002b) for all-thin-film Josephson junctions, in which controllably a high-T_c cuprate superconductor (such as YBCO or $Nd_{2-x}Ce_xCuO_4$ (NCCO)) is connected with a low-T_c material (Nb) along freely chosen directions of the high-T_c cuprates. A thin interlayer is incorporated in the junctions to obtain an increased transparency. The interlayer restores the surface damaged by ion milling and has the advantage of an in-situ barrier deposition between the two superconductors, leading to clean and well-defined interfaces. To avoid oxygen migration at the interface, a thin but chemically closed barrier separating the materials is used, for which Au is found to be the most suitable material (Smilde et al., 2001a; Smilde et al., 2001b).

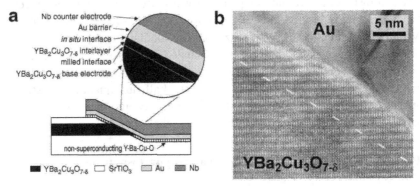

Figure 2. (a) Schematic cross section of the $YBa_2Cu_3O_{7-\delta}$/Au/Nb ramp-type junction including the interlayer. (b) Bright-field transmission electron microscopy image of the $YBa_2Cu_3O_{7-\delta}$/Au interface at the ramp-edge area, including an interlayer of 6 nm deposited $YBa_2Cu_3O_{7-\delta}$. Crystalline $YBa_2Cu_3O_{7-\delta}$ material is observed up to the Au interface, while no clear interface is observed between the base electrode and the interlayer (dashed line) [from (Smilde et al., 2002b)].

For the preparation, first a [001]-oriented high-T_c cuprate and a $SrTiO_3$ insulator layer are epitaxially grown by pulsed laser deposition (PLD) on [001]-oriented $SrTiO_3$ single crystal substrates. In these films, beveled edges (ramps) are etched by Ar-ion milling under an angle of 45° using a photoresist stencil, yielding ramps with an angle of ~20° with the substrate plane (Blank and Rogalla, 1997). In order to facilitate a good alignment of the junction with the ⟨100⟩-axes of the high-T_c film, edge-aligned substrates are used, with an alignment accuracy better than 1°. After stripping of the photoresist, a low-voltage ion mill step is applied to clean the surface, in-situ followed by an annealing step and the deposition of a thin interlayer of 5 – 7 nm YBCO at a condition similar to the deposition of the YBCO base-electrode. The annealing procedure is introduced to recrystallize residual amorphous material present at the ramp edge. This is followed by an in-situ deposition of Au-barrier at room temperature. After deposition of the Au-barrier layer with a thickness ranging from 6 to 120 nm, a photoresist lift-off stencil is applied to define the junction area. Before Nb deposition, maximally 2 nm of the Au layer is removed by rf-sputter etching, followed in situ by dc-sputter deposition of 150 nm Nb. After lift-off, the redundant uncovered Au is removed using ion milling.

A schematic of the junction obtained in this way is presented in Fig. 2a. The interlayer concept employs the difference in homoepitaxial and heteroepitaxial growth of high-T_c material. The thin interlayer is anticipated to be superconducting only if deposited on the YBCO ramp area, whereas on the SrTiO$_3$ substrate and isolation layer it is anticipated not to become superconducting (Smilde et al., 2002b). Figure 2b presents a TEM micrograph of the ramp edge area near the YBCO/Au interface. Because of the application of the thin YBCO interlayer, crystalline high-T_c material extends up to the Au barrier, and an amorphous layer was never observed. The interface between the base electrode and the interlayer could not be distinguished by TEM, indicating nearly perfect homoepitaxial growth.

Following this procedure results in normal state resistance $R_n A$ values of ∼10^{-12} Ωm^2 at liquid helium temperature. By adapting the Au-barrier thickness d_{Au}, the junction critical current density can be tuned in a wide range from 10^5 A/m^2 for d_{Au} ∼ 120 nm, up to values approaching 10^9 A/m^2 for d_{Au} ∼ 7 nm. In the following, we will review recent experiments based on high-T_c versus low-T_c contacts prepared using the procedure that has been described in this section.

3. Pairing symmetry test experiments

Understanding the nature of the ground state and its low-lying excitations in the copper oxide superconductors is a prerequisite for determining the origin of high temperature superconductivity. A superconducting order parameter (that is, the energy gap) with a predominantly $d_{x^2-y^2}$ symmetry is well-established (van Harlingen, 1995; Tsuei and Kirtley, 2000a). There are, however, several important issues that remain highly controversial. For example (in hole-doped compound such as YBCO) various deviations from a pure d-wave pair state, such as the possibility of Cooper pairing with broken time-reversal symmetry (BTRS) or an admixed $d_{x^2-y^2} + s$ pair state, have been theoretically predicted (Laughlin, 1988; Varma, 1999; Sigrist, 1998; Lofwander et al., 2001) and actively sought in numerous experimental studies (Spielman et al., 1992; Lawrence et al., 1992; Kaminski et al., 2002; Varma, 2002; Fauque et al., 2005; Covington et al., 1997; Dagan and Deutscher, 2001; Sharoni et al., 2002; Mathai et al., 1995; Schulz et al., 2000). Furthermore, a transition of the pairing symmetry from d-wave behavior to s-wave-like behavior was also suggested as a function of doping (Skinta et al., 2002; Biswas et al., 2002; Qazilbash et al., 2003) and temperature (Balci and Greene, 2004) in various electron doped compounds.

In view of this ongoing discussion, there is a need for further phase-sensitive experiments as a function of doping and temperature, and specifically for studying the possible existence of BTRS states. Many phase-sensitive experiments have been performed to look for evidence of BTRS (van Harlingen, 1995; Tsuei and Kirtley, 2000a; Mathai et al., 1995; Schulz et al., 2000). From these it has been concluded that, if present, the imaginary component must be quite small for high-T_c cuprates over a broad range of doping (Tsuei et al., 2004). Tsuei and Kirtley succeeded in performing phase-sensitive measurements for various compounds and temperatures based on grain boundary junctions and half-flux quantum effects (Tsuei and Kirtley, 2000a). Geometrical restrictions of the grain boundaries makes such experiments very challenging, especially for investigations as a function of momentum. In this section we describe various phase-sensitive experiments that have been performed based on high-T_c/Nb Josephson contacts.

3.1. LOW-INDUCTANCE-SQUID INTERFEROMETRY

Superconducting quantum interference devices present excellent tools to perform phase-sensitive experiments on the order parameter symmetry in superconductors (Wollman et al., 1993; Brawner and Ott, 1994a; Mathai et al., 1995; van Harlingen, 1995; Schulz et al., 2000). In a dc SQUID in which an isotropic s-wave superconductor contacts the superconductor to be studied in two crystal orientations, the critical current versus the applied magnetic flux dependence contains information about the relative phase and magnitude of the order parameter wave function at both contacts. Previous SQUID experiments on the order parameter symmetry of the high-T_c cuprates have been performed using, for example, Pb as a counter electrode. In these experiments, single crystals (Brawner and Ott, 1994a) as well as twinned and untwinned thin films (Wollman et al., 1993; Mathai et al., 1995) of $YBa_2Cu_3O_7$ have been used. These SQUIDs generally had inductances of $LI_c \approx \Phi_0$ or larger.

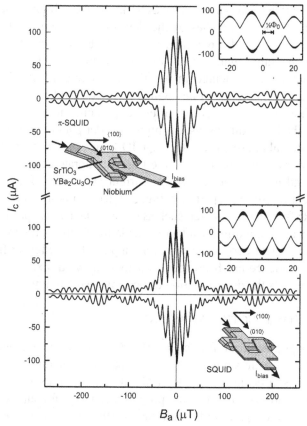

Figure 3. (Color) Critical current as a function of the applied magnetic field at $T = 4.2$ K of a π-SQUID (top) and standard SQUID (bottom). The insets present schematically the corresponding configuration and an enlargement of the $I_c(B)$ dependence near zero field. Both SQUIDs are in the low inductance limit [from (Smilde et al., 2004a)].

Low-inductance SQUIDs enable a more precise analysis (Schulz et al., 2000; Smilde et al., 2004a). Low-inductance all-high-T_c π-SQUIDs have been prepared using tetra-crystal substrates, providing clear evidence for predominant $d_{x^2-y^2}$-wave order parameter symmetry (Schulz et al., 2000). These dc π-SQUIDs were based on symmetric 45° [001]-tilt grain-boundary junctions. By definition, grain-boundary junctions are subject to limitations with respect to the orientation of the superconductors on both sides of the junction interface. Additionally, a complicating factor is presented by the fact that both electrodes are characterized by the order parameter symmetry under investigation. Junctions combining a well-characterized isotropic superconductor with the superconductor to be studied provide the ability to probe the order parameter in any desired orientation.

We have fabricated low-inductance SQUIDs based on high-T_c versus low-T_c Josephson contacts. Sketches of the devices are shown in the insets of Fig. 3. If the high-T_c order parameter is probed by an isotropic superconductor along the same main crystal orientation in a dc SQUID (bottom), i.e., both junctions are oriented in parallel, a maximum critical current is then observed in the absence of an applied magnetic field as shown in Fig. 3 (bottom). When the junctions are oriented perpendicular with respect to each other in the [100]-directions, the high-T_c order parameter symmetry induces an additional phase-shift in the SQUID-loop. A time reversal invariant $d_{x^2-y^2}$ order parameter symmetry of the high-T_c cuprate corresponds to a π phase-shift, and maxima in the critical current are observed at a magnetic field equivalent to $\frac{1}{2}\Phi_0$ as depicted in Fig. 3 (top). A time-reversal symmetry breaking (TRSB) order parameter, such as predominant $d_{x^2-y^2}$ pairing with an imaginary s-wave admixture, results in a deviation from π, and consequently the maximum critical current occurs at an applied flux differing from $\frac{1}{2}\Phi_0$. Therefore no evidence has been found for imaginary admixtures. Twinning of the film prevents the determination of possible real admixtures.

3.2. JOSEPHSON JUNCTION MODULATION

Multiple π-loops placed controllably at arbitrary positions would enable more detailed and systematic studies of the order parameter symmetry and its effects on Josephson devices (Smilde et al., 2002a; Ariando et al., 2005), as well as the realization of theoretically proposed elements for superconducting (quantum) electronics (Terzioglu and Beasley, 1998; Ioffe et al., 1999; Blatter et al., 2001).

We have fabricated well-defined zigzag-shaped ramp-type Josephson junctions between various high-T_c superconductors and Nb, and used such structures to test the order parameter symmetry of these high-T_c compounds. The zigzag configuration is depicted in Fig. 4a. In this structure, all interfaces are aligned along one of the $\langle 100 \rangle$ directions of the cuprate, and are designed to have identical critical current density (J_c) values. If the high-T_c cuprate were an s-wave superconductor, there would be no significant difference between a zigzag and a straight junction, aligned along one of the facet's direction. With the high-T_c superconductor having a $d_{x^2-y^2}$-wave symmetry, the facets oriented in one direction experience an additional π-phase difference compared to those oriented in the other direction. For a given number of facets, the characteristics of these zigzag structures then depend on the ratio of the facet length a and the Josephson penetration depth λ_J; see, e.g., (Zenchuk and Goldobin, 2004). In the small facet limit, $a \ll \lambda_J$, the zigzag structure can be envisaged as a one-dimensional array of Josephson contacts with an alternating sign of J_c, leading to anomalous magnetic field dependencies of

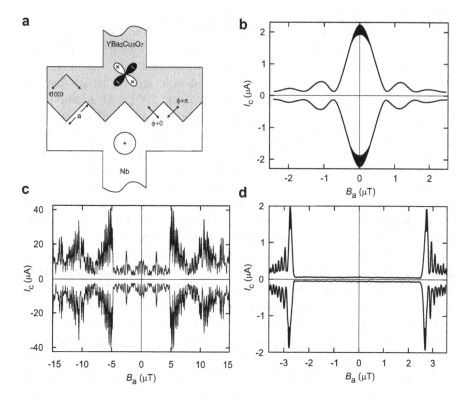

Figure 4. (a) A schematic representation of a zigzag junction. Critical current I_c as a function of applied magnetic field B_a for (b) a straight junction, and (c) a YBCO/Nb and (d) a NCCO/Nb zigzag array comprised of 80 facets of 5 μm width (T = 4.2 K) [from (Smilde et al., 2002a) and (Ariando et al., 2005)].

the critical current. In the large facet limit, the energetic ground state includes the spontaneous formation of half-integer magnetic flux quanta at the corners of the zigzag structures, as seen in (Hilgenkamp et al., 2003). All experiments described in this section are in the small facet limit.

Figures 4c and 4d shows the $I_c(B_a)$ dependence for a YBCO/Nb and a NCCO/Nb zigzag array with 80 facets having a facet length of 5 μm, respectively. The $I_c(B_a)$ dependencies of these zigzag structures clearly exhibit the characteristic features with an absence of a global maximum at $B_a(0)$ and the sharp increase in the critical current at a given applied magnetic field, resembling in their basic features the ones observed for asymmetric 45° [001]-tilt grain boundary junctions. This behavior can only be explained by the facets being alternatingly biased with and without an additional π-phase change (Copetti et al., 1995; Hilgenkamp et al., 1996; Mannhart et al.,

1996; Mints and Kogan, 1997). This provides a direct evidence for a π-phase shift in the pair wave function for orthogonal directions in momentum space and thus for a predominant $d_{x^2-y^2}$ order parameter symmetry. If the order parameter were to comprise an imaginary s-wave admixture, the $I_c(B_a)$ dependencies for the zigzag junctions would be expected to display distinct asymmetries, especially for low fields (Smilde et al., 2002a). In addition, the critical current at zero applied field is expected to increase with the fraction of s-wave admixture. From the high degree of symmetry of the measured characteristics of Figs. 4c and 4d and the very low zero field I_c, an upper limit of an imaginary s-wave symmetry admixture to the predominant $d_{x^2-y^2}$ symmetry of 1% for YBCO can be set and no indication for subdominant symmetry components for NCCO can be distinguished. The d-wave result for the electron-doped superconductor corroborates the results obtained using grain boundary junctions (Tsuei and Kirtley, 2000b; Chesca et al., 2003)

To investigate a possible change of the order parameter symmetry with doping (Biswas et al., 2002; Qazilbash et al., 2003), we have fabricated similar zigzag structures using $Nd_{1.835}Ce_{0.165}CuO_4$/Nb junctions. The results also indicated a predominant $d_{x^2-y^2}$-wave symmetry. When cooling the samples to $T = 1.6$ K all the basic features displayed by the structures at $T = 4.2$ K remain unaltered. We thus see no indication for an order parameter symmetry crossover for $Nd_{1.835}Ce_{0.165}CuO_4$ in this temperature range, as was reported for $Pr_{2-x}Ce_xCuO_{4-y}$ (Balci and Greene, 2004). Similar results were obtained for optimally doped samples upon cooling to $T = 1.6$ K.

3.3. ANGLE-RESOLVED ELECTRON TUNNELING

The upper limit of an imaginary s-wave symmetry admixture can be determined using the experiments described above. However, these experiments used geometries in which the junction normals are perpendicular to the a- and b- axes. They are, therefore, insensitive to an imaginary d_{xy} component which has nodes along the a- and b-axes. Another possible modification of the pure d-wave gap parameter is an admixed $d_{x^2-y^2}+s$ pair state, where s is the s-wave real component of the gap, as required by group theory for the in-plane CuO lattice symmetry of orthorhombic cuprate superconductors such as YBCO (Tsuei and Kirtley, 2000a). Although a $d + s$ pair state in YBCO is established (Polturak et al., 1993; Basov et al., 1995; Rykov and Tajima, 1998; Lu et al., 2001; Engelhardt et al., 1999), reports of the magnitude of the s-wave admixture vary. Twinning of the YBCO thin films used in the experiments discussed above prevents the determination of possible real admixtures. To further examine the d-wave order parameter symmetry, it is therefore important to perform pairing symmetry tests as a function of in-plane momentum using untwinned films. We have performed such experiments on untwinned

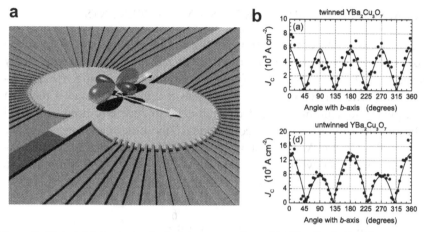

Figure 5. (Color) (a) Angle-resolved electron tunneling with YBa$_2$Cu$_3$O$_7$/Au/Nb ramp-type junctions oriented every 5° over 360°. The arrows indicate the main crystal orientations in the *ab* plane of the high-T_c superconducting material. (b) Critical current densities J_c as a function of the junction orientation with respect to the YBa$_2$Cu$_3$O$_7$ crystal for (top) twinned and (bottom) untwinned YBa$_2$Cu$_3$O$_7$ at T = 4.2 K and in zero magnetic field [from (Smilde et al., 2005)].

optimally-doped YBCO films based on YBCO/Nb contacts (Smilde et al., 2005; Kirtley et al., 2006).

The experimental layout is summarized in Fig. 5a. Basically, the YBCO base electrode is patterned into a nearly circular polygon, changing the orientation from side to side by 5°. A Au barrier and Nb counterelectrode contact each side. In this way, the angle with respect to the (010)-orientation is varied as a single parameter. Figure 5b presents the electrical characterization of the twinned base-electrode sample (top), and the untwinned one (bottom). The superconducting properties of the Au/Nb bilayer are independent of the orientation. Therefore, J_c depends on the in-plane orientation θ with respect to the *b*-axis of the YBCO crystal only, and presents four maxima for both samples, approaching zero in between. This is in agreement with predominant $d_{x^2-y^2}$-wave symmetry of the superconducting wave function in one electrode only. In closer detail, the nodes of the untwinned YBCO sample are found at 5° from the diagonal between the *a*- and the *b*-axis. This presents direct evidence for a significant real isotropic *s*-wave admixture. An estimate for the *s*- over $d_{x^2-y^2}$-wave gap ratio is 17% for a node angle θ_0 = 50°, resulting in a gap amplitude 50% higher in the *b* (Cu-O chain) direction than in the *a* direction. For the twinned base electrode, the nodes are found at the diagonal, which is expected if all twin orientations are equally present, and contributions of subdominant components average to zero.

3.4. ANGLE-RESOLVED PHASE-SENSITIVE EXPERIMENTS

Measurements of the critical currents of single YBCO-Nb junctions described above resulted in a gap 50% larger in the b than the a direction for optimally doped YBCO. To further quantify the deviations from a pure $d_{x^2-y^2}$ symmetry, in particular BTRS states, phase-sensitive experiments as a function of in-plane momentum are needed. Such experiments have been suggested to be sensitive to an imaginary component to the order parameter (Beasley et al., 1994; Ng and Varma, 2004). We have performed phase-sensitive experiments based on two-junction rings connecting YBCO and Nb to be able to accurately determine the in-plane pairing symmetry (Kirtley et al., 2006).

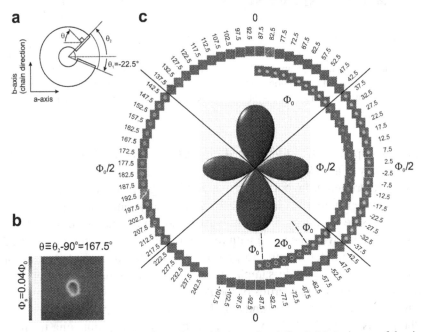

Figure 6. (color) (a) Schematic of one ring, with the angles defined. (b) An image of the ring with second junction normal angle $\theta = 167.5°$ relative to the majority twin a-axis direction, cooled and imaged in nominally zero field. (c) Images for all of the rings labelled by θ and arranged in a polar plot. Rings cooled in zero field were either in the $n = 0$ or the $n = 1/2$ flux quantum states (outer circle); the rings cooled in 0.2 μT were in the $n = 1/2$, $n = 1$ or $n = 2$ states (inner circle) [from (Kirtley et al., 2006)].

A schematic of the rings for this experiment is shown in Fig. 6a. Each ring has one junction with a fixed angle of $-22.5°$ relative to the majority twin a-axis direction in the YBCO, with the second junction angle varying in intervals of 5°. The ring geometries were optimized with several considerations in mind. For this measurement it is desirable to have large rings, because this leads to large $I_c L$ products, so that the spontaneous magnetization is not reduced from the asymptotic $\Phi_0/2$ value. Large rings are also easier to resolve with the SQUID microscope. It is important that the junction interfaces are well-defined, clean and have a sufficiently high critical current density.

Figure 6 shows a SQUID microscope image of one of the rings, cooled and imaged in zero field. The spontaneous flux is clearly visible in the center of the ring; the ring walls are nearly invisible. Figure 6b shows images of the rings in this sample, cooled and imaged in zero field, depicted as a polar plot. This underlines the nearly four-fold symmetry of the data, with the transitions between 0-rings and π-rings occurring at angle of the variable junction normal θ values close to $(2m + 1)45°$, m being an integer. The deviations from four-fold symmetry are systematic and are consistent with the gap being larger in the b-axis direction than in the a-axis direction. The outer circle of images are of the sample cooled in zero field, in which case the rings either have zero spontaneous flux (fluxoid number $n = 0$) or spontaneous flux close to $\Phi_0/2$ ($n = 1/2$). To test whether all of the rings had sufficiently large junction critical currents to sustain an appreciable circulating supercurrent, we recooled the sample in a field of 0.2 μT, with the resulting images shown in the inner semi-circle of Fig. 6c. In this case the 0-rings were either in the $n = 1$ or 2 state, whereas the π-rings remained in the $n = 1/2$ state. This shows that the nodal direction in YBCO films with a predominant twin orientation is shifted by at least a few degrees from the $(2m + 1)45°$ angles expected for a pure $d_{x^2-y^2}$ superconductor. Furthermore, the spontaneous fluxes which do not deviate from $\Phi_0/2$ or zero underline the fact that BTRS is not associated with high temperature superconductivity in optimally doped YBCO.

4. Coupling of half-flux quanta

Using grain boundary junctions, it has only been possible to controllably generate individual π-ring (Tsuei et al., 1994; Tsuei and Kirtley, 2000a). Such ring has a doubly degenerate ground state in zero applied flux. This is of interest as model systems for studying magnetic phenomena −including frustration effects− in Ising antiferromagnets (Aeppli and Chandra, 1997; Moessner and Sondhi, 2001; Chandra and Doucot, 1988; Davidovic et al., 1996; Pannetier et al., 1984; Lerch et al., 1990). Furthermore, studies of coupled π-loops can be useful for designing quantum computers based on flux-qubits (Mooij et al., 1999; van der Wal et al., 2000; Friedman et al., 2000; Ioffe et al., 1999; Blais and Zagoskin, 2000) with viable quantum error correction capabilities (Preskill, 1998; Bennet and DiVincenzo, 2000). However, these require a large number of rings. In this section we concentrate on the realization of large-scale coupled π-loop arrays based on YBCO/Nb

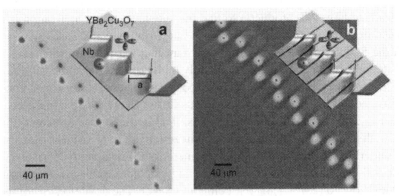

Figure 7. (color) Scanning SQUID microscope images of (a) continous zigzag junction with 40 μm facet lengths, and (b) electrically disconnected zigzag junction with 40 μm between facet corners [from (Hilgenkamp et al., 2003)].

Josephson contacts. Scanning SQUID microscopy (Kirtley et al., 1995b; Kirtley et al., 1995a) has been used to study the ordering of half-flux quanta in these structures.

We have first investigated the generation and coupling of half-integer flux quanta in the zigzag array that have been discussed in Sec. 3.2, shown schematically in Fig. 7 insets. In the scanning SQUID microscopy image presented in Fig. 7a, a spontaneously induced magnetic flux is clearly seen at every corner of the zigzag structure. For this sample $a = 40$ μm, which implies that the facets are well within the wide limit. The observed corner fluxons are arranged in an antiferromagnetic fashion. This antiferromagnetic ordering was found to be very robust, occurring for many cool-downs and for different samples with comparable geometries. Deviations from an antiferromagnetic arrangement were only observed when a magnetic field was applied during cool-down, or when an Abrikosov vortex was found trapped in (or near) the junction interface.

In the zigzag configuration, all the half-fluxons are generated in a singly connected superconducting structure; the question therefore arises as to whether the antiferromagnetic ordering is due to a magnetic interaction between the fractional fluxons, or to an interaction via the superconducting connection between the corners. To investigate this, we have also fabricated arrays of corner junctions, in a similar configuration as the zigzag arrays but with 2.5 μm-wide slits etched halfway between the corners, as schematically shown in Fig. 7b inset. In this situation there is no superconducting connection between the separate flux-generating corner junctions. For a distance between the corners equal to the facet length in the connected array, $a = 40$ μm, a ferromagnetic arrangement of the fractional flux quanta was observed (Fig. 7b). The magnetic interaction between the half-flux quanta at these distances is expected to be very weak, and alignment along minute spurious background fields in the scanning SQUID microscope is anticipated to be the dominating mechanism for their parallel arrangement. When the distance was decreased to about 20 μm, with a slit width of 1.5 μm, a tendency towards an antiferromagnetic coupling was observed.

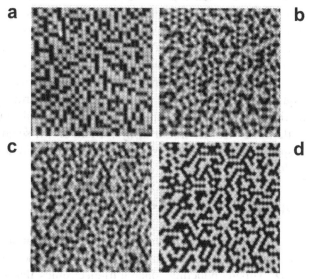

Figure 8. SQUID microscopy images of four electrically disconnected arrays of π-rings with 2.7 μm junctions and 11.5 μm ring to ring spacings for (a) square, (b) honeycomb, (c) kagomé, and (d) triangle lattices [from (Kirtley et al., 2005)].

We also have investigated 2-dimensional π-ring arrays made up of individual rings with various nearest-neighbor distances (25 μm or closer) (Hilgenkamp et al., 2003; Kirtley et al., 2005). The rings were arranged into arrays with 4 different geometries: square, honeycomb, triangular, and kagomé. The square and honeycomb arrays are geometrically unfrustrated, as their magnetic moments can be arranged so that all nearest neighbors have opposite spins, and the ground state of these lattices are only doubly degenerate. In contrast, the triangle and kagomé lattices are geometrically frustrated, since it is impossible for all of the rings to have all nearest neighbors anti-ferromagnetically aligned, and the ground states are highly degenerate. Figure 8 shows examples of scanning SQUID microscope images of the arrays after cooling in nominally zero field. Although regions of antiferromagnetic ordering are seen in the unfrustrated arrays Figs. 8a and 8b, anti-ferromagnetic ordering beyond a few lattice distances was never observed. Nevertheless, antiferromagnetic correlations were seen in all the 2D π-ring arrays.

We have shown that it is possible to realize large arrays of photolithographically patterned π-rings. Half-fluxon Josephson vortices in electrically connected zigzag junctions order with strongly anti-parallel half-fluxon vortices through the superconducting order parameter phase. Electrically isolated 1D and 2D arrays order much less strongly. The 2D π-ring arrays show stronger anti-ferromagnetic correlations than reported previously for 0-ring arrays (Davidovic et al., 1997; Davidovic et al., 1996), but do not order beyond a few lattice constants. One possibility is that our arrays correspond to a spin glass (Kirtley et al., 2005). However, in the absence of some hidden symmetry breaking, it appears that our rings are simply doubly degenerate. Furthermore, qualitatively similar results are obtained for arrays with and without geometrical frustration. Our experiments with repeated cooling show that there is little fixed disorder in our arrays. Therefore, it seems unlikely that we have a spin glass, unless there is some form of frozen-in disorder that varies from cooldown to cooldown.

5. Possible applications of π-phase-shifting elements

Besides holding a clue to the mechanism of high-T_c superconductivity, the unconventional d-wave symmetry in high-T_c cuprates provides unique possibilities to realize superconducting (quantum)-electronics that exploit the associated sign-changes in the order parameter for orthogonal directions in k-space. An intriguing consequence of this sign-change is the spontaneous generation of fractional magnetic flux quanta in superconducting rings incorporating a d-wave induced π-phase shift. In the following, we discuss some possible applications of π-phase-shifting elements for novel (quantum)-electronics based on high-T_c versus low-T_c Josephson contacts.

5.1. COMPLEMENTARY JOSEPHSON ELECTRONICS

For a built-in phase shift of $k\pi$, the $I_c(B)$ characteristics of π-SQUIDs are shifted by $\frac{1}{2}k\Phi_0$ compared to a standard SQUID. Two $I_c(B)$ dependencies are possible, differing with the polarity of the built-in phase shift. By constructing a phase-shifting element in which the polarity can be switched, a bistable superconducting device can be realized, with SQUID characteristics shifted by $+\frac{1}{2}k\Phi_0$ or $-\frac{1}{2}k\Phi_0$ compared to the standard case. This bistability can be used to construct superconducting memory elements, like flip-flops or programmable logic. Furthermore, Ioffe et al. (1999) proposed to design qubits for quantum computation based on the energetically degenerate ground states. They suggested a possible practical realization with $k = \pm\frac{1}{2}$ (Fig. 9a). It is based on a superconducting ring containing four identical Josephson junctions. In this ring a π–phase shift is incorporated using one of the concepts described earlier. It invokes a persistent circulating supercurrent I_{circ} relating to a phase drop of practically $\frac{1}{4}\pi$ per junction, with a slight deviation proportional to the magnetic flux in the ring, which is equal to the product of I_{circ} and the inductance of the phase-shifting element $L_{shifter}$.

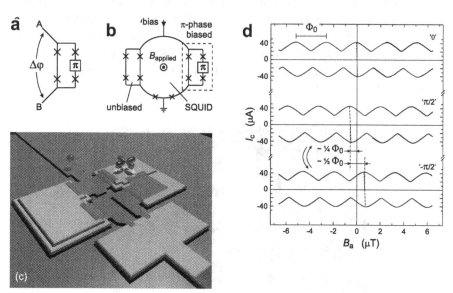

Figure 9. (Color) Schematic of (a) a $\pi/2$ phase-shifting element consisting of four Josephson junctions and a built-in π phase shift, and (b) a switchable $\pi/2$ SQUID, in which a SQUID loop incorporates the $\pi/2$ phase shifter (dashed rectangle) and an unbiased four-junction loop. (c) Implementation of the switchable $\pi/2$ SQUID. (d) Critical current versus applied magnetic field for a reference SQUID incorporating two unbiased four-junction loops (top), and for the two complementary states of the switchable $\pi/2$ SQUID (middle and bottom) [from (Smilde et al., 2004b)].

We have realized a superconducting bistable device by incorporating a $\pi/2$ phase-shifting element into a dc SQUID. Incorporating this element in a larger SQUID structure, as shown in Fig. 9b, a phase shift over the terminals A and B of $\pm\pi/2$ is obtained whose polarity depends on the direction of I_{circ}. With the I_c's of the junctions in the overall SQUID being much smaller than those in the phase shifter, the final device is bistable, with the $I_c(B)$ characteristics shifted by $+\frac{1}{4}\Phi_0$ or $-\frac{1}{4}\Phi_0$ compared to a standard SQUID, as shown in Fig. 9d. Notice that the characteristics of these two states are complementary to each other, which makes these structures suitable for the construction of (programmable) complementary Josephon electronics.

5.2. HALF-FLUX QUANTA AS INFORMATION CARRIERS

The doubly degenerate groundstates of π-rings allow their use for information storage. As previously discussed, under specific circumstances these groundstates are characterized by the spontaneous formation of half a quantum of magnetic flux. Their polarity can be set by the action of a logic (quantum-)gate, making them potentially useful elements for a random access memory. This is demonstrated in Fig. 10, showing a scanning SQUID microscopy image of an array of half-integer flux quanta forming the characters 'IBM+UT'. The fluxes were all set to have the same polarity by cooling in a modest externally applied magnetic field. The polarity of selected elements was reversed by applying pulses of control current to a single turn coil incorporated into the SQUID measurement chip.

Figure 10. Scanning SQUID microscopy image of an array of half-integer flux quanta forming the characters 'IBM+UT'.

5.3. π-SHIFTS IN RAPID SINGLE FLUX QUANTUM-LOGIC

In isolated π-rings, fabricated with high-T_c grain boundaries (Tsuei et al., 1994) or with connections between high-T_c and low-T_c superconductors (Hilgenkamp et al., 2003), the generation and manipulation of fractional flux quanta has already been demonstrated using scanning SQUID microscopy. An important step towards its application in electronic circuitry is the incorporation of such π-loops in superconducting logic gates, in which a controlled operation on an electronically applied input signal leads to a predefined output signal (Terzioglu and Beasley, 1998; Ustinov and Kaplunenko, 2003). We have performed experiments realizing this idea (Ortlepp et al., 2006), and showed the first realization of a Toggle Flip-Flop (TFF) based on Josephson contacts between high-T_c and low-T_c superconductors, in which the polarity of the fractional flux quantum provides the internal memory.

The spontaneous generation of fractional flux in the π-shift device eliminates the need for the asymmetrically injected bias current, which reduces the amount of connections to external control-electronics and allows for symmetry in the design parameters. This greatly benefits the design-process and fabrication and also leads to denser circuitry; our first realization needed only a quarter of the size of a standard Toggle Flip-Flop in established Niobium technology with the same feature size of 2.5 μm.

5.4. π-SHIFT QUBIT CONCEPTS

Over the last decades quantum-computation has received much attention because of its potential to solve mathematical problems, which classical computers cannot solve in acceptable times (DiVincenzo, 1995; DiVincenzo and Loss, 1998). As usual, it is a long way from basic principles to their practical implementation. A central aspect in this context is the availability of appropriate qubit elements, which allow integration into logic gates. A qubit is a quantum-system which, similar to a bit in a conventional computer, is characterized by two states but which can exist in arbitrary quantum-superpositions of these states.

While techniques from quantum-optics (cold trapped atoms, photons in cavities) or from molecular physics (nuclear magnetic resonance methods) appear to provide promising technologies for the realization of individual and few qubits, freedom in the design-parameters of the qubits and upscaling to real computing devices will require the scalability and variability of solid-state implementations. Superconductors appear to provide the best starting point to achieve this goal.

Proposed superconducting qubits operate either with charge (Schnirman et al., 1997; Averin, 1998; Bouchiat et al., 1998; Nakamura et al., 1999), or flux-/phase-states (Bocko et al., 1997; Mooij et al., 1999; Ioffe et al., 1999; Feigel'man et al., 2000; Blatter et al., 2001). In the phase-qubit, the information is mainly expressed through the phase degree of freedom, with only a minor coupling to the environment or to other qubits. The phase-state itself carries no charge and no current, rendering the device electrically and magnetically robust. A first proposal that was made by the ETH Zürich group for a superconducting phase-qubit involved mesoscopic junctions combining s- and d-wave superconductors, so called SDS'-junctions (Ioffe et al., 1999). Subsequently, a more easily realizable scheme was proposed involving superconducting loops with five Josephson junctions, four conventional ones and a π-junction (Blatter et al., 2001). In the design, the quantum degrees of freedom involve only the conventional junctions, while π-junction acts primarily as a phase-shifter. A very appealing aspect hereby is the fact that the qubits do not require a flux-bias, as is the case for all-low-T_c concepts.

Acknowledgements

We would like to thank A. Brinkman, M. Dekkers, A. Golubov, S. Harkema, T. Ortlepp, and F. Roesthuis for discussions. This work was supported by the Dutch Foundation for Research on Matter (FOM), the Netherlands Organization for Scientific Research (NWO), the Dutch STW NanoNed programme, and the European Science Foundation (ESF) PiShift programme.

References

Aeppli, G. and Chandra, P. (1997) Seeking a simple complex system, *Science* **275**, 177.

Akoh, H., Camerlingo, C., and Takada, S. (1990) Anisotropic Josephson junctions of Y-Ba-Cu-O/Au/Nb film sandwiches, *Appl. Phys. Lett.* **56**, 1487.

Akoh, H., Shinoki, F., Takahashi, M., and Takada, S. (1988) S-N-S Josephson junction consisting of Y-Ba-Cu-O/Au/Nb thin-films, *Jpn. J. Appl. Phys.* **27**, L519.

Akoh, H., Shinoki, F., Takahashi, M., and Takada, S. (1989) Fabrication of S-N-S Josephson-junctions of Y-Ba-Cu-O/Au/Nb sandwices, *IEEE Trans. Magn.* **25**, 795.

Ariando, Darminto, D., Smilde, H.-J., Rogalla, H., and Hilgenkamp, H. (2005) Phase-sensitive order parameter symmetry test experiments utilizing $Nd_{2-x}Ce_xCuO_{4-y}$/Nb zigzag junctions, *Phys. Rev. Lett.* **94**, 167001.

Averin, D. V. (1998) Adiabatic quantum computation with Cooper pairs, *Sol. State Commun.* **105**, 659.

Balci, H. and Greene, R. L. (2004) Anomalous change in the field dependence of the electronic specific heat of an electron-doped cuprate superconductor, *Phys. Rev. Lett.* **93**, 067001.

Basov, D. N., Liang, R., Bonn, D. A., Hardy, W. N., Dabrowski, B., Quijada, M., Tanner, D. B., Rice, J. P., Ginsberg, D. M., and Timusk, T. (1995) In-plane anisotropy of the penetration depth in $YBa_2Cu_3O_{7-x}$ and $YBa_2Cu_4O_8$ superconductors, *Phys. Rev. Lett.* **74**, 598.

Beasley, M. R., Lew, D., and Laughlin, R. B. (1994) Time-reversal symmetry breaking in superconductors: a proposed experimental test, *Phys. Rev. B* **49**, 12330.

Bennet, C. H. and DiVincenzo, D. P. (2000) Quantum information and computation, *Nature* **404**, 247.

Biswas, A., Fournier, P., Qazilbash, M. M., Smolyaninova, V. N., Balci, H., and Greene, R. L. (2002) Evidence of a $d-$ to s-wave pairing symmetry transition in the electron-doped cuprate superconductor $Pr_{2-x}Ce_xCuO_4$, *Phys. Rev. Lett.* **88**, 207004.

Blais, A. and Zagoskin, A. M. (2000) Operation of universal gates in a solid-state quantum computer based on clean Josephson junctions between d-wave superconductors, *Phys. Rev. A* **61**, 042308.

Blank, D. H. A. and Rogalla, H. (1997) Effect of ion milling on the morphology of ramp-type Josephson junctions, *J. Mater. Res.* **12**, 2952.

Blatter, G., Geshkenbein, V. B., and Ioffe, L. B. (2001) Design aspects of superconducting-phase quantum bits, *Phys. Rev. B* **63**, 174511.

Bocko, M. F., Herr, A. M., and Feldman, M. J. (1997) Prospects for quantum coherent computation using superconducting electronics, *IEEE Trans. Appl. Supercond.* **7**, 3638.

Bouchiat, V., Vion, D., Joyez, P., Esteve, D., and Devoret, M. H. (1998) Quantum coherence with a single Cooper pair, *Physica Scripta* **T76**, 165.

Brawner, D. A. and Ott, H. R. (1994)a Evidence for an unconventional superconducting order-parameter in $YBa_2Cu_3O_6$.9, *Phys. Rev. B* **50**, 6530.

Brawner, D. A. and Ott, H. R. (1994)b Evidence for unconventional superconductivity in $YBa_2Cu_3O_6$.9, *Physica C* **235**, 1867.

Brawner, D. A. and Ott, H. R. (1996) Evidence for a non-*s*-wave superconducting order parameter in $YBa_2Cu_3O_6$.6 with T_c = 60 K, *Phys. Rev. B* **53**, 8249.

Bulaevskii, L. N., Kuzii, V. V., and Sobyanin, A. A. (1977) Superconducting system with weak coupling to current in ground-state, *JETP Lett.* **25**, 290.

Chandra, P. and Doucot, B. (1988) Possible spin-liquid state at large S for the frustrated square Heisenberg lattice, *Phys. Rev. B* **38**, 9335.

Chesca, B., Doenitz, D., Dahm, T., Huebener, R. P., Koelle, D., Kleiner, R., Ariando, Smilde, H. J. H., and Hilgenkamp, H. (2006) Observation of Andreev bound states in $YBa_2Cu_3O_{7-x}$/Au/Nb ramp-type Josephson junctions, *Phys. Rev. B* **73**, 014529.

Chesca, B., Ehrhardt, K., Mossle, M., Straub, R., Koelle, D., Kleiner, R., and Tsukada, A. (2003) Magnetic-field dependence of the maximum supercurrent of $La_{2-x}Ce_xCuO_{4-y}$ interferometers: Evidence for a predominant $d_{x^2-y^2}$ superconducting order parameter, *Phys. Rev. Lett.* **90**, 057004.

Chesca, B., Seifried, M., Dahm, T., Schopohl, N., Koelle, D., Kleiner, R., and Tsukada, A. (2005) Observation of Andreev bound states in bicrystal grain-boundary Josephson junctions of the electron-doped superconductor $La_{2x}Ce_xCuO_{4y}$, *Phys. Rev. B* **71**, 104504.

Copetti, C. A., Rüders, F., Oelze, B., Buchal, C., Kabius, B., and Seo, J. W. (1995) Electrical-properties of 45° grain-boundaries of epitaxial YBaCuO, dominated by crystalline microstructure and *d*-wave-symmetry, *Physica C* **253**, 63.

Covington, M., Aprili, M., Paraoanu, E., , Greene, L. H., Xu, F., Zhu, J., , and Mirkin, C. A. (1997) Observation of surface-induced broken time-reversal symmetry in $YBa_2Cu_3O_7$ tunnel junctions, *Phys. Rev. Lett.* **79**, 277.

Dagan, Y. and Deutscher, G. (2001) Doping and magnetic field dependence of in-plane tunneling into $YBa_2Cu_3O_{7-x}$: possible evidence for the existence of a quantum critical point, *Phys. Rev. Lett.* **87**, 177004.

Davidovic, D., Kumar, S., Reich, D. H., Siegel, J., Field, S. B., Tiberio, R. C., Hey, R., and Ploog, K. (1996) Magnetic correlations, geometrical frustration, and tunable disorder in arrays of superconducting rings, *Phys. Rev. B* **55**, 6518.

Davidovic, D., Kumar, S., Reich, D. H., Siegel, J., Field, S. B., Tiberio, R. C., Hey, R., and Ploog, K. (1997) Correlations and disorder in arrays of magnetically coupled superconducting rings, *Phys. Rev. Lett.* **76**, 815.

DiVincenzo, D. P. (1995) Quantum Computation, *Science.* **270**, 255.

DiVincenzo, D. P. and Loss, D. (1998) Quantum computation is physical, *Superlattices Microstruct.* **23**, 419.

Engelhardt, A., Dittmann, R., and Braginski, A. I. (1999) Subgap conductance features of $YBa_2Cu_3O_{7-\delta}$ edge Josephson junctions, *Phys. Rev. B* **59**, 3815.

Fauque, B., Sidis, Y., Hinkov, V., Pailhes, S., Lin, C., Chaud, X., and Bourges, P. (2005) Magnetic order in the pseudogap phase of high-T_c superconductors, *Preprint at http://arxiv.org/abs/cond-mat/0509210.*

Feigel'man, M. V., Ioffe, L. B., Geshkenbein, V. B., and Blatter, G. (2000) Andreev spectroscopy for superconducting phase qubits, *J. Low Temp. Phys.* **118**, 805.

Foote, M. C., Hunt, B. D., and Bajuk, L. J. (1991) $YBa_2Cu_3O_{7-\delta}$/Au/Nb sandwich geometry SNS weak links on *c*-axis oriented $YBa_2Cu_3O_{7-\delta}$, *IEEE Trans. Magn.* **27**, 1335.

Friedman, J. R., Patel, V., Chen, W., Tolpygo, S. K., and Lukens, J. E. (2000) Quantum superposition of distinct macroscopic states, *Nature* **406**, 43.

Fujimaki, A., Takai, Y., and Hayakawa, H. (1991) Experimental-analysis of superconducting properties of Y-Ba-Cu-O/Ag proximity interfaces, *IEEE Trans. Magn.* **27**, 1353.

Fujimaki, A., Tamaoki, T., Hidaka, T., Yanagase, M., Shiota, T., Takai, Y., and Hayakawa, H. (1990) Experimental-analysis of $YBa_2Cu_3O_x$/Ag proximity interfaces, *Jpn. J. Appl. Phys.* **29**, L1659.

Gao, J., Aarnink, W. A. M., Gerritsma, G. J., and Rogalla, H. (1990) Controlled preparation of all high-T_c SNS-type edge junctions and dc squids, *Physica C* **171**, 126.

Gao, J., Boguslavskij, Y., Klopman, B. B. G., Terpstra, D., Gerritsma, G. J., and Rogalla, H. (1991) Characteristics of advanced $YBa_2Cu_3O_x$ $PrBa_2Cu_3O_x$ $YBa_2Cu_3O_x$ edge type junctions, *Appl. Phys. Lett.* **59**, 2754.

Geshkenbein, V. B. and Larkin, A. I. (1986) The Josephson effect in superconductors with heavy fermions, *JETP Lett.* **43**, 395.

Geshkenbein, V. B., Larkin, A. I., and Barone, A. (1987) Vortices with half magnetic flux quanta in heavy-fermion superconductors, *Phys. Rev. B* **36**, 235.

Gim, Y., Mathai, A., Black, R. C., Amar, A., and Wellstood, F. C. (1996) Symmetry of the phase of the order parameter in $YBa_2Cu_3O_{7-\delta}$, *Journal de Physique I* **6**, 2299.

Goldobin, E., Sterck, A., Gaber, T., Koelle, D., and Kleiner, R. (2004) Dynamics of semifluxons in Nb long Josephson 0-π junctions, *Phys. Rev. Lett.* **92**, 057005.

Golubov, A. A., Kupriyanov, M. Y., and Il'ichev, E. (2004) The current-phase relation in Josephson junctions, *Rev. Mod. Phys.* **76**, 411.

Hilgenkamp, H., Ariando, Smilde, H. J. H., Blank, D. H. A., Rijnders, G., Rogalla, H., Kirtley, J. R., and Tsuei, C. C. (2003) Ordering and manipulation of the magnetic moments in large-scale superconducting π-loop arrays, *Nature* **422**, 50.

Hilgenkamp, H. and Mannhart, J. (2002) Grain boundaries in high-T_c superconductors, *Rev. Mod. Phys.* **74**, 485.

Hilgenkamp, H., Mannhart, J., and Mayer, B. (1996) Implications of $d_{x^2-y^2}$ symmetry and faceting for the transport properties of grain boundaries in high-T_c superconductors, *Phys. Rev. B* **53**, 14586.

Hu, C. R. (1994) Midgap surface states as a novel signature for $d_{x^2-y^2}$-wave superconductivity, *Phys. Rev. Lett.* **72**, 1526.

Hunt, B. D., Foote, M. C., and Bajuk, L. J. (1991) Edge-geometry $YBa_2Cu_3O_{7-x}$/Au/Nb SNS devices, *IEEE Trans. Magn.* **27**, 848.

Hunt, B. D., Foote, M. C., and Vasquez, R. P. (1990) Electrical characterization of chemically modified $YBa_2Cu_3O_{7-x}$ surfaces, *Appl. Phys. Lett.* **56**, 2678.

Ioffe, L. B., Geshkenbein, V. B., Feigel'man, M. V., Fauchere, A. L., and Blatter, G. (1999) Environmentally decoupled sds-wave Josephson junctions for quantum computing, *Nature* **398**, 679.

Kaminski, A., Rosenkranz, S., Fretwell, H. M., Campuzano, J. C., Li, Z., Raffy, H., Cullen, W. G., You, H., Olson, C. G., Varma, C. M., and Hochst, H. (2002) Spontaneous breaking of time reversal symmetry in the pseudogap state of a high-T_c superconductor, *Nature* **416**, 610.

Kashiwaya, S. and Tanaka, Y. (2000) Tunnelling effects on surface bound states in unconventional superconductors, *Rep. Prog. Phys.* **63**, 1641.

Kirtley, J. R., Ketchen, M. B., Stawiasz, K. G., Sun, J. Z., Gallagher, W. J., Blanton, S. H., and Wind, S. J. (1995)a High-resolution scanning SQUID microscope, *Appl. Phys. Lett.* **66**, 1138.

Kirtley, J. R., Ketchen, M. B., Tsuei, C. C., Sun, J. Z., Gallagher, W. J., Yu-Jahnes, L. S., Gupta, A., Stawiasz, K. G., and Wind, S. J. (1995)b Design and applications of a scanning SQUID microscope, *IBM J. Res. Develop.* **39**, 655.

Kirtley, J. R., Moler, K. A., and Scalapino, D. J. (1997) Spontaneous flux and magnetic-interference patterns in 0-π Josephson junctions, *Phys. Rev. B* **56**, 886.

Kirtley, J. R., Tsuei, C. C., Ariando, Smilde, H. J. H., and Hilgenkamp, H. (2005) Antiferro-magnetic ordering in arrays of superconducting π-rings, *Phys. Rev. B* **72**, 214521.

Kirtley, J. R., Tsuei, C. C., Ariando, A., Verwijs, C. J. M., Harkema, S., and Hilgenkamp, H. (2006) Angle-resolved phase-sensitive determination of the in-plane gap symmetry in $YBa_2Cu_3O_{7-\delta}$, *Nature Physics*.

Laughlin, R. B. (1988) The relationship between high-temperature superconductivity and the fractional quantum hall effect, *Science* **242**, 525.

Lawrence, T. W., Szoker, A., and Laughlin, R. B. (1992) Absence of circular dichroism in high-temperature superconductors, *Phys. Rev. Lett.* **69**, 1439.

Lerch, P., Leemann, C., Theron, R., and Martinoli, P. (1990) Dynamics of the phase-transition in proximity-effect arrays of Josephson-junctions at full frustration, *Phys. Rev. B* **41**, 11579.

Lofwander, T., Shumeiko, V. S., and Wendin, G. (2001) Andreev bound states in high-T_c superconducting junctions, *Supercond. Sci. Technol.* **14**, R53.

Lu, D. H., Feng, D. L., Armitage, N. P., Shen, K. M., Damascelli, A., Kim, C., Ronning, F., Bonn, D. A., Liang, R., Hardy, W. N., Rykov, A. I., and Tajima, S. (2001) Superconducting gap and strong in-plane anisotropy in untwinned $YBa_2Cu_3O_{7-\delta}$, *Phys. Rev. Lett.* **86**, 4370.

Majer, J. B., Butcher, J. R., and Mooij, J. E. (2002) Simple phase bias for superconducting circuits, *Appl. Phys. Lett.* **80**, 3638.

Mannhart, J., Mayer, B., and Hilgenkamp, H. (1996) Anomalous dependence of the critical current of 45° grain boundaries in $YBa_2Cu_3O_{7-x}$ on an applied magnetic field, *Z. Phys. B* **101**, 175.

Mathai, A., Gim, Y., Black, R. C., Amar, A., and Wellstood, F. C. (1995) Experimental proof of a time-reversal-invariant order parameter with a π shift in $YBa_2Cu_3O_{7-\delta}$, *Phys. Rev. Lett.* **74**, 4523.

Mints, R. G. and Kogan, V. G. (1997) Josephson junctions with alternating critical current density, *Phys. Rev. B* **55**, R8682.

Mints, R. G., Papiashvili, I., Kirtley, J. R., Hilgenkamp, H., Hammerl, G., and Mannhart, J. (2002) Observation of splintered Josephson vortices at grain boundaries in $YBa_2Cu_3O_{7-\delta}$, *Phys. Rev. Lett.* **89**, 067004.

Moessner, R. and Sondhi, S. L. (2001) Ising models of quantum frustration, *Phys. Rev. B* **63**, 224401.

Mooij, J. E., Orlando, T. P., Levitov, L., Tian, L., van der Wal, C. H., and Lloyd, S. (1999) Josephson persistent-current qubit, *Science* **285**, 1036.

Nakamura, Y., Pashkin, Y. A., and Tsai, J. S. (1999) Coherent control of macroscopic quantum states in a single-Cooper-pair box, *Nature* **398**, 786.

Ng, T. K. and Varma, C. M. (2004) Experimental signatures of time-reversal-violating superconductors, *Phys. Rev. B* **70**, 054514.

Ortlepp, T., Ariando, Mielke, O., Verwijs, C. J. M., Foo, K. F. K., Rogalla, H., Uhlmann, F. H., and Hilgenkamp, H. (2006) Flip-flopping fractional flux quanta, *Science* **312**, 1495.

Pannetier, B., Chaussy, J., Rammal, R., and Villegier, J. C. (1984) Experimental fine tuning of frustration: Two-dimensional superconducting network in a magnetic Field, *Phys. Rev. Lett.* **53**, 1845.

Polturak, E., Koren, G., Cohen, D., and Aharoni, E. (1993) Measurements of the anisotropy and temperature dependence of the in-plane energy gap in $YBa_2Cu_3O_{7-\delta}$ using Andreev reflections, *Phys. Rev. B* **47**, 5270.

Preskill, J. (1998) Reliable quantum computers, *Proc. R. Soc. Lond. A* **454**, 385.

Qazilbash, M. M., Biswas, A., Dagan, Y., Ott, R. A., and Greene, R. L. (2003) Point-contact spectroscopy of the electron-doped cuprate superconductor $Pr_{2x}Ce_xCuO_4$: The dependence of conductance-voltage spectra on cerium doping, barrier strength, and magnetic field, *Phys. Rev. B* **68**, 024502.

Rykov, M. F. L. A. I. and Tajima, S. (1998) Raman scattering study on fully oxygenated $YBa_2Cu_3O_7$ single crystals: xy anisotropy in the superconductivity-induced effects, *Phys. Rev. Lett.* **80**, 825.

Schnirman, A., Schon, G., and Hermon, Z. (1997) Quantum manipulations of small Josephson junctions, *Phys. Rev. Lett.* **79**, 2371.

Schulz, R. R., Chesca, B., Goetz, B., Schneider, C. W., Schmehl, A., Bielefeldt, H., Hilgenkamp, H., Mannhart, J., and Tsuei, C. C. (2000) Design and realization of an all d-wave dc π-superconducting quantum interference device, *Appl. Phys. Lett.* **76**, 912.

Sharoni, A., Millo, O., Kohen, A., Dagan, Y., Beck, R., Deutscher, G., and Koren, G. (2002) Local and macroscopic tunneling spectroscopy of $Y_{1-x}Ca_xBa_2Cu_3O_{7-d}$ films: evidence for a doping-dependent *is* or *idxy* component in the order parameter, *Phys. Rev. B* **65**, 134526.

Sigrist, M. (1998) Time-reversal symmetry breaking states in high-temperature superconductors, *Prog. Theor. Phys.* **99**, 899.

Sigrist, M. and Rice, T. M. (1992) Paramagnetic effect in high-T_c superconductors - a hint for d-wave superconductivity, *J. Phys. Soc. Jpn.* **61**, 4283.

Skinta, J. A., Kim, M. S., Lemberger, T. R., Greibe, T., and Naito, M. (2002) Evidence for a transition in the pairing symmetry of the electron-doped cuprates $La_{2x}Ce_xCuO_{4y}$ and $Pr_{2x}Ce_xCuO_{4y}$, *Phys. Rev. Lett.* **88**, 207005.

Smilde, H. J. H., Ariando, Blank, D. H. A., Gerritsma, G. J., Hilgenkamp, H., and Rogalla, H. (2002)a d-waveinduced Josephson current counterflow in $YBa_2Cu_3O_7$/Nb Zigzag Junctions, *Phys. Rev. Lett.* **88**, 057004.

Smilde, H. J. H., Ariando, Blank, D. H. A., Hilgenkamp, H., and Rogalla, H. (2004)a π-SQUIDs based on Josephson contacts between high-T_c and low-T_c superconductors, *Phys. Rev. B* **70**, 024519.

Smilde, H. J. H., Ariando, Rogalla, H., and Hilgenkamp, H. (2004)b Bistable superconducting quantum interference device with built-in switchable $\pi/2$ phase shift, *Appl. Phys. Lett.* **85**, 4091.

Smilde, H. J. H., Golubov, A. A., Ariando, Rijnders, G., Dekkers, J. M., Harkema, S., Blank, D. H. A., Rogalla, H., and Hilgenkamp, H. (2005) Admixtures to d-wave gap symmetry in untwinned $YBa_2Cu_3O_7$ superconducting films measured by angle-resolved electron tunneling, *Phys. Rev. Lett.* **95**, 257001.

Smilde, H. J. H., Hilgenkamp, H., Gerritsma, G. J., Blank, D. H. A., and Rogalla, H. (2001)a Realization and properties of ramp-type $YBa_2Cu_3O_{7-\delta}$/Au/Nb junctions, *Physica C* **350**, 269.

Smilde, H. J. H., Hilgenkamp, H., Gerritsma, G. J., Blank, D. H. A., and Rogalla, H. (2001)b Y-Ba-Cu-O/Au/Nb ramp-type Josephson junctions, *IEEE Trans. Appl. Supercond.* **11**, 501.

Smilde, H. J. H., Hilgenkamp, H., Rijnders, G., Rogalla, H., and Blank, D. H. A. (2002)b Enhanced transparency ramp-type Josephson contacts through interlayer deposition, *Appl. Phys. Lett.* **80**, 4579.

Spielman, S., Dodge, J. S., Lombardo, L. W., Eom, C. B., Fejer, M. M., Geballe, T. H., and Kapitulnik, A. (1992) Measurement of the spontaneous polar Kerr effect in $YBa_2Cu_3O_7$ and $Bi_2Sr_2CaCu_2O_8$, *Phys. Rev. Lett.* **68**, 3472.

Tafuri, F. and Kirtley, J. R. (2005) Weak links in high critical temperature superconductors, *Rep. Prog. Phys.* **68**, 2573.

Terai, H., Fujimaki, A., Takai, Y., and Hayakawa, H. (1993) Magnetic-field dependence of high J_c $YBa_2Cu_3O_{7-x}$/Au/Nb junctions using alpha-axis-oriented $YBa_2Cu_3O_{7-x}$ thin-films, *Jpn. J. Appl. Phys.* **32**, L901.

Terai, H., Fujimaki, A., Takai, Y., and Hayakawa, H. (1995) Electrical interface structure $YBa_2Cu_3O_{7-x}$ metal contact, *IEEE Trans. Appl. Supercond.* **5**, 2408.

Terzioglu, E. and Beasley, M. R. (1998) Complementary Josephson junction devices and circuits: A possible new approach to superconducting electronics, *IEEE Trans. Appl. Sup.* **8**, 48.

Tsuei, C. C. and Kirtley, J. R. (2000)a Pairing symmetry in cuprate superconductors, *Rev. Mod. Phys.* **72**, 969.

Tsuei, C. C. and Kirtley, J. R. (2000)b Phase-sensitive evidence for d-wave pairing symmetry in electron-doped cuprate superconductors, *Phys. Rev. Lett.* **85**, 182.

Tsuei, C. C., Kirtley, J. R., Chi, C. C., Yu-jahnes, L. S., Gupta, A., Shaw, T., Sun, J. Z., and Ketchen, M. B. (1994) Pairing symmetry and flux quantization in a tricrystal superconducting ring of $YBa_2Cu_3O_{7-\delta}$, *Phys. Rev. Lett.* **73**, 593.

Tsuei, C. C., Kirtley, J. R., Hammerl, G., Mannhart, J., Raffy, H., and Li, Z. Z. (2004) Robust $d_{x^2-y^2}$ pairing symmetry in hole-doped cuprate superconductors, *Phys. Rev. Lett.* **93**, 187004.

Usagawa, T., Wen, J. G., Ishimaru, Y., Koyama, S., Utagawa, T., and Enomoto, Y. (1998) Stability of ultrasmooth surface morphology of (110) $YBa_2Cu_3O_{7-\delta}$ homoepitaxial films and Nb/Au/(110) $YBa_2Cu_3O_{7-\delta}$ junctions, *Appl. Phys. Lett.* **72**, 3202.

Ustinov, A. V. and Kaplunenko, V. K. (2003) Rapid single-flux quantum logic using π-shifters, *J. Appl. Phys.* **94**, 5405.

van der Wal, C. H., ter Haar, A. C. J., Wilhelm, F. K., Schouten, R. N., and Mooij, C. J. P. M. H. T. P. O. S. L. J. E. (2000) Quantum superposition of macroscopic persistent-current states, *Science* **290**, 773.

van Harlingen, D. J. (1995) Phase-sensitive tests of the symmetry of the pairing state in the high-temperature superconductorsEvidence for $d_{x^2-y^2}$ symmetry, *Rev. Mod. Phys.* **67**, 515.

Varma, C. M. (1999) Pseudogap phase and the quantum-critical point in copper-oxide metals, *Phys. Rev. Lett.* **83**, 3538.

Varma, M. E. S. C. M. (2002) Detection and implications of a time-reversal breaking state in underdoped cuprates, *Phys. Rev. Lett.* **89**, 247003.

Verhoeven, M. A. J., Moerman, R., Bijlsma, M. E., Rijnders, A. J. H. M., Blank, D. H. A., Gerritsma, G. J., and Rogalla, H. (1996) Nucleation and growth of $PrBa_2Cu_3O_{7-\delta}$ barrier layers on ramps in $DyBa_2Cu_3O_{7-\delta}$ studied by atomic force microscopy, *Appl. Phys. Lett.* **68**, 1276.

Wen, J. G., Koshizuka, N., Tanaka, S., Satoh, T., Hidaka, M., and Tahara, S. (1999) Atomic structure and composition of the barrier in the modified interface high-T_c Josephson junction studied by transmission electron microscopy, *Appl. Phys. Lett.* **75**, 2470.

Wollman, D. A., van Harlingen, D. J., Giapintzakis, J., and Ginsberg, D. M. (1995) Evidence for $d_{x^2-y^2}$ Pairing from the Magnetic Field Modulation of $YBa_2Cu_3O_7$-Pb Josephson Junctions, *Phys. Rev. Lett.* **74**, 797.

Wollman, D. A., van Harlingen, D. J., Lee, W. C., Ginsberg, D. M., and Leggett, A. J. (1993) Experimental determination of the superconducting pairing state in YBCO from the phase coherence of YBCO-Pb dc SQUIDs, *Phys. Rev. Lett.* **71**, 2134.

Zenchuk, A. and Goldobin, E. (2004) Analysis of ground states of 0-π long Josephson junctions, *Phys. Rev. B* **69**, 024515.

ANISOTROPIC RESONANCE PEAK IN ORTHORHOMBIC SUPERCONDUCTORS

Dirk Manske[1] (manske@itp.phys.ethz.ch), Ilya Eremin[2,3]
[1] *Institut für Theoretische Physik, ETH Zürich, Hönggerberg,
CH-8093 Zürich, Switzerland*
[2] *Max-Planck-Institut für Physik komplexer Systeme, D-01187 Dresden,
Germany*
[3] *Institute für Mathematische und Theoretische Physik, Technische Universität Carolo-Wilhelmina zu Braunschweig, D-38106 Braunschweig, Germany*

Abstract. Extending our previous studies employing a generalized RPA-type theory we calculate the in-plane anisotropy of the magnetic excitations in hole-doped high-T_c superconductors. Using an effective two-dimensional one-band Hubbard model we consider anisotropic hopping matrix elements ($t_x \neq t_y$) and a mixing of d- and s-wave symmetry of the superconducting order parameter in order to describe orthorhombic superconductors. We compare our calculations with new experimental data on fully untwinned $YBa_2Cu_3O_{6.85}$ and find good agreement. Our results are in contrast to earlier interpretations on the in-plane anisotropy in terms of stripes.

Key words: Spin susceptibility, High-T_C superconductivity, Orthorhombicity

1. Introduction

Since the discovery of high-T_c superconductors, its mechanism is still under debate. Several theoretical scenarios have proposed a mechanism for superconductivity in the high-T_c cuprates originating from magnetism. It has been argued that superconductivity occurs from an exchange of AF spin fluctuations between Fermi-like quasiparticles [1, 2] or from a recombination in momentum space of holons and spinons in a spin-charge-separated normal state [3]. In the Fermi-liquid picture, an Eliashberg-like formalism is applied in order to calculate properties of the normal and superconducting states. In particular, a so-called resonance peak develops below T_c which is often attributed to a feedback effect of superconductivity [2]. In the stripe scenario [4], strong electronic interactions result in normal and superconducting states in which spin and charge are separated in predominantly one-dimensional region, called stripes, of the CuO_2 planes. In order to distinguish between

K. Scharnberg and S. Kruchinin (eds.),
Electron Correlation in New Materials and Nanosystems, 175–186.
© 2007 *Springer.*

both pictures, a detailed analysis of untwinned cuprates is necessary. Recently, an inelastic neutron scattering (INS) study in the fully untwinned high-temperature superconductor $YBa_2Cu_3O_{6.85}$ reveals the two-dimensional character of the magnetic fluctuations [5], in contrast to the previous conclusions from measurements in the partially untwinned samples [6].

2. Generalized Eliashberg theory

In this paper, we employ an effective one-band Hubbard Hamiltonian for each CuO_2-plane

$$H = H_0 + H_1, \tag{1}$$

$$H_0 = -\sum_{\langle ij\rangle\sigma} t_{ij}c^{\dagger}_{i\sigma}c_{j\sigma} - \mu\sum_{i\sigma}n_{i\sigma} \quad , \quad H_1 = U\sum_i n_{i\uparrow}n_{i\downarrow}. \tag{2}$$

Here, $c^{\dagger}_{i\sigma}$ is the creation operator of a quasiparticle with spin σ on site i, $n_{i\sigma} = c^{\dagger}_{i\sigma}c_{i\sigma}$ is the spin-dependent local number operator, t_{ij} is a hopping matrix element in the CuO_2-plane, μ is the chemical potential, and U denotes a residual on-site (i.e., intra-orbital) Coulomb repulsion. We assume that a rigid-band approximation, in which all effects of doping are incorporated into a doping dependent chemical potential, can be used. Because of the translational symmetry of the underlying crystal lattice, H_0 can be diagonalized by the usual transformation to momentum space:

$$H_0 = -\sum_{k\in BZ}\varepsilon_k\sum_{\sigma}c^{\dagger}_{k\sigma}c_{k\sigma} \tag{3}$$

The band energy

$$\varepsilon_k = \sum_{\ell} t_{\ell}\, e^{k\cdot R_{\ell}}, \tag{4}$$

with R_{ℓ} a lattice vector and k from the first Brillouin zone, is evaluated by extending the summation up to the 5th neighbor shell:

$$\begin{aligned}
\varepsilon_k &= \frac{t_1}{2}(1+\delta_0)\cos k_x + \frac{t_1}{2}(1-\delta_0)\cos k_y + t_2\cos k_x\cos k_y \\
&+\frac{t_3}{2}(1+\delta_0)\cos 2k_x + \frac{t_3}{2}(1-\delta_0)\cos 2k_y \\
&+\frac{t_4}{2}\cos 2k_x\cos k_y + \frac{t_4}{2}\cos k_x\cos 2k_y + t_5\cos 2k_x\cos 2k_y + \mu.
\end{aligned} \tag{5}$$

We shall choose the parameters $t_1,\ldots,t_5$ so as to fit qualitatively the Fermi surface as measured by angle-resolved photoemission spectroscopy (ARPES) on twinned crystals [7]. The parameter $\delta_0 \neq 0$ breaks the tetragonal symmetry

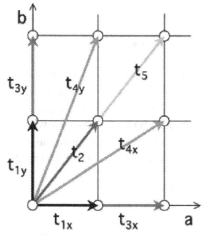

Figure 1. The hopping parameters used in the tight-binding dispersion (5) are $t_1 = -588.1$ meV, $t_2 = 146.1$ meV, $t_3 = 9.5$ meV, $t_4 = -129.8$ meV, and $t_5 = 6.9$ meV throughout this paper. Note that only t_1 and t_3 (along the bonds) are changed due to orthorhombicity.

as the k_x and k_y directions in the BZ of the square lattice are not equivalent. This is illustrated in Fig. 1. An orthorhombic symmetry implies that rotation symmetry by $\pi/2$ is broken, i.e., that $t_{1x} \equiv t_1(1 + \delta_0)/4$ is not equal to $t_{1y} \equiv t_1(1 - \delta_0)/4$ and that $t_{3x} \equiv t_3(1 + \delta_0)/4$ is not equal to $t_{3y} \equiv t_3(1 - \delta_0)/4$. In this one-band model we assume that the *same* electrons (holes) are participating in the formation of antiferromagnetic fluctuations and in Cooper-pairing arising from the exchange of these fluctuations. Thus, all self-energy parts as well as the charge and spin susceptibilities must be calculated self-consistently. This is possible in the FLEX approximation [8, 9, 10, 11, 12]. Note that the FLEX approximation ignores the fact that cuprates are doped Mott insulators and, secondly, that it is based on perturbation theory. On the other hand, in defense of this approach, we argue that for optimally doped or overdoped cuprates the Mott physics is no longer dominant.

Like in conventional Eliashberg theory it is convenient to formulate the theory in terms of self-energy components. In a short-hand notation, the Eliashberg equations, generalized to a repulsive spin fluctuation-mediated pairing interaction, can be written in terms of the self-energy components Σ_ν ($\nu = 0, 3, 1$), which are the expansion coefficients with respect to Pauli matrices τ_ν of the 2×2 matrix self-energy in the Nambu representation[14, 15]. The only non-vanishing components are $\Sigma_0 = \omega(1 - Z)$ (mass renormalization), $\Sigma_3 = \xi$ (energy shift), and the off-diagonal element $\Sigma_1 = \phi = Z\Delta$. This last equality defines the superconducting gap function $\Delta(\mathbf{k}, \omega)$. These self-energy components have to be determined from the equations

$$\Sigma_\nu(\mathbf{k}, \omega) = \sum_{\mathbf{k}'} \int_0^\infty d\Omega \, V_{\text{eff}}(\mathbf{k} - \mathbf{k}', \Omega) \int_{-\infty}^{+\infty} d\omega' I(\omega, \Omega, \omega') A_\nu(\mathbf{k}', \omega') \quad (6)$$

with the effective pairing interaction

$$V_{\text{eff}} = [P_s(\mathbf{k} - \mathbf{k}', \Omega) - (\delta_{\nu 1} - \delta_{\nu 0} - \delta_{\nu 3}) P_c(\mathbf{k} - \mathbf{k}', \Omega)]. \qquad (7)$$

P_s and P_c denote the spectral densities of the spin and charge excitations, respectively, and are treated within RPA, $P_s = (2\pi)^{-1} U^2 \operatorname{Im}(3\chi_s - \chi_{s0})$ with $\chi_s = \chi_{s0}(1 - U\chi_{s0})^{-1}$ and $P_c = (2\pi)^{-1} U^2 \operatorname{Im}(3\chi_c - \chi_{c0})$ with $\chi_c = \chi_{c0}(1 + U\chi_{c0})^{-1}$. Note that P_s and P_c might reveal strong changes below T_c which are calculated self-consistently. In particular, a pole in χ_s can occur as we will discuss below. The kernel I and the spectral functions A_ν in Eq. (6) read

$$I(\omega, \Omega, \omega') = \frac{f(-\omega') + b(\Omega)}{\omega + i\delta - \Omega - \omega'} + \frac{f(\omega') + b(\Omega)}{\omega + i\delta - \Omega - \omega'}, \qquad (8)$$

$$A_\nu(\mathbf{k}, \omega) = -\pi^{-1} \operatorname{Im}[a_\nu(\mathbf{k}, \omega)/D(\mathbf{k}, \omega)], \qquad (9)$$

with

$$D = [\omega Z]^2 - [\epsilon_\mathbf{k}^0 + \xi]^2 - \phi^2, \qquad (10)$$

$$a_0 = \omega Z(\mathbf{k}, \omega), \quad a_3 = \epsilon_\mathbf{k}^0 + \xi(\mathbf{k}, \omega), \quad a_1 = \phi(\mathbf{k}, \omega). \qquad (11)$$

In Eq. (8), f and b are the Fermi and Bose distribution function, respectively.

The dressed one-electron Green's functions obtained from our generalized Eliashberg equations are used to calculate the bare charge and spin susceptibilities χ_{c0} and χ_{s0}, which appear in the expressions for P_s and P_c above. Hence, no input parameters in an Ornstein-Zernicke ansatz are needed because the magnetic correlation length ξ and the characteristic energy scale of the spin fluctuations ω_{sf} (roughly the peak in $\operatorname{Im}\chi(\mathbf{q} = \mathbf{Q} = (\pi, \pi), \omega)$ in the normal state) are calculated self-consistently for fixed values of $U, t_1, ..., t_5$, and μ. In Eq. (7) an effective Berk-Schrieffer-like [13] pairing interaction V_{eff}, describing the exchange of charge and spin fluctuations, has been constructed from these susceptibilities. More details on this theory can be found in Ref. [8].

What is the solution of the FLEX equations below T_c? For tetragonal symmetry, the resulting superconducting order parameter has $d_{x^2-y^2}$-wave symmetry. However, in presence of orthorhombicity the superconducting order parameter changes, since s-wave and d-wave symmetries belong now to the same irreducible representation of the point group symmetry. Since we discuss later our results in the limit $T = 0$, it is instructive to analyze the weak-coupling limit ($Z \to 1$). Furthermore, for studying the dispersion of the magnetic excitations below T_c, it turns out that the ω-dependence of the gap function is not important and will be neglected from now on. Then, the total superconducting gap has the form

$$\Delta(\mathbf{k}) = g(\delta_0)\Delta_s(\mathbf{k}) + f(\delta_0)\Delta_d(\mathbf{k}), \tag{12}$$

where, for simplicity, we employ $g(\delta_0) = \delta_0$, $f(\delta_0) = 1 - \delta_0$ and $\Delta_d = \Delta_0(\cos k_x - \cos k_y)/2$, $\Delta_s = \Delta_0$. For the set of parameters described above, we get $\Delta_0 = 26 \text{meV}$.

How do we calculate the resonance peak below T_c? In a simple view $(T = 0)$ Within the generalized RPA, the imaginary part of the dynamical spin susceptibility is given by

$$\text{Im}\chi(\mathbf{q}, \omega) = \frac{\text{Im}\chi_0(\mathbf{q}, \omega)}{(1 - U\text{Re}\chi_0(\mathbf{q}, \omega))^2 + U^2\text{Im}\chi_0^2(\mathbf{q}, \omega)}, \tag{13}$$

where χ_0 is the BCS Lindhard response function[16]. Without orthorhombicity, the $d_{x^2-y^2}$ superconducting gap opens rapidly due to a feedback effect on the elementary excitations [17] yielding at $T = 0$ a jump at $\omega = 2\Delta_0$ in $\text{Im}\chi_0(\mathbf{q} = \mathbf{Q}, \omega)$. Thus, the corresponding real part of χ_0 reveals a logarithmic singularity at $\omega = 2\Delta_0$ and the resonance condition [18, 19, 20]

$$1 - U\text{Re}\,\chi_0(\mathbf{q} = \mathbf{Q}, \omega = \omega_{res}) = 0 \tag{14}$$

can be fulfilled. Due to the fact that many inelastic neutron scattering experiments are performed at low temperature, the occurence of a resonance peak has been interpreted as a fingerprint for a $d_{x^2-y^2}$-wave gap in cuprates. Since at $T = 0$ $\text{Im}\chi_0$ is zero below $2\Delta_0$, the resonance condition (14) reveals a strong delta-like peak in $\text{Im}\chi$ which occurs only below T_c. Note that its position is mainly determined by the maximum of the d-wave superconducting gap Δ_0 and also by the proximity to an antiferromagnetic instability described by the characteristic energy scale ω_{sf}. Then, the resonance peak scales with the maximum of the d-wave superconducting gap in optimally doped and over-doped compounds. On the other hand, in the underdoped cuprates it rather scales with ω_{sf}[18, 19, 20] due to stronger antiferromagnetic fluctuations. Thus, one finds for the whole doping range $\omega_{res}/k_BT_c \approx const$ [18, 19, 20] in good agreement with experiments [21]. In the orthorhombic case, in which d- and s-wave symmetries are mixed, many of the above arguments still hold as we will discuss below.

3. Results and discussion

First, we analyze the dispersion of the resonance peak for the tetragonal case. To a good approximation the RPA expression for $\text{Im}\chi(\mathbf{Q}, \omega)$ (see Eq. (13))

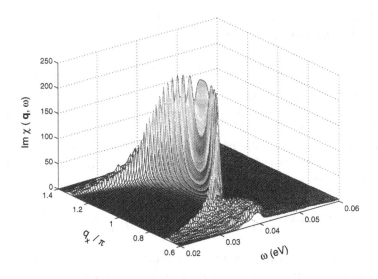

Figure 2. The dispersion of the resonance peak using $q_y = \pi$ and the bandstructure of Eq. (5) for $\delta_0 = 0$.

can be expanded around its resonance condition yielding a quadratic dispersions of the resonance peak. Our results are shown in Fig. 2 as a function of q_x and fixed $q_y = \pi$. One clearly sees a parabola-like envelope which is in good agreement with experiment [21]. Closer inspection shows that for frequencies smaller than $\omega \simeq 35$meV a strong decrease of the resonance peak occurs. In such a situation, the numerator of Eq. (13) may become dominant yielding four peaks around (π, π). This seems to be case in La$_{2-x}$Sr$_x$CuO$_4$ (LSCO) in which the gap is too small to generate a resonance peak. Recently, Tranquada *et al.* measured $\chi(\omega)$ at one of these **q**-points and found only a rearrangement of spectral weight, but no resonance peak [22].

Before analyzing the superconducting state for the orthorhombic case it is instructive to understand how the normal state properties and the electronic structure of a CuO$_2$ plane are affected by the presence of the orthorhombic distortions. In Fig. 3 we show the calculated density of states (DoS) and the Fermi surface topology for different orthorhombic distortions, as quantified by the parameter δ_0. With increasing orthorhombicity the Fermi surface closes around the $(0, \pi)$ and $(0, -\pi)$. One of the immediate consequence is the splitting of the corresponding singularities. In the inset of Fig. 3 we show the calculated changes of the topology of the Fermi surface. Note that for $\delta_0 < 0$ (and $\mu \simeq 120$meV) the distorted Fermi surface is quasi one-dimensional along the k_x-direction. This has to be compared with available ARPES data [23, 24] which, however, are under discussion [25]. It is interesting to note that this

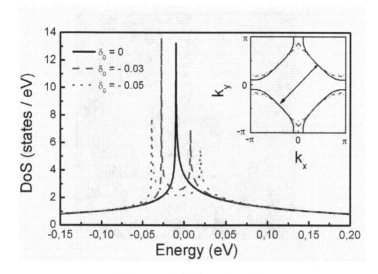

Figure 3. Calculated density of states with and without orthorhombicity. For a comparison the corresponding changes of the Fermi surface topology are shown in the inset. The arrows refer to the two $2\mathbf{k}_F$ instabilities.

Fermi surface deformation breaks the point-group symmetry and looks similar to what is expected for the case of a $d_{x^2-y^2}$-wave Pomeranchuk instability due to strong electron-electron interactions [26].

What is happening below the resonance threshold ($\omega < \omega_{res}$) in fully untwinned YBCO for constant energy scans as a function of the momenta q_x and q_y? In Fig. 4 we show the calculated projected momentum dependence of Im $\chi(\mathbf{q}, \omega = 35\text{meV})$ without (a) and with (b) orthorhombicity. In accordance with *ab-initio* calculations, we have chosen $\delta_0 = -0.03$. In the tetragonal case one sees that the spin excitations form a ring around (π, π) with four pronounced peaks at $(\pi \pm q_0, \pi)$ and $(\pi, \pi \pm q_0)$. In a simple view, away from $\mathbf{Q}$ we are connecting points at the Fermi surface which lie closer to the diagonal of the BZ. The superconducting gap tends to zero there and thus the position and the intensity of the resonance peak are decreasing. However, for the diagonal wave vectors $(\pi \pm q_0, \pi \pm q_0)$ it happens faster than for the vector $(\pi \pm q_0, \pi)$ or $(\pi, \pi \pm q_0)$. Therefore, effectively the latter peaks are 'closer' to the resonance condition at $\mathbf{Q} = (\pi, \pi)$; their intensities are higher than those for the other wave vectors. This explains the observed symmetry of the dominant spin excitations for $\omega < \omega_{res}$ shown in Fig. 4(a). For the orthorhombic case the situation is different. The ring of the excitations becomes distorted and, most importantly, there are only two well pronounced peaks. The latter is a result of strongly changed electronic properties, in

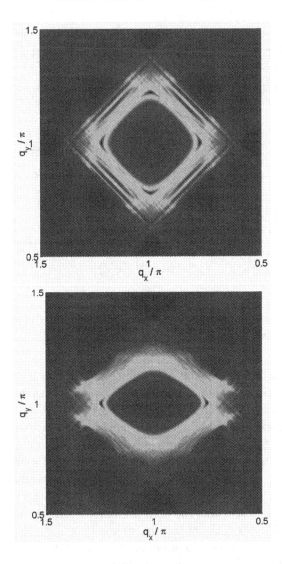

Figure 4. Calculated normalized two-dimensional intensity plot for a constant energy of $\hbar\omega = 35$ meV, (a) without, (b) with inclusion of orthorhombicity ($\delta_0 = -0.03$).

particular, the topology of the quasi one-dimensional Fermi surface. Our main result is that, despite the fact that there are only two pronounced peaks, the resonant spin excitations remains basically two-dimensional. This is in good agreement with recent experiments [5]. Furthermore, this result, based on a standard Fermi-liquid approach, is in contrast to the stripe scenario of the resonance peak [27]. Another interesting observation is that the dispersion

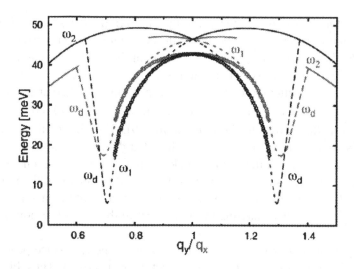

Figure 5. Momentum dependence of the threshold frequencies ω_1, ω_2, and ω_d calculated from $\min_k E_2(\mathbf{q}, \mathbf{k})$ and $\delta_0 = -0.03$. The open diamonds represent the position of the resonance peak which is different for the q_x- and q_y-direction, respectively.

will have a different slope along q_x- and q_y-direction, respectively. This can be further tested experimentally.

Finally, we explain the qualitative behavior of the imaginary part of the RPA spin susceptibility $\chi''_{RPA}(\mathbf{q}, \omega)$ for an orthorhombic superconductor in terms of the properties of $\chi''_0(\mathbf{q}, \omega)$ and the two-particle energy $E_2(\mathbf{q}, \mathbf{k})$. We recall that in the limit of $T = 0$ and for positive frequencies the imaginary part of the noninteracting BCS-Lindhard response function $\chi_0(\mathbf{q}, \omega)$ simplifies to [28]

$$\chi''_0(\omega, \mathbf{q}) = \frac{\pi}{N} \sum_{\mathbf{k}} C^{+,-}_{\mathbf{q},\mathbf{k}} \delta(\omega - E_2(\mathbf{q}, \mathbf{k})), \tag{15}$$

$$C^{+,-}_{\mathbf{q},\mathbf{k}} = \frac{1}{4}\left(1 - \frac{\varepsilon_{\mathbf{k}+\mathbf{q}}\varepsilon_{\mathbf{k}} + \Delta_{\mathbf{k}+\mathbf{q}}\Delta_{\mathbf{k}}}{E_{\mathbf{k}+\mathbf{q}}E_{\mathbf{k}}}\right), \quad E_2(\mathbf{q}, \mathbf{k}) = E_{\mathbf{k}+\mathbf{q}} + E_{\mathbf{k}}, \tag{16}$$

where $E_{\mathbf{k}} = \sqrt{\varepsilon_{\mathbf{k}}^2 + \Delta_{\mathbf{k}}^2}$ denotes the dispersion of the quasiparticles in the superconducting state. At a fixed wave vector $\mathbf{q}$ the imaginary part of the non-interacting spin susceptibility $\chi''_0(\mathbf{q}, \omega)$ vanishes below the threshold frequency

$$\omega_c(\mathbf{q}) = \min_{\mathbf{k} \in BZ} E_2(\mathbf{q}, \mathbf{k}) \tag{17}$$

that defines the border to a continuum of particle-hole excitations.

What is the dispersion of the various threshold frequencies? For a d-wave superconductor the low-energy border of the continuum has a nontrivial form (see Fig. 5). It is bounded by several segments of different curves along each of which $\chi_0''(\mathbf{q}, \omega)$ exhibits either a jump (ω_1 and ω_2 in Fig. 5) or a kink (ω_d in Fig. 5) as a function of frequency, depending on whether the coherence factor $C_{\mathbf{q,k}}^{+,-}$ in Eq. 15 is vanishing for the wave vectors $\mathbf{k}$ contributing to $\chi_0''(\mathbf{q}, \omega)$ at the border to the continuum [19]. The size of the jump in $\chi_0''(\mathbf{q}, \omega)$ is controlled by two criteria: (i) how flat the two-particle dispersion at the corresponding minimum in $E_2(\mathbf{q}, \mathbf{k})$ is, and (ii) by the degeneracy of the minimum itself. As explained in Ref. [28] the degeneracy of the minima $\min_{\mathbf{k}} E_2(\mathbf{q}, \mathbf{k})$ is increased for $\mathbf{q}$ on a high symmetry axes of the magnetic BZ, i.e., on the k_x- or k_y-axes passing through (π, π) in the case of orthorhombic symmetry.

The open diamonds represent the dispersion of the resonance peak (see ω_1) that is different along the q_x- and q_y-direction, respectively. Thus, Fig. 5 is a generalization of Fig. 2 for the tetragonal case in which no difference along both directions are found. The prediction of the anisotropic dispersion of the resonance peak in $YBa_2Cu_3O_{6.85}$ can be tested experimentally by constant energy scans, for example.

4. Conclusions

In summary we have presented a theory for spin excitations in orthorhombic superconductors. In particular, we have considered ansiotropic hopping matrix elements and a mixing of d- and s-wave symmetry of the superconducting order parameter for calculating various constant energy scans. We have contrasted our results with those for the tetragronal case. Finally, employing a generalized RPA-like description, we predict two different dispersions along the q_x- and q_y-direction, respectively which can be tested experimentally. We believe that our analysis and a detailed comparison with experimental data will help to distinguish the Fermi liquid-like scenario from theoretical approaches based on stripes.

Acknowledgements

We wish to thank A. Schnyder for numerical help and C. Mudry, M. Sigrist, H. Yamase, V. Hinkov, and B. Keimer for various discussions. D.M. thanks the Alexander von Humboldt-foundation for financial support.

References

1. Scalapino, D.J. (1995) The case for $d_{x^2-y^2}$ pairing in the cuprate superconductors, Phys. Rep. **250**, 329.

2. Manske, D., *Theory of Unconventional Superconductors* (Springer, Heidelberg, 2004).

3. Lee, P.A., Nagaosa, N., Wen, X.-G. (2006) Doping a Mott insulator: Physics of high temperature superconductivity, Rev. Mod. Phys. **78**, 17.

4. Kivelson, S.A., Bindloss, I.P., Fradkin, E., Oganesyan, V. Tranquada, J.M., Kapitulnik, A., and Howald, C. (2003) How to detect fluctuating stripes in the high-temperature superconductors, Rev. Mod. Phys. **75**, 1201.

5. Hinkov, V., Pailhes, S., Bourges, P., Sidis, Y., Ivanov, A., Kulakov, A., Lin, C. T., Chen, D. P., Bernhard, C., and Keimer, B. (2004) Two-dimensional geometry of spin excitations in the high-transition-temperature superconductor $YBa_2Cu_3O_{6+x}$, *Nature* **430**, 650-654.

6. Mook, H.A., Dai, P., Dogan, F., and Hunt, R.D. (2000) One-dimensional nature of the magnetic fluctuations in $YBa_2Cu_3O_{6.6}$, *Nature* **404**, 729-731.

7. Norman, M.R. (2001) Magnetic collective mode dispersion in high-temperature superconductors, Phys. Rev. B **63**, 092509.

8. Manske, D., Eremin, I., and Bennemann, K.H. (2003) Renormalization of the elementary excitations in hole- and electron-doped high-temperature superconductors, Phys. Rev. B **67**, 134520.

9. Bickers, N.E., Scalapino, D.J., and White, S.R. (1989) Conserving approximations for strongly correlated electron systems: Bethe-Salpeter equation and dynamics for the two-dimensional Hubbard model, Phys. Rev. Lett. **62**, 961; Monthoux, P., and Scalapino, D.J. (1995) Self-consistent $d_{x^2-y^2}$ pairing in a two-dimensional Hubbard model, Phys. Rev. Lett. **72**, 1874.

10. Dahm, T., and Tewordt, L. (1995) Quasiparticle and spin excitations spectra in the normal and d-wave superconducting state of the two-dimensional Hubbard model, Phys. Rev. Lett. **74**, 793.

11. Langer, M., Schmalian, J., Grabowski, S., and Bennemann, K.H. (1995) Theory for the excitation spectrum of high-T_c superconductors: Quasiparticle dispersion and shadows of the Fermi surface, Phys. Rev. Lett. **75**, 4508.

12. Lenck, St., Carbotte, J.P., and Dynes, R.C. (1994) Self-consistent calculation of superconductivity in a nearly antiferromagnetic Fermi liquid, Phys. Rev. B **50**, 10149.

13. Berk, N.F., and Schrieffer, J.R. (1966) Effect of ferromagnetic spin correlations on superconductivity, Phys. Rev. Lett. **17**, 433.

14. Nambu, Y. (1960) Quasi-particles and gauge invariance in the theory of superconductvity, Phys. Rev. **117**, 648.

15. Schrieffer, J.R., *Theory of Superconductivity*, Addison-Wesley (Redwood City, 1964).

16. Tchernyshyov, O., Norman, M.R., and Chubukov, A.V. (2001) Neutron resonance in high-T_c superconductors is not the π particle, Phys. Rev. B **63**, 144507.

17. Manske, D., Eremin, I., and Bennemann, K.H. (2001) Analysis of the elementary excitations in high-T_c cuprates: Explanation of the new energy scale observed by angle-resolved photoemission spectroscopy, Phys. Rev. Lett. **87**, 177005.

18. Manske, D., Eremin, I., and Bennemann, K.H. (2001) Analysis of the resonance peak and magnetic coherence seen in inelastic neutron scattering of cuprate superconductors: A consistent picture with tunneling and conductivity data, Phys. Rev. B **63**, 054517.

19. Onufrieva, F. and Pfeuty, P. (2002) Spin dynamics of a two-dimensional metal in a superconducting state: Application to the high-T_c cuprates, Phys. Rev. B **65**, 054515.

20. Abanov, Ar., Chubukov, A.V., Eschrig, M., Norman, M.R., and Schmalian, J. (2002) Neutron resonance in the cuprates and its effect on fermionic excitations, Phys. Rev. Lett. **89**, 177002.

21. Bourges, P., Sidis, Y., Fong, H.F., Regnault, L.P., Bossy, J., Ivanov. A., Keimer, B. (2000) The spin excitation spectrum in superconducting $YBa_2Cu_3O_{6.85}$, Science **288**, 1234.

22. Tranquada, J.M., Lee, C.H., Yamada, K., Lee, Y.S., Regnault, L.P., and Ronnow, H.M. (2004) Evidence for an incommensurate magnetic resonance in $La_{2-x}SrCuO_4$, Phys. Rev. B **69**, 174507.

23. Schabel, M.C., Park, C.-H., Matsuura, A., Shen, Z.-X., Bonn, D.A., Liang, R., Hardy, W.N. (1998) Angle-resolved photoemission on untwinned $YBa_2Cu_3O_{6.95}$. I. Electronic structure and dispersion relations of surface and bulk bands, Phys. Rev. B **57**, 6090; Schabel, M.C., Park, C.-H., Matsuura, A., Shen, Z.-X., Bonn, D.A., Liang, R., Hardy, W.N. (1998) Angle-resolved photoemission on untwinned $YBa_2Cu_3O_{6.95}$. II. Determination of Fermi surfaces, Phys. Rev. B **57**, 6107.

24. Lu, D.H., Feng, D.L., Armitage, N.P., Shen, K.M., Damascelli, A., Kim, C., Ronning, F., Shen, Z.-X., Bonn, D.A., Liang, R., Hardy, W.N., Rykov, A.I., Tajima, S. (2001) Superconducting gap and strong in-plane anisotropy in untwinned $YBa_2Cu_3O_{7-\delta}$, Phys. Rev. Lett. **86**, 4370.

25. Borisenko, S., private communication.

26. Metzner, W., Rohe, D. and Andergassen, S. (2003) Soft Fermi surfaces and breakdown of Fermi-liquid behavior, Phys. Rev. Lett. **91**, 066402.

27. Vojta, M., and Ulbricht,T. (2004) Magnetic excitations in a bond-centered stripe phase: Spin waves far from the semiclassical limit, Phys. Rev. Lett. **93**, 127002; Uhrig, G.S., Schmidt, K.P., and Grüninger, M. (2004) Unifying magnons and triplons in stripe-ordered cuprate superconductors, Phys. Rev. Lett. **93**, 267003.

28. Schnyder, A.P., Bill, A., Mudry, C., Gilardi, R., Ronnow, H.M., and Mesot, J. (2004) Influence of higher d-wave gap harmonics on the dynamical magnetic susceptibility of high-temperature superconductors, Phys. Rev. B **70**, 214511.

DYNAMICAL SPIN SUSCEPTIBILITY IN THE UNDERDOPED

CUPRATE SUPERCONDUCTORS:

DDW STATE AND INFLUENCE OF ORTHORHOMBICITY

J.-P. Ismer[1], I. Eremin[1,2] (ieremin@mpipks-dresden.mpg.de),
D. K. Morr[3,4]
[1] *Max-Planck-Institut für Physik komplexer Systeme, D-01187 Dresden, Germany*
[2] *Institute für Mathematische und Theoretische Physik, Technische Universität Carolo-Wilhelmina zu Braunschweig, 38106 Braunschweig, Germany*
[3] *Department of Physics, University of Illinois at Chicago, Chicago, IL 60607*
[4] *Institut für Theoretische Physik, Freie Universität Berlin, D-14195 Berlin, Germany*

Abstract. We present a study of the dynamical spin susceptibility in the so-called id-density wave (DDW) state in application to the pseudogap phase of underdoped cuprates. In particular we analyze the structure of the dynamical spin susceptibility in the DDW phase at the antiferromagnetic wave vector $\mathbf{Q} = (\pi, \pi)$. We find that similar to the superconducting state a resonance peak forms. However, away from $\mathbf{Q}$ it shows nearly no dispersion. We also analyze the spin response in the coexisting DDW and $d_{x^2-y^2}$-wave superconducting (DSC) states and discuss the peculiar features of the resonance peak dispersion. Furthermore, we investigate the influence of various tight-binding parameters and the orthorhombic distortions on the robustness of the features discussed.

Key words: Spin susceptibility, Charge density wave, High-T_C superconductivity

1. Introduction

The phenomenon of pseudogap formation in underdoped high-T_c-cuprates is far from being completely understood (1). Among various scenarios proposed so far (2), the so-called id-density wave (DDW) ordered phase (3) is attractive due to its relative simplicity and, at the same time, an ability to cover some important features of the underdoped cuprates in the pseudogap regime. In particular, within this scenario the pseudogap formation temperature, T^*, corresponds to a real phase transition into the DDW-state. Furthermore, below the superconducting transition temperature both states (DDW and d-wave superconducting) coexist. Therefore, within this scenario the superconducting

K. Scharnberg and S. Kruchinin (eds.),
Electron Correlation in New Materials and Nanosystems, 187–197.

state in the underdoped cuprates will be significantly different from that in optimally and overdoped cuprates.

Here, we analyze the behavior of the dynamical spin susceptibility in the pseudogap region of underdoped cuprates. In particular, we discuss the formation of the so-called resonance peak, seen by inelastic neutron scattering(INS) (4), and its dispersion in the DDW-state and the combined DDW+ DSC state. Furthermore, we investigate the sensitivity of our results to variations in the band structure parameters and also to the orthorhombic distortions on the spin excitations in these phases. The latter is, in particular, interesting in view of the recent INS studies (9; 10) on the twin-free $YBa_2Cu_3O_{7-\delta}$ compound.

2. Pure DDW-state

We start our analysis from the effective mean field Hamiltonian

$$H_{CDW} = \sum_{k,\sigma} \varepsilon_k c^\dagger_{k,\sigma} c_{k,\sigma} + \sum_{k,\sigma} W_k c^\dagger_{k,\sigma} c_{k+Q,\sigma} \quad (1)$$

with $Q = (\pi, \pi)$ being an antiferromagnetic wave vector. Here, we choose the charge density wave (CDW) order parameter in the form $W_k = iW^d_k + W^s$ with $W^d_k = \frac{W_0}{2}\left(\cos k_x - \cos k_y\right)$ being the id-charge density wave (DDW) gap while W^s is an admixture of a real s-wave component to the CDW order parameter due to orthorhombic distortions. Concerning the relative ratio of the s- and d-wave components of the gap, it is to be noted that in the superconducting state one finds an anisotropy of the Josephson current in junctions between $YBa_2Cu_3O_7$ and s-wave Nb, which is well fitted using a gap with 83 % d-wave and 17 % s-wave symmetry (5). We will assume a similar situation for the distorted DDW state. We use the normal state tight-binding energy dispersion

$$
\begin{aligned}
\varepsilon_k = {} & \mu + \frac{t_1}{2}\left((1 + \delta_0)\cos k_x + (1 - \delta_0)\cos k_y\right) + t_2 \cos k_x \cos k_y \\
& + \frac{t_3}{2}\left((1 + \delta_0)\cos 2k_x + (1 - \delta_0)\cos 2k_y\right) \\
& + \frac{t_4}{2}\left(\cos 2k_x \cos k_y + \cos k_x \cos 2k_y\right) + t_5 \cos 2k_x \cos 2k_y \quad (2)
\end{aligned}
$$

where $t_1, t_2, \ldots$ are the hopping integrals to nearest, next nearest neighbors and so on, while μ is the chemical potential. The parameter δ_0 accounts for the orthorhombicity. Please note that we have assumed the hopping along the CuO chains to be larger than perpendicular to them. This follows from an analysis of the experimental transport data (6).

TABLE I. The values of the tight-binding parameters and of W_0 used in the calculations.

	disp1	disp2
μ/eV	0.2707	0.084
t_1/eV	-1	-0.5547
t_2/eV	0.4	0.1327
t_3/eV	0	0.0132
t_4/eV	0	-0.1849
t_5/eV	0	0.0265
W_0/eV	0.042	0.042

After diagonalizing the Hamiltonian, Eq. (1), one finds the excitation spectrum:

$$E_{\mathbf{k}}^{\pm} = \varepsilon_{\mathbf{k}}^{+} \pm \sqrt{\left(\varepsilon_{\mathbf{k}}^{-}\right)^2 + \left(W_{\mathbf{k}}^{d}\right)^2 + (W^s)^2} \tag{3}$$

with $\varepsilon_{\mathbf{k}}^{\pm} = (\varepsilon_{\mathbf{k}} \pm \varepsilon_{\mathbf{k+Q}})/2$ (7). The sets of chemical potential, hopping matrix elements, gap anisotropy, and orthorhombic distortions used in the calculations are given in Table I. In the following we use a simple tight-binding fit including hopping between nearest and next nearest neighbors (7) and, for comparison, a six-parameter fit explored by Norman (8). In order to keep the same doping level in the DDW state, a slight change of the chemical potential occurs. An inclusion of the orthorhombicity is done in a similar way as in Ref. (12).

In Fig. 1 we show the calculated changes of the Fermi surface topology and of the electronic density of states (DOS) in the DDW state. Due to the doubling of the unit cell, the Fermi surface consists of hole pockets centered around $(\pm\pi/2, \pm\pi/2)$ and electron pockets around $(\pm\pi, 0)$ and $(0, \pm\pi)$ (7). We note, however, that for certain parameter sets, e.g. $disp2$ (see Fig. 1(a)) one of the bands does not cross the Fermi level and, thus, the Fermi surface consists only of the hole pockets centered around $(\pm\pi/2, \pm\pi/2)$. The occurrence of the electron pockets is determined by the relative strength of t_2: in particular, they form if $t_2 \geq t_3 + t_5 + W_0 + \mu$. Therefore, the orthorhombic distortions which affect mostly the Fermi surface close to the $(\pm\pi, 0)$ and $(0, \pm\pi)$ points of the first Brillouin Zone (BZ),(12) do not influence the present Fermi surface topology in the DDW state significantly. The fact that the Fermi surface reduces to hole pockets has a strong effect on the electronic density of states. In particular, in Fig. 1 we show the results for the electronic density of states in the normal (b) and the DDW state (d). As one could see in the case of $disp1$,

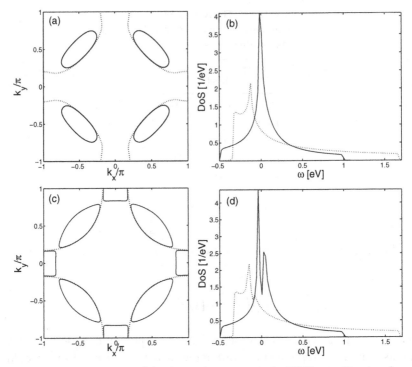

Figure 1. Calculated changes of the electronic structure in the DDW state: Fermi surfaces for the normal (dotted curve) and DDW state (solid curve) using *disp2* (a) and *disp1* (c) from Table I and electronic density of states in the normal (b) and DDW (d) phase for two various sets of parameters *disp1* (dotted curve) and the *disp2* (solid curve).

the DOS possesses almost no gap structure due to the fact that both bands cross the Fermi level and therefore the total DOS remains mainly unchanged. At the same time in the other case (*disp2*) the pseudogap forms at the Fermi level.

The bare part of the dynamical spin susceptibility is given by

$$\chi_0(\mathbf{q}, i\Omega_m) = -\frac{1}{2\beta} \sum_{\mathbf{k},\sigma,n} \text{Tr}\left[G(\mathbf{k}, i\omega_n)G(\mathbf{k}+\mathbf{q}, i\omega_n - i\Omega_m)\right]. \quad (4)$$

Performing the sum over Matsubara frequencies and continuing the result to the real frequency axis one obtains (7)

$$\chi_0(\mathbf{q}, \omega) = \frac{1}{2} \sum_{\mathbf{k}} \left(\left(1 + \frac{\varepsilon_{\mathbf{k}}^- \varepsilon_{\mathbf{k}+\mathbf{q}}^- + W_{\mathbf{k}}^d W_{\mathbf{k}+\mathbf{q}}^d + (W^s)^2}{\sqrt{\left(\varepsilon_{\mathbf{k}}^-\right)^2 + \left(W_{\mathbf{k}}^d\right)^2 + (W^s)^2}\sqrt{\left(\varepsilon_{\mathbf{k}+\mathbf{q}}^-\right)^2 + \left(W_{\mathbf{k}+\mathbf{q}}^d\right)^2 + (W^s)^2}}\right)\right.$$

$$\left.\times \left(\frac{f(E_{\mathbf{k}+\mathbf{q}}^+) - f(E_{\mathbf{k}}^+)}{\omega + i0^+ - E_{\mathbf{k}+\mathbf{q}}^+ + E_{\mathbf{k}}^+} + \frac{f(E_{\mathbf{k}+\mathbf{q}}^-) - f(E_{\mathbf{k}}^-)}{\omega + i0^+ - E_{\mathbf{k}+\mathbf{q}}^- + E_{\mathbf{k}}^-}\right)\right.$$

$$+\left(1-\frac{\varepsilon_{\mathbf{k}}^-\varepsilon_{\mathbf{k+q}}^- + W_{\mathbf{k}}^d W_{\mathbf{k+q}}^d + (W^s)^2}{\sqrt{\left(\varepsilon_{\mathbf{k}}^-\right)^2 + \left(W_{\mathbf{k}}^d\right)^2 + (W^s)^2}\sqrt{\left(\varepsilon_{\mathbf{k+q}}^-\right)^2 + \left(W_{\mathbf{k+q}}^d\right)^2 + (W^s)^2}}\right)$$

$$\times \left(\frac{f(E_{\mathbf{k+q}}^-) - f(E_{\mathbf{k}}^+)}{\omega + i0^+ - E_{\mathbf{k+q}}^- + E_{\mathbf{k}}^+} + \frac{f(E_{\mathbf{k+q}}^+) - f(E_{\mathbf{k}}^-)}{\omega + i0^+ - E_{\mathbf{k+q}}^+ + E_{\mathbf{k}}^-}\right)\right) \tag{5}$$

where $f(\varepsilon)$ is the Fermi function and the **k**-summation covers the reduced Brillouin-zone defined by $|k_x| + |k_y| \le \pi$. We note, that the first two terms describe intraband scattering, while the last two terms describe interband scattering .

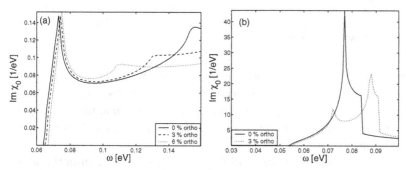

Figure 2. Imaginary part of the bare susceptibility in the DDW state as a function of frequency at $q = Q$ for *disp*1 (a) and *disp*2 (b) for various orthorhombic distortions. 3% ortho corresponds to $\delta_0 = -0.03$ and $W^s = 6meV$, 6% ortho corresponds to $\delta_0 = -0.06$, $W^s = 12meV$ and $W^d = 40.7meV$.

Let us first discuss the imaginary part of the bare susceptibility shown in Fig. 2(a). Since $E_{\mathbf{k+Q}}^\pm = E_{\mathbf{k}}^\pm$ the intraband scattering terms do not contribute to $\mathrm{Im}\chi_0$ at **Q** and the main contribution to $\mathrm{Im}\chi_0$ for the positive frequencies comes from the fourth term of Eq. (5) (Note the third term in Eq. (5) contributes mainly for negative frequencies). Furthermore, one finds that for $\mathbf{q} = \mathbf{Q}$ the Im χ_0 remains gapped up to a certain threshold frequency of about $2W_{\mathbf{k}_0}$ where $\mathbf{k}_0$ is the momentum at which the hole pocket around $(\pi/2, \pi/2)$ is intersected by the DDW Brillouin zone boundary. Above the threshold frequency $\mathrm{Im}\chi_0$ shows a square-root-like increase (7). The latter is determined by the difference in the population of the states that are involved in the scattering process, and not by energy conservation as in the superconducting state. At higher energies (of about $\omega_p \approx 150meV$ for *disp*1 and $\omega_p \approx 76meV$ for *disp*2) Im χ_0 shows an additional peak, which is due to a nearly flat part of $E_2 = E_{\mathbf{k+q}}^+ - E_{\mathbf{k}}^-$ close to $(\pi, 0)$ points of the BZ. In the first case the peak is less pronounced due to smaller regions of the flat dispersion.

As soon as one gets away from $\mathbf{q} = \mathbf{Q}$, the fourth term in Eq. (5) is no longer the only one that contributes. The first two terms describing intraband

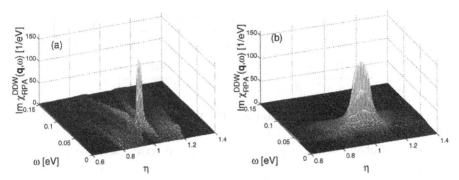

Figure 3. Imaginary part of the RPA spin susceptibility in the DDW state as a function of frequency and momentum along the diagonal of the first BZ for $U_0 = 0.56$eV with *disp*1 (a) and $U_0 = 0.138$eV with *disp*2 (b).

scattering now also make an important contribution because $E_{\mathbf{k}}^{\pm} = E_{\mathbf{k+q}}^{\pm}$ no longer holds. Furthermore these terms contribute already at small frequencies resulting in the gapless Im χ_0 away from (π, π). At the same time, at larger frequencies the square-root-like jump arising from interband scattering splits into three jumps (7).

In order to see how the susceptibility is affected by the orthorhombic distortions we show in Fig. 2 Im χ_0 as a function of frequency for different orthorhombic distortions. As one can see the square-root-like jumps are affected only slightly. The reason for that is obvious. In contrast to the superconducting state the gap in Im χ_0 is not directly related to the DDW gap but rather to the difference in the population of the two bands. The latter is not strongly affected by the orthorhombic distortions. At the same time, the main difference occurs for the high-energy peak in Im χ_0 which splits into two peaks separated by the energies of about 20 meV (*disp*2) and 50 meV (*disp*1). This splitting reflects the difference in the flat bands positions along the k_x and k_y directions. Note, for *disp*1 the second peak occurs at higher energies than are shown in the picture.

To analyze the formation of the resonance peak we include the interaction between the quasiparticles. Within the random phase approximation (RPA) the susceptibility is given by

$$\chi_{RPA} = \frac{\chi_0(\mathbf{q}, \omega)}{1 - U_{\mathbf{q}}\chi_0(\mathbf{q}, \omega)}, \tag{6}$$

where we use $U_{\mathbf{q}} = U_0 - 0.1U_0\left(\cos q_x + \cos q_y\right)$ as the fermionic four-point vertex (11). For a resonance, the conditions $U_{\mathbf{q}}\mathrm{Re}\chi_0 = 1$ and Im $\chi_0 = 0$ have to be satisfied for a certain pair $(\mathbf{q}, \omega)$. For the DDW state, Imχ_0 is only gapped for $\mathbf{q} = \mathbf{Q}$. Therefore one observes a resonance which is rapidly

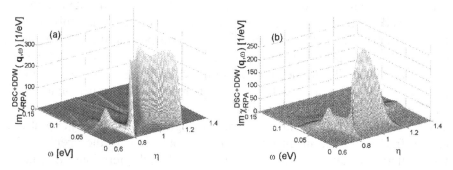

Figure 4. RPA susceptibility of the combined DSC + DDW state as a function of frequency and momentum along the diagonal for *disp*1, $U_0 = 0.583eV$ (a) and for *disp*2, $U_0 = 0.19eV$ (b).

damped away from **Q** as shown in Fig. 3. In particular, since Im χ_0 exhibits a square-root-like increase at the critical frequency Re χ_0 is enhanced but does not show a singularity as in the superconducting state. Therefore, the resonance occurs only if U is larger than a certain U_{cr}. Away from **Q** the mode becomes rapidly damped due to the opening of the scattering channel for intraband transitions. Due to much weaker features of the $p-h$ continuum for *disp*2 the structure of Im χ_{RPA} is much less pronounced away from (π, π). The orthorhombic distortions influence the resonance peak in the DDW state much less than in the superconducting state, because the states in the BZ responsible for the formation of the resonance peak are far from the $(\pi, 0)$ points and thus less affected by the orthorhombicity.

3. Combined DSC+DDW-state

Within DDW scenario the superconducting state in the underdoped cuprates is a combined state where both DDW and d-wave superconductivity coexist. To describe DDW+DSC we add the superconducting pairing term into the Hamiltonian (1):

$$H_{DSC} = \sum_{\mathbf{k}} \Delta_{\mathbf{k}} c_{\mathbf{k},\uparrow}^{\dagger} c_{-\mathbf{k},\downarrow}^{\dagger} + \text{h.c.} \qquad (7)$$

Here, $\Delta_{\mathbf{k}} = \Delta_0(\cos k_x - \cos k_y)/2 + \Delta^s$ is the superconducting gap with a $d_{x^2-y^2}$-wave and a s-wave components and we use $\Delta_0 = W_0$ and $\Delta^s = W^s$. The new Hamiltonian can be diagonalized using standard Bogolyubov-Valatin transformations and the new energy bands are $\Omega_{\mathbf{k}}^{\pm} = \sqrt{\left(E_{\mathbf{k}}^{\pm}\right)^2 + \Delta_{\mathbf{k}}^2}$. The dynamical spin susceptibility can be calculated in a usual way using Eq. (4) with the only difference that the Green's functions are now 4×4 matrices.

The explicit expression for the susceptibility in the combined state is rather lengthy and the details can be found in Ref. (7). We note, that only three terms contribute at small temperatures which correspond to the Cooper-pair creation like in a pure d-wave superconductor, but multiplied with DDW coherence factors. Then, the situation in the combined state resembles the pure superconducting state with the only difference that the superconductivity occurs for the DDW Fermi surface and two band system. In particular, at $Q = (\pi, \pi)$ ($\eta = 1.0$), Imχ_0 exhibits a single *discontinuous* jump at a critical frequency. The magnetic scattering associated with the opening of this scattering channel connects the "hot spots" in the fermionic BZ, i.e. those momenta $\mathbf{k}$ and $\mathbf{k} + \mathbf{Q}$ for which $\varepsilon_\mathbf{k} = \varepsilon_{\mathbf{k}+\mathbf{Q}} = 0$. Correspondingly, the critical frequency is given by $\Omega_{cr}^{coex} = 2\sqrt{\Delta^2(\mathbf{k}_{hs}) + W^2(\mathbf{k}_{hs})}$, where $\mathbf{k}_{hs}$ is the momentum of the hot spots. In contrast, away from $Q = (\pi, \pi)$, one finds that Imχ_0 exhibits 5 discontinuous jumps at critical frequencies, $\Omega_{cr}^{(i)}$ with $i = 1, .., 5$ indicating the opening of new scattering channels (7).

Similarly to the pure DSC state, the discontinuous jump in Im χ_0 in the coexistence phase is accompanied by a logarithmic divergence in Re χ_0, which in turn gives rise to a resonance peak below the particle-hole continuum for an arbitrary small fermionic interaction. In Fig. 4 we show the RPA susceptibility of the combined state as a function of frequency and momentum along the diagonal for two dispersions. Note, the difference in the dispersion of the resonance peaks in the pure DSC state and the coexistent DDW+DSC state is particularly pronounced around Q ($0.95 \lesssim \eta \lesssim 1.05$), with a more cusp-like dispersion in the coexistence phase. This cusp follows the form of the particle-hole continuum in the vicinity of Q (7). Thus, the dispersion of the resonance peak directly reflects the different momentum dependence of the particle-hole continuum in the vicinity of Q in the coexisting DDW and DSC state and the pure DSC state. However, away from Q, the particle-hole continuum, as well as the dispersion of the resonance peak are quite similar in both phases. One finds that these feature are quite robust with respect to the choice of different dispersions.

An influence of orthorhombic distortions on the resonance peak formation has been recently studied by various groups (12; 13; 14). In particular, it has been found that weak orthorhombic distortions do not influence the position of the resonance peak, but induce an anisotropy in the intensity pattern away from it. Namely, due to the slight change of the Fermi surface topology the ring of the resonant excitations at $\omega < \omega_{res}$ becomes distorted and two peaks along q_y-direction are suppressed with respect to their q_x counterparts. In the coexistent DDW+DSC phase the situation remains the same. It is illustrated in Fig. 5 for two different dispersions. Interestingly, for *disp*1 the ring of the excitations is distorted opposite to *disp*2 which depends on the details

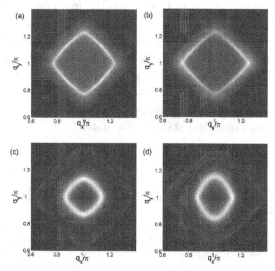

Figure 5. Imaginary part of the RPA susceptibility in the coexistent DDW+DSC state as a function of momentum for $\omega = 32meV$ for *disp2* with (b) and without (a) orthorhombicity using $U_0 = 0.19eV$ and for *disp1* with (d) and without (c) orthorhombicity using $U_0 = 0.583eV$. Here we employ $\delta_0 = -0.03$, $W^s = \Delta^s = 6meV$.

of the particle-hole continuum. At the same time in both cases the peaks are suppressed along q_y-direction and this is a result of the reduced phase-space in that direction.

4. Conclusions

In summary we have analyzed the momentum and frequency dependence of the dynamical spin susceptibility in the pure DDW and the coexistent DSC+DDW state for two different sets of tight-binding parameters and also examined the influence of the orthorhombic distortions on the resonance excitations in these phases. We find that in the DDW state due to the formation of a spin gap in the imaginary part of the bare susceptibility a resonance peak below the particle-hole continuum emerges in the RPA susceptibility at the antiferromagnetic wave vector $\mathbf{Q} = (\pi, \pi)$. Furthermore, the resonance excitations become strongly damped away from $\mathbf{Q}$ due to rapid closing of the gap in the $p - h$ continuum due to intraband transitions. Moreover, in contrast to the superconducting state the influence of the orthorhombicity on the resonant excitations in the DDW state is much less pronounced. In the coexistent DSC+DDW state the behavior of Im χ_0 resembles the superconducting state for DDW Fermi surface. The spin gap is mainly determined by the superconducting order parameter and therefore the resonance peak shows similar behavior to that of a pure DSC state. The main difference to the pure DSC state exists only in the close vicinity of **Q**. There the low frequency structure determined by the DSC order is suppressed by the DDW coherence factors. Thus, the interband transitions of the DDW state gain influence on the structure of the resonance which results in the cusp of the resonance peak dispersion. Similar to the pure superconducting state an inclusion of the orthorhombic distortions introduces some anisotropy in the structure of the resonance in the coexistent DSC+DDW state.

References

Norman, M. R., Pines, D., and Kallin, C. (2005) The pseudogap: friend or foe of high T_c?, *cond-mat*/0507031 (unpublished).

Anderson, P.W., *et al.* (2004) The physics behind high-temperature superconducting cuprates: the 'plain vanilla' version of RVB, *J. Phys. Condens. Mat.* **16**, R755-R769; Ivanov, D. A. , Lee, P. A., and Wen, X.-G. (2000) Staggered-Vorticity Correlations in a Lightly Doped t-J Model: A Variational Approach, *Phys. Rev. Lett.* **84**, 3958-3961; Varma, C. M. (1999) Pseudogap Phase and the Quantum-Critical Point in Copper-Oxide Metals, *Phys. Rev. Lett* **83**, 3538-3541; Emery, V. J., Kivelson, S. A., and Zachar, O. (1997) Spin-gap proximity effect mechanism of high-temperature superconductivity, *Phys. Rev. B* **56**, 6120-6147; Benfatto, L., Caprara, S., and Castro, C. Di (2000) Gap and pseudogap evolution within the charge-ordering scenario for superconducting cuprates, *Eur. Phys. Jour. B* **17**. 95-102 ; Schmalian, J., Pines, D., and Stojkovic, B. (1998) Weak Pseudogap Behavior in the Underdoped Cuprate Superconductors, *Phys. Rev. Lett.* **80**, 38393842; Engelbrecht, J.R., Nazarenko, A., Randeria, M., and Dagotto, E. (1998) Pseudogap above T_C in a model with $d_{x^2-y^2}$ pairing, *Phys. Rev. B* **57**, 1340613409; Chen, Q., Kosztin, I., Janko, B., and Levin, K. (1999) Superconducting transitions from the pseudogap state: d-wave symmetry, lattice, and low-dimensional effects, *Phys. Rev. B* **59**, 70837093; Zhang, S. C. (1997) A Unified Theory Based on SO(5) Symmetry of Superconductivity and Antiferromagnetism, *Science* **275**, 1089-1096.

Chakravarty, S., Laughlin, R. B., Morr, D. K., and Nayack, C. (2001) Hidden order in the cuprates, *Phys. Rev. B* **63**, 094503.

See for review P. Bourges, in "The gap Symmetry and Fluctuations in High Temperature Superconductors" edited by J. Bok, G. Deutscher, D. Pavuna and S.A. Wolf (Plenum Press, 1998).

Smilde, H.J.H., Golubov, A.A., Ariando, Rijnders, G., Dekkers, J.M., Harkema, S., Blank, D.H.A., Rogalla, H., and Hilgenkamp, H. (2005) Admixtures to d-wave gap symmetry in untwinned $YBa_2Cu_3O_7$ superconducting films measured by angle-resolved electron tunneling, *Phys. Rev. Lett.* **95**, 257001.

Ito, T., Takenaka, K., and Uchida, S. (1993) Systematic deviation from T-linear behavior in the in-plane resistivity of $YBa_2Cu_3O_{7-y}$: Evidence for dominant spin scattering, *Phys. Rev. Lett.* **70**, 3995-3998.

Ismer, J.-P., Eremin, I., and Morr, D. K. (2006) Dynamical spin susceptibility and the resonance peak in the pseuogap region of the underdoped cuprate superconductors, *cond-mat*/0601173 (*unpublished*).

Norman, M. R. (2001) Magnetic collective mode dispersion in high-temperature supercon-ductors, *Phys. Rev. B* **63**, 092509.

Mook, H. A., Dai, P. C., Dogan, F., and Hunt, R. D. (2000) One-dimensional nature of the magnetic fluctuations in $YBa_2Cu_3O_{6.6}$, *Nature* **404**, 729-731.

Hinkov, V., Pailhes, S., Bourges, P., Sidis, Y., Ivanov, A., Kulakov, A., Lin, C. T., Chen, D. P., Bernhard, C., and Keimer, B. (2004) Two-dimensional geometry of spin excitations in the high-transition-temperature superconductor $YBa_2Cu_3O_{6+x}$, *Nature* **430**, 650-654.

Eremin, I., Morr, D.K., Chubukov, A.V., Bennemann, K.-H., Norman, M.R. (2005) Novel neutron resonance mode in $d_{x^2-y^2}$-wave superconductors, *Phys. Rev. Lett.* **94**, 147001.

Schnyder, A.P., Manske, D., Mudry, C., and Sigrist, M. (2005) Theory for Inelastic Neutron Scattering in Orthorhombic High-T_C Superconductors, *cond-mat*/0510790; Eremin, I., and Manske, D. (2005) Fermi-liquid based theory for the in-plane magnetic anisotropy in untwinned High-T_C superconductors, *Phys. Rev. Lett.* **94**, 067006.

Kao, Y.-J., and Kee, H.-Y. (2005) Anisotropic spin and charge excitations in superconductors: Signature of electronic nematic order, *Phys. Rev. B* **72**, 024502.

Bascones, E., and Rice, T. M. (2005) Spin susceptibility of underdoped cuprates: the case of Ortho-II $YBa_2Cu_3O_{6.5}$, *cond-mat*/0511661.

DISORDER EFFECTS IN d-WAVE SUPERCONDUCTORS

C. T. Rieck, K. Scharnberg (scharnbe@physnet.uni-hamburg.de),
and S. Scheffler
I. Institut für Theoretische Physik, Universität Hamburg
20355 Hamburg, Germany

Abstract. In the theoretical analyses of impurity effects in superconductors the assumption is usually made that all quantities, except for the Green functions, are slowly varying functions of energy. When this so-called Fermi Surface Restricted Approximation is combined with the assumption that impurities can be represented by δ-function potentials of arbitrary strength, many reasonable looking results can be obtained. The agreement with experiments is not entirely satisfactory and one reason for this might be the assumption that the impurity potential has zero range. The generalization to finite range potentials appears to be straightforward, independent of the strength of the potential. However, the selfenergy resulting from scattering off finite range impurities of infinite strength such as hard spheres, diverges in this approximation at frequencies much larger than the gap amplitude! To track down the source of this unacceptable result we consider the normal state. The elementary results for scattering off a hard sphere, including the result that even an infinitely strong δ-function potential does not lead to scattering at all in systems of two and more dimensions, are recovered only when the energy dependencies of all quantities involved are properly taken into account. To obtain resonant scattering, believed to be important for the creation of mid-gap states, the range of the potential is almost as important as its strength.

Key words: Unconventional superconductivity, Disorder, Non-s-wave scattering, Quasi-classical approximation, Particle-hole symmetry

1. Introduction

Scattering of a particle by a potential is a time-honored problem of quantum mechanics. (Taylor, 1972) It is straightforward to formulate although a microscopic derivation of the scattering potential requires very intricate considerations. Most often the scattering potential is modelled by some plausible function in real space. This is the route we shall follow here.

Even though potential scattering is elastic, calculation of the scattered wave function requires consideration of all virtual states, up to infinitely high energies. When the scattered wave function is calculated for piecewise constant potentials by imposing boundary conditions, the need to include high energy virtual states is not apparent. We shall not solve the Schrödinger

K. Scharnberg and S. Kruchinin (eds.),
Electron Correlation in New Materials and Nanosystems, 199–221.

equation directly but rather use the Green function formalism, because this allows a straightforward generalization to scattering in a metal containing an ensemble of defects, even when the metal goes superconducting. However, in order to check the approximations usually made in the application of this formalism we shall revisit the simplest problems of scattering theory and reproduce known results.

The driving force behind the study of disorder effects in superconductors is the hope that such investigations will give information on the pairing state and the pairing interaction. Indeed, qualitative differences are expected between conventional and unconventional superconductors, the latter being defined by a vanishing Fermi surface average of the order parameter. Potential scattering in an anisotropic conventional superconductor would at high enough concentration lead to a finite, isotropic gap, while unconventional superconductors are expected to acquire midgap states before superconductivity is destroyed. Conventional superconductors show this kind of behavior in the presence of spin-flip scattering. (Abrikosov and Gor'kov, 1961) Because of the innate magnetism in high temperature superconductors, non-magnetic impurities can induce local moments and thus blur the seemingly clear distinction between potential and spin-flip scattering.

Here, we shall assume that we are dealing with d-wave superconductors. Since for unconventional superconductors there is no qualitative difference between these two types of scattering, we shall confine ourselves to the study of potential scattering. Even with this limitation there is a wide range of theoretical predictions as regards T_c-suppression, density of states, transport properties etc, depending on the electronic structure, that is assumed, the way disorder is modelled and depending on the analytical and numerical approximations employed. (Atkinson et al., 2000; Hirschfeld and Atkinson, 2002; Balatsky et al., 2006)

In a solid the scattering potential is due to defects of the crystalline lattice which are distributed more or less randomly over the whole sample. For large systems, an average over defect configurations has to be taken, not only because the problem would be untractable otherwise but also because the defect configuration is unknown and, except at very low temperatures, changes with time. Thus the problem of treating disorder in solids is usually broken down into two parts: scattering off a single defect and averaging with respect to such individual scattering events. As model for disorder we shall use an alloy model in which some of the host atoms are randomly replaced by some other kind of atom. Averaging independently with respect to all possible defect positions, which limits the applicability of this theory to small concentration of defects, leads to the selfconsistent T-matrix approximation (SCTMA).

We shall use the T-matrix equation (Lippmann-Schwinger equation (Taylor, 1972)) to describe an individual scattering event. This is a two-dimensional Fredholm integral equation of the second kind with a singular kernel. When the scattering potential is so weak that this equation can be solved by iteration up to 2nd order (Born approximation), integrals are not actually evaluated but are parametrized through quantities like lifetimes, transport times, and shifts in the chemical potential. Then, a detailed knowledge of the potential is necessary only when a microscopic calculation of these parameters is undertaken. We are interested in the case of strong scatterers, including infinitely high potentials, for which the Born series diverges. The limiting case might seem unphysical but we know from elementary quantum mechanics that scattering off a hard sphere is described by perfectly well-behaved wave functions. The reason for the wide spread interest in the strong scattering limit is the realisation that in systems with energy gaps strong repulsive potentials might create states inside the gaps. (Joynt, 1997; Balatsky et al., 2006)

From this T-matrix the local density of states (LDOS) can be obtained. The selfenergy of the configuration averaged and hence translationally invariant Green function is given by a different but closely related quantity which might be called generalized T-matrix. (Taylor, 1972; Mahan, 1981) The Green function appearing in the kernel of the integral equation which determines this generalized T-matrix depends on some arbitrary energy ω. Only when this is set equal to the energy of the scattered particle does one recover the T-matrix known from scattering theory.

The non-iterative solution of the two-dimensional integral equation, which gives the desired T-matrix, is a formidable problem which we have not yet solved!

One frequently used simplification, especially in the theory of superconductivity, is to fix all momentum variables at the Fermi energy except in the Green function which is the only quantity integrated with respect to energy. (Fermi surface restricted approach) However, this energy integral diverges! This is related to the fact that the real part of the Green function in position space $G_\omega(r, r')$ diverges for $r' \to r$ in two and three dimensions. The divergence can be removed by invoking particle-hole symmetry which eliminates the offensive term, (Flatté and Byers, 1999; Salkola et al., 1996) by averaging the Green function with respect to r' over a small volume around r, (Flatté and Byers, 1999) or by restricting consideration to a single band of finite width. Omitting the divergent term is central to the immensely successful quasi-classical theory of superconductivity. (Eilenberger, 1968; Larkin and Ovchinnikov, 1969; Serene and Rainer, 1983) The quite convincing argument for this approach is that the divergence comes from energies far removed from the Fermi energy were the modifications of the electronic states due to the onset of superconductivity are negligible. So, when differences between superconducting and normal state properties are calculated one expects such contributions to cancel. In many cases, however, no such differences are calculated: one does expect the results for the superconducting state to reduce to their normal state equivalents as $T \to T_c$.

In much of the published work, the theory is further simplified by assuming δ-function potentials. With this assumption, which has been made by many authors including ourselves (Hensen et al., 1997), the (generalized) T-matrix equation is no longer an integral equation and even for a superconductor the solution is trivial. Then a variety of interesting results can be derived, most of which are in reasonable agreement with experiment. Modelling the scattering potential by a δ-function potential actually reduces the complexity of the problem to such an extent that it is possible to drop the assumption of scattering events at different defects being independent. (Atkinson et al., 2003) We should note, however, that according to elementary quantum mechanics a δ-function potential in two and three dimensions does not scatter at all.

Even for defects within the CuO_2-planes it seems rather doubtful that their effect is limited to a single site. Defects due to oxygen nonstoichiometry and cation disorder, which reside on lattice sites away from the conducting CuO_2-planes, are only poorly screened and hence are certainly long ranged. One conceivable consequence of the finite range of the defect potentials is a mitigation of the T_c suppression by potential scattering in unconventional superconductors. Balian et al. (Balian and Werthamer, 1963) already noted that forward scattering would reduce the deleterious effect of potential scattering on T_c, while Foulkes and Gyorffy presented a detailed calculation for p-wave superconductors in the Born approximation, showing that it is the transport time that controls the T_c-suppression. (Foulkes and Gyorffy, 1977) Millis et al. seem to have been the first to apply these ideas to d-wave superconductors, again using the Born approximation. (Millis et al., 1988)

We tried to generalize the treatment of angle dependent scattering to arbitrarily large potentials, still using the Fermi surface restricted approach so that only one-dimensional integral equations had to be solved. (Rieck et al., 2005) As model potential we used a Gaussian for computational convenience. We shall show below that in the unitary limit the normal state selfenergy calculated within the SCTMA is proportional to the number of scattering channels considered. For the pair breaking parameter a similarly unphysical result is found when the unitary limit is taken.

In Section 2 we develop the general theory of potential scattering in a d-wave superconductor within a continuum description of the electronic structure. In the following section we introduce the widely used Fermi surface restricted approximation and calculate the selfenergy and the pair breaking parameter using the selfconsistent T-matrix approximation. We show that these results in the strong scattering limit must be wrong and deduce from that the inadequacy of the Fermi surface restricted approximation, at least in the context of potential scattering in a metal. In Section 4 we study scattering off a single impurity and demonstrate what needs to be done in order to get reliable results.

2.   Basic theory

The Gor'kov equations for the retarded Green functions describing a weak coupling spin singlett superconductor in the absence of magnetic interactions can be written in Nambu space as

$$\left(\omega_+\hat{\sigma}_0 - h(r, \nabla^2)\,\hat{\sigma}_3\right)\hat{G}(r, r'; \omega_+) - \int d^2\rho\Delta(r, \rho)\hat{\sigma}_1\hat{G}(\rho, r'; \omega_+)$$
$$= \delta(r - r')\hat{\sigma}_0 \qquad (1)$$

The $\hat{\sigma}$'s are Pauli matrices and $\hat{G}$ is a 2×2 matrix with only two independent elements. Though not strictly necessary, the introduction of a matrix Green function proves to be very useful. The Hamiltonian in (1) is given by

$$h(r, \nabla^2) = -\frac{1}{2\mu}\nabla^2 - \epsilon_F + V(r) + V_{\text{lattice}}(r). \qquad (2)$$

$V(r)$ is a defect potential to be specified later and $V_{\text{lattice}}(r)$ is the periodic lattice potential. A complete solution for these Gor'kov equations is not in sight. One either uses a nearly free electron model, i.e. essentially ignores $V_{\text{lattice}}(r)$, or one uses a localized description requiring numerical calculations on a lattice of finite size. The relation between these two limiting cases will be discussed elsewhere. In an intermediate model some tight-binding dispersion relation leading to a band of finite width is introduced, but the difference between plane waves and Bloch functions is ignored. In this paper we treat the charge carriers as a two-dimensional (nearly) free electron gas. The Fermi surface in this model is circular and the band width is infinite and without $V(r)$ the system is translationally invariant. In particular, the order parameter in (1) will then depend only on $r - \rho$, so that (1) can easily be solved by Fourier transformation:

$$\hat{G}^0(k; \omega_+) = \frac{\omega\hat{\sigma}_0 + \varepsilon(k)\hat{\sigma}_3 + \Delta(k)\hat{\sigma}_1}{\omega_+^2 - \varepsilon^2(k) - \Delta^2(k)} = \sum_i G^{0i}(k, \omega_+)\hat{\sigma}_i. \qquad (3)$$

True to our assumptions we should have $\varepsilon(k) = \frac{1}{2\mu}k^2 - \epsilon_F$ with μ some effective carrier mass, but one could also insert any (model) dispersion relation.

The order parameter Δ is assumed to have d-wave symmetry with respect to its dependence on k. Using the Green function $\hat{G}^0$ one can rewrite the integro-differential equation (1) as inhomogeneous integral equation. However, the perturbation $V(r)$ affects the off-diagonal elements of $\hat{G}$ which, via the weak-coupling selfconsistency equation, change the order parameter, even when we ignore the possibility that the pairing interaction is modified by the presence of the defect(s). We, therefore, write the order parameter as

$$\Delta(r,\rho) = \Delta(r - \rho) + \delta\Delta(r,\rho). \tag{4}$$

The equation for $\hat{G}$ then has the general form of a Fredholm integral equation of the second kind

$$\hat{G}(r,r') = \hat{G}^0(r,r') + \int d^2\rho\, \hat{K}(r,\rho)\, \hat{G}(\rho,r'). \tag{5}$$

With the order parameter fluctuations taken into account, the kernel

$$\hat{K}(r,\rho) = \int d^2\rho'\hat{G}^0(r,\rho')\,\{V(\rho)\delta(\rho - \rho')\hat{\sigma}_3 + \delta\Delta(\rho',\rho)\hat{\sigma}_1\} \tag{6}$$

is itself an integral. Since the frequency only appears as parameter we have suppressed it for clarity. For an isotropic s-wave superconductor one has $\delta\Delta(\rho',\rho) = \delta\Delta(\rho)\delta(\rho - \rho')$. Then the suppression of the order parameter near a defect does not complicate the problem significantly.

For such a (3D) superconductor containing a single spherical impurity, Fetter (Fetter, 1965) has calculated the Friedel oscillations of the electron density and the order parameter amplitude in the asymptotic regime. The assumption of a zero-range pairing interaction is justified for conventional superconductors because the range of the pairing interaction is certainly much less than the coherence length considered to be the shortest length relevant for superconductivity. For cuprate superconductors even this assumption can be called into question because the coherence length is very short and the pairing interaction must have a finite range to allow for d-wave pairing. Starting from a weak-coupling version of the spin-fluctuation exchange (Dahm et al., 1993) one finds that the pairing interaction extends over several lattice constants and hence is comparable with the in-plane coherence length. Friedel oscillations occur on a length scale given by the Fermi wavelength. So, both for conventional and for cuprate superconductors the assumption, that the range of the pairing interaction is short compared with the length scale of the defect induced order parameter fluctuations, seems to be hard to justify.

Since in our view there are even more important shortcomings in the description of defects in unconventional superconductors, we shall neglect these order parameter fluctuations for the time being. Then Eqs. (5) and (6) reduce to

$$\hat{G}(r,r';\omega)=\hat{G}^0(r-r';\omega) + \int d^2\rho\, \hat{G}^0(r-\rho;\omega)\, V(\rho)\hat{\sigma}_3\, \hat{G}(\rho,r';\omega) \quad (7)$$

Since we are interested in very strong potentials for which the Born series does not converge, the solution of this equation for general $V(\rho)$ is still quite a tall order, especially as this (screened) potential ought to be calculated selfconsistently taking into account the Friedel oscillations it induces in the charge density. Again, we neglect this effect and use some model potential.

The most popular model is a δ-function potential

$$V(\rho) = V\,\delta(\rho), \quad (8)$$

because for this potential the task of solving (7) is trivial:

$$\hat{G}(r,r';\omega)=\hat{G}^0(r-r';\omega)+V\,\hat{G}^0(r;\omega)\,\hat{\sigma}_3\left(\hat{\sigma}_0 - V\,\hat{G}^0(0;\omega)\,\hat{\sigma}_3\right)^{-1}\hat{G}^0(-r';\omega)$$
$$(9)$$

Unless regularized as described in the introduction, the diagonal components of $\hat{G}^0(0;\omega) = \int \frac{d^D k}{(2\pi)^D}\, \hat{G}^0(k;\omega_n)$ diverge in two and three dimensions. Hence

$$\hat{G}(r,r';\omega)=\hat{G}^0(r-r';\omega) \quad (10)$$

and we have to conclude that δ-function potentials, no matter how strong, have no effect on the properties of a superconductor. This is in accord with the results, obtainable by elementary quantum mechanics, for the scattering phase shifts of sphere- or disk-shape potentials (see Figs. 3 and 4). When the divergence is removed by hand, $\hat{G}^0(r - r';\omega)$ no longer solves the original equation of motion (1).

The order of the divergence can be seen from the explicit forms of the Green's functions in position space, which are easily calculated using an integral representation for the Hankel function (Watson, 1952), when the order parameter in (3) is momentum independent. Since the result for the two-dimensional case is not readily available (Scheffler, 2004), we shall give it here for $\omega > 0$:

$$G^{00}(r,\omega_+) = -\frac{m}{4}\frac{\omega_+}{\sqrt{\omega_+^2-\Delta^2}}[iJ_0(r\Omega_+)+iJ_0(r\Omega_-)-Y_0(r\Omega_+)+Y_0(r\Omega_-)]\,(11)$$

$$G^{03}(r,\omega_+) = -\frac{m}{4}[iJ_0(r\Omega_+) - iJ_0(r\Omega_-) - Y_0(r\Omega_+) - Y_0(r\Omega_-)] \quad (12)$$

with $\quad \Omega_{\pm} = k_F^2 \pm 2m\sqrt{\omega_+^2 - \Delta^2} \quad (13)$

J_0 and Y_0 are Bessel functions of the first and second type. For small argument, the leading term in $Y_0(z)$ is $0.5\pi \ln(0.5z)$. Hence, $\mathcal{R}eG^{03}(r, \omega_+)$ diverges as $\ln r$ for $r \to 0$, while the divergent terms in G^{00} cancel.

If corresponding expressions were available for d-wave superconductors it would probably be easiest to calculate the local density of states in the vicinity of a single defect for given $V(\rho)$ directly from (7). Since this is not the case and in view of the interest in properties of superconductors containing random ensembles of defects, we use an eigenfunction represention for $\hat{G}^0(r - r'; \omega)$ and introduce a generalized T-matrix to rewrite (7) as:

$$\hat{G}(r, r', \omega) = \hat{G}^0(r - r', \omega) +$$
$$\int \frac{d^2k}{(2\pi)^2} \int \frac{d^2k'}{(2\pi)^2} e^{ikr} \hat{G}^0(k, \omega) \hat{T}(k, k'; \omega) \hat{G}^0(k', \omega) e^{-ik'r'}$$

$$(14)$$

$$\hat{T}(k, k'; \omega) = V(k - k')\hat{\sigma}_3 + \int \frac{d^2p}{(2\pi)^2} V(k - p)\,\hat{\sigma}_3\,\hat{G}^0(p, \omega)\,\hat{T}(p, k'; \omega) \quad (15)$$

One possible description of an ensemble of (identical) impurities is the socalled alloy model

$$V(r) = \sum_{i=1}^{N} v(r - R_i) \qquad N \gg 1. \tag{16}$$

Taking an average with respect to the random impurity sites R_i, neglecting interference between scattering processes at different sites, consistent with assuming a small concentration n_{imp} of impurities, gives

$$\hat{t}(k, k'; \omega) = v(k - k')\hat{\sigma}_3 + \int \frac{d^2p}{(2\pi)^2} v(k - p)\,\hat{\sigma}_3\,\hat{G}(p, \omega)\,\hat{t}(p, k'; \omega) \quad (17)$$

This equation is very similar to (15): $\hat{t}$ is now the $\hat{T}$-matrix for a single defect and $\hat{G}^0(k, \omega)$ has been replaced by

$$\hat{G}(k, \omega) = \left[\omega\hat{\sigma}_0 - \varepsilon(k)\hat{\sigma}_3 - \Delta(k)\hat{\sigma}_1 - \hat{\Sigma}(k, \omega) \right]^{-1} = \sum_i G^i(k, \omega)\,\hat{\sigma}_i. \tag{18}$$

which, like $\hat{G}^0(k, \omega)$, describes a translationally invariant system.

$$\hat{\Sigma}(k, \omega) = n_{imp}\,\hat{t}(k, k; \omega) \tag{19}$$

is a selfenergy which has to be calculated selfconsistently. When $\hat{t}(k, k'; \omega)$, together with the selfenergy, are expanded in terms of Pauli matrices:

$$\hat{t} = t^0\hat{\sigma}_0 + t^1\hat{\sigma}_1 + it^2\hat{\sigma}_2 + t^3\hat{\sigma}_3 \tag{20}$$

one obtains four coupled 2D integral equations for t^ℓ, $\ell = 0, \ldots, 4$

$$t^0 = \int \frac{d^2p}{(2\pi)^2} v \left[\frac{\omega - \Sigma^0}{D} t^3 + \frac{\varepsilon + \Sigma^3}{D} t^0 - \frac{\Delta + \Sigma^1}{D} t^2 \right] \tag{21}$$

$$t^1 = \int \frac{d^2p}{(2\pi)^2} v \left[\frac{\omega - \Sigma^0}{D} t^2 + \frac{\varepsilon + \Sigma^3}{D} t^1 - \frac{\Delta + \Sigma^1}{D} t^3 \right] \tag{22}$$

$$t^2 = \int \frac{d^2p}{(2\pi)^2} v \left[\frac{\omega - \Sigma^0}{D} t^1 + \frac{\varepsilon + \Sigma^3}{D} t^2 + \frac{\Delta + \Sigma^1}{D} t^0 \right] \tag{23}$$

$$t^3 = v + \int \frac{d^2p}{(2\pi)^2} v \left[\frac{\omega - \Sigma^0}{D} t^0 + \frac{\varepsilon + \Sigma^3}{D} t^3 + \frac{\Delta + \Sigma^1}{D} t^1 \right] \tag{24}$$

with

$$D(p, \omega) = \left(\omega - \Sigma^0 \right)^2 - \left(\varepsilon + \Sigma^3 \right)^2 - \left(\Delta + \Sigma^1 \right)^2 \tag{25}$$

These equations are obviously very difficult to solve in all generality and nobody has yet succeeded in deriving a complete solution! They have been solved by a large number of authors, including the present ones, (Hensen et al., 1997; Rieck et al., 2005; Scheffler, 2004; Balatsky et al., 2006) using a variety of approximations. We shall show here that some approximations, while leading to seemingly reasonable results, cannot be trusted.

3. Selfconsistent $\hat{T}$-matrix approximation (SCTMA) in the Fermi surface restricted approximation

In the quasiclassical approximation, the energy integration is performed assuming particle-hole symmetry. Terms in Eqs. (21)–(24) with $\varepsilon + \Sigma^3$ in the numerator, which are responsible for the divergence of the real part of $\hat{G}^0(0; \omega)$, then vanish. This quasiclassical approximation is justified (Eilenberger, 1968; Larkin and Ovchinnikov, 1969) with the argument that only differences between the superconducting and the normal state need to be considered. In Eqs. (21)–(24), subtracting the corresponding normal state equations does not seem to improve the convergence of the ε-integral.

In the present case of a translationally invariant system, the quasiclassical approximation reduces to the omission of terms odd in energy, putting momenta equal to their values on the Fermi surface (line) except when they appear as argument of ε, and then integrating with respect to the ε-dependence of the Denominator (25). For a circular Fermi surface all momenta are of the form $k_F(\cos \varphi, \sin \varphi)$ so that the set of equations (21) - (24) reduces to

$$t^0(\varphi, \phi) = \pi N_F \int_0^{2\pi} \frac{d\psi}{2\pi} v(\varphi - \psi) \left[g^0(\psi) \, t^3(\psi, \phi) - g^1(\psi) \, t^2(\psi, \phi) \right] \tag{26}$$

$$t^1(\varphi,\phi) = \pi N_F \int_0^{2\pi} \frac{d\psi}{2\pi} v(\varphi-\psi) \left[g^0(\psi)\, t^2(\psi,\phi) - g^1(\psi)\, t^3(\psi,\phi) \right] \qquad (27)$$

$$t^2(\varphi,\phi) = \pi N_F \int_0^{2\pi} \frac{d\psi}{2\pi} v(\varphi-\psi) \left[g^0(\psi)\, t^1(\psi,\phi) + g^1(\psi)\, t^0(\psi,\phi) \right] \qquad (28)$$

$$t^3(\varphi,\phi) = v(\varphi-\phi) + \pi N_F \int_0^{2\pi} \frac{d\psi}{2\pi} v(\varphi-\psi) \left[g^0\, t^0(\psi,\phi) + g^1\, t^1(\psi,\phi) \right] \qquad (29)$$

$g^0(\psi,\omega_+)$ and $g^1(\psi,\omega_+)$ are the energy integrated normal and ano malous retarded Green functions

$$g^0(\psi,\omega_+) = -\frac{\omega - \Sigma^0(\psi,\omega_+)}{\sqrt{[\Delta(\psi) + \Sigma^1(\psi,\omega_+)]^2 - [\omega - \Sigma^0(\psi,\omega_+)]^2}} \qquad (30)$$

$$g^1(\psi,\omega_+) = -\frac{\Delta(\psi) + \Sigma^1(\psi,\omega_+)}{\sqrt{[\Delta(\psi) + \Sigma^1(\psi,\omega_+)]^2 - [\omega - \Sigma^0(\psi,\omega_+)]^2}} \qquad (31)$$

Since particle-hole symmetry is assumed, g^0 and g^1 are independent of $t^3(\psi,\psi)$ and g^3 vanishes. However, all four components $t^\ell(\varphi,\phi)$ are required for the calculation of $\Sigma_{0,1}$.

If the defect potentials are taken to be δ-functions in real space, $v = v_0$ is independent of angle and so are the t^ℓ. Then one has to take the Fermi surface average of $g^1(\psi)$, which vanishes. Hence $t^1 = t^2 = 0$ and

$$t^0 = \frac{\pi N_F v_0^2 < g^0 >}{1 - (\pi N_F v_0)^2 < g^0 >^2}, \qquad t^3 = \frac{v_0}{1 - (\pi N_F v_0)^2 < g^0 >^2} \qquad (32)$$

for arbitrarily large v_0. These are standard results derived and used by many authors, including the present ones (Hensen et al., 1997), to describe (transport) properties of cuprate superconductors.

Here, we shall show that a straightforward extension of this theory to defects with finite range leads to unacceptable results!

For specific examples of impurity potentials with finite range we consider a Gaussian and a disk:

$$v_G(r) = \bar{v}\, \frac{1}{\pi a^2}\, e^{-r^2/a^2} \qquad v_D(r) = \bar{v} \begin{cases} \frac{1}{\pi a^2} & \text{for} \quad r < a \\[2mm] 0 & \text{for} \quad r > a \end{cases} \qquad (33)$$

Both potentials reduce to δ-functions when the range a goes to zero while the spatial average $\bar{v} = v_{\max}\pi a^2$ is kept constant. Fourier transformation gives:

$$v_G(\mathbf{k}' - \mathbf{k}) = \bar{v}\, e^{-(\mathbf{k}'-\mathbf{k})^2 a^2/4} \qquad v_D(\mathbf{k}' - \mathbf{k}) = \bar{v}\, \frac{2}{|\mathbf{k}'-\mathbf{k}|a}\, J_1\left(|\mathbf{k}'-\mathbf{k}|a\right)$$

$$(34)$$

Because of the symmetry assumed for the normal state electronic structure, both functions can be expanded into a cosine series with respect to the scattering angle φ between $\boldsymbol{k}$ and $\boldsymbol{k}'$. For brevity, we write the series in complex form:

$$v(k', k, \cos\varphi) = \sum_{n=-\infty}^{+\infty} v_n(k', k)\, e^{in\varphi} \qquad \text{with}$$

$$v_n(k', k) = \bar{v} e^{-\frac{1}{4}(k'^2 + k^2)a^2} I_n\left(\frac{k' k a^2}{2}\right), \tag{35}$$

where the explicit expression for $v_n(k', k)$ applies to the Gaussian potential. The I_n's are modified Bessel functions. For the disk-shaped potential, these coefficients have to be evaluated numerically.

Within the Fermi surface restricted approach, $k' = k = k_F$. Hence, it is convenient to redefine the above expansion in terms of new parameters:

$$v(k_F, \cos\varphi) = v_0 \sum_{n=-\infty}^{+\infty} u_n e^{in\varphi} \qquad \text{with} \tag{36}$$

$$v_0 = \bar{v} e^{-\frac{1}{2}k_F^2 a^2} I_0(\gamma), \quad u_n = \frac{I_n(\gamma)}{I_0(\gamma)} \quad \text{and} \quad \gamma = \frac{1}{2} k_F^2 a^2. \tag{37}$$

These expansion coefficients are often treated as free parameters.

Just like the defect potential the t^ℓ's are expanded into Fourier series' and the coefficients are collected in the form of matrices $\hat{t}^\ell$ with elements:

$$t_{nm}^\ell = \frac{\pi N_F}{\sin^2 \delta_s} \int_0^{2\pi} \frac{d\varphi}{2\pi} \int_0^{2\pi} \frac{d\phi}{2\pi} t^\ell(\varphi, \phi)\, e^{-in\varphi + im\phi}. \tag{38}$$

For later convenience we have defined t_{nm}^ℓ with a factor depending on an s-wave scattering phase shift δ_s, which characterizes the strength of the potential (37)

$$\pi N_F v_0 = \tan \delta_s. \tag{39}$$

It follows from symmetry consideration (Klemm et al., 2000) that the most general real order parameter transforming as $k_x^2 - k_y^2$ can be written in a Fermi surface restricted approach as

$$\Delta(T, \varphi) = \sum_{m=-\infty}^{\infty} a_{4m-2} \cos[(4m - 2)\varphi]. \tag{40}$$

Then $g^\ell(\psi; \omega)$ has the general form

$$g^\ell(\psi; \omega) = \sum_{m=-\infty}^{+\infty} g_{q, q+|4m-2\ell|}^\ell(\omega) \cos[(4m - 2\ell)\psi] \qquad \ell = 0, 1. \tag{41}$$

It is convenient to write the expansion coefficients in the form of symmetric matrices $\tilde{g}^\ell = \left(g^\ell_{q,k}\right)$ with the additional property $g^\ell_{q,k} = g^\ell_{-q,-k}$, even though $g^\ell_{q,q+m}$ is actually independent of q. Since the matrix elements are complex, these matrices are not hermitian. When the Fourier coefficients u_n of the potential (35) are written in the form of a diagonal matrix $\tilde{u}$, the four integral equations (26)–(29) are transformed to four equations for $\tilde{t}^\ell$. From these, $\tilde{t}^2$ and $\tilde{t}^3$ can be eliminated so that we finally obtain

$$\tilde{t}^0 \pm \tilde{t}^1 = \left[\cos^2\delta_s - \sin^2\delta_s\, \tilde{u}\,(\tilde{g}^0 \mp \tilde{g}^1)\,\tilde{u}\,(\tilde{g}^0 \pm \tilde{g}^1)\right]^{-1} \tilde{u}\,(\tilde{g}^0 \mp \tilde{g}^1)\,\tilde{u}. \quad (42)$$

This is the central equation that has to be solved numerically for various choices of the scattering potential (35). Because of the selfconsistency requirement the solution depends also on n_{imp} through the parameter Γ^{el}_N (49). The $t^\ell(\varphi, \phi)$ have the following symmetries

$$t^{0,1,3}(\varphi, \phi) = t^{0,1,3}(\phi, \varphi); \quad t^2(\varphi, \phi) = -t^2(\phi, \varphi); \quad t^\ell(\varphi, \phi) = t^\ell(-\varphi, -\phi),$$
$$(43)$$

from which we derive

$$t^{0,1}_{m,n} = t^{0,1}_{n,m} = t^{0,1}_{-m,-n}. \quad (44)$$

It follows from (41) and (42) that

$$t^0_{q,q'} = t^0_{q,q+4m}\, \delta_{q',q+4m} \quad (45)$$

$$t^1_{q,q'} = t^1_{q,q+4m-2}\, \delta_{q',q+4m-2} \quad (46)$$

In contrast to (41), $t^{0,1}_{q,q'}$ does depend on q. Hence, the selfenergy (19) has the same structure (41) as $g^{0,1}(\psi; \omega)$:

$$\Sigma^1(\psi; \omega) = \Gamma^{el}_N \sum_{\ell=-\infty}^{+\infty} \left\{\sum_q t^1_{q,q+4\ell-2}\right\} \cos\left[(4\ell-2)\psi\right] \equiv \sum_{\ell=-\infty}^{+\infty} \Sigma^{1\ell}\cos\left[(4\ell-2)\psi\right] (47)$$

$$\Sigma^0(\psi; \omega) = \Gamma^{el}_N \sum_{\ell=-\infty}^{+\infty} \left\{\sum_q t^0_{q,q+4\ell}\right\} \cos\left[4\ell\psi\right] \equiv \sum_{\ell=-\infty}^{+\infty} \Sigma^{0\ell} \cos\left[4\ell\psi\right] \quad (48)$$

where the parameter

$$\Gamma^{el}_N = \frac{n_{imp}}{\pi N_F}\, \sin^2\delta_s \quad (49)$$

has been introduced.

3.1. RESULTS FOR SELFENERGIES AND PAIR BREAKING PARAMETERS

Results for the selfenergies Σ^{00} and Σ^{01} are shown in Figure (1) for a Gaussian potential with $\gamma = 5$. The OP has been taken to be $\Delta(T; \varphi) = \Delta_{max}(T) \cos 2\varphi$ with a low temperature amplitude $\Delta_{max} = 16 \text{ meV}$. The exact numerical value is of no physical significance here. We focus our attention on the limiting behavior for $\omega \gg \Delta_{max}$. In this limit, $\Sigma^{0\ell}$ reduces to the normal state result: $\Sigma^{0\ell} = 0$ for $\ell \neq 0$ because in the system considered, rotational symmetry is broken only by the superconducting order parameter. The purely imaginary isotropic contribution

$$\Sigma^{00}(\omega_+) = -i \Gamma_N^{el} \sum_{\ell=-\infty}^{\infty} \frac{u_\ell^2}{\cos^2 \delta_s + \sin^2 \delta_s u_\ell^2} \tag{50}$$

is easily obtained because the Green function (30) reduces to $g^0(\psi, \omega_+) = -i$ while $\hat{\tau}^0$ in (42) is diagonal and both $\hat{\tau}^1$ and $\tilde{g}^1$ can be neglected. For δ-function potentials $u_\ell = 0$ for $\ell \neq 0$ and (50) reduces to $\Sigma^{00}(\omega_+) = -i \Gamma_N^{el}$, which elucidates the physical significance of the parameter Γ_N^{el} introduced above. In the Born limit $\delta_s \ll \pi/2$, the denominator in (50) can be replaced by 1. The remaining sum is related to the Fermi surface average of the squared potential: $\langle v^2(\varphi) \rangle = \langle v(\varphi) \rangle^2 \sum_{\ell=-\infty}^{\infty} u_\ell^2$. For a Gaussian potential (37) the sum can easily be performed yielding $\Sigma^{00}(\omega_+) = -i \Gamma_N^{el} I_0(2\gamma)/I_0^2(\gamma)$. For $\gamma = 5$, the correction factor is 3.8. In the unitary limit $\delta_s = 0.5\pi$ the sum in (50) diverges because u_ℓ^2 cancels, no matter how small. The numerical results shown in Fig. (1) were obtained with a cut-off $\ell_{max} = 10$, hence the limiting value of 4.2 meV. Whether the set of u_ℓ's represent a Gaussian or any other potential is obviously immaterial for this argument.

One might argue that an infinitely high potential is unphysical. However, as known from elementary quantum mechanics, a hard disk or sphere simply presents a part of space which the particles cannot enter and the difference between a large and an infinite potential is, in fact, not substantial. In Fig. (1) we also show results for $\delta_s = 0.49\pi$. In this case, contributions from terms $\ell > 7$ are negligible. However, the limiting value is still much larger than for point-like scatterers or weak scatterers. This appears to be an artefact of the Fermi surface restricted approximation, which goes unnoticed when one starts with considering scattering only in very few angular momentum channels.

This theory has been used by us to calculate the T_c-degradation of an unconventional superconductor with an order parameter of the general form (40) (Rieck et al., 2005). In this context one requires $\Sigma^1(\psi, i\omega_n)$ which also diverges for $\delta_s = \pi/2$. For one component order parameters $\Delta(\varphi) = \Delta_{max} \cos$

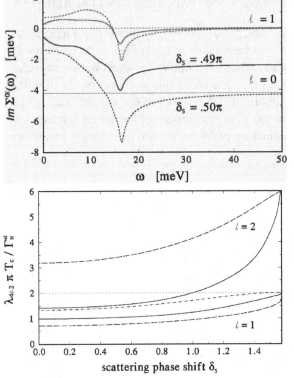

Figure 1. Imaginary part of the selfenergy defined in (48) for Gaussian defect potentials (34) with $\gamma = 5$ (37) as function of frequency for $\Gamma_N^{el} = 0.2$ meV and defect potentials at or near the unitary limit. The parameter δ_s is defined in (39) Extrema occur at $\omega = \Delta_{max}$.

Figure 2. λ_2 and λ_6 for Gaussian potentials with widths $\gamma = 1$ (solid) and $\gamma = 5$ (dashed) as function of potential strength, parametrized by $\delta_s = \tan^{-1}(\pi N_F v_0)$. The dot-dashed line shows the result for λ_2 when only s-, p-, and d-wave scattering is included. For δ-function scatterers the reduced pairbreaking parameters plotted in this figure are identically 1.

$(4\ell - 2)\varphi$ we find the standard Abrikosov-Gor'kov formula

$$\ln \frac{T_c}{T_{c0}} = \psi\left(\frac{1}{2}\right) - \psi\left(\frac{1}{2} + \frac{\lambda_{4\ell-2}}{2}\right) \tag{51}$$

with pair breaking parameters

$$\lambda_{4\ell-2} = \frac{\Gamma_N^{el}}{\pi T_c} \frac{1}{2} \sum_{m=-\infty}^{\infty} \frac{(u_m - u_{m+4\ell-2})^2}{(\cos^2 \delta_s + u_m^2 \sin^2 \delta_s)(1 + u_{m+4\ell-2}^2 \tan^2 \delta_s)} \tag{52}$$

Taking the unitary limit in (52) it would appear as if every term in the series vanishes. However, when the series is terminated at $m = \pm m_0$ (Kulić and Dolgov, 1999), one finds $\lambda_{4\ell-2} = \frac{1}{\pi T_c} \Gamma_N^{el} (4\ell - 2)$, i.e. 2 and 6 for the examples shown in Fig. (2). independent of m_0. These results are clearly also artefacts of the Fermi surface restricted approximation and have nothing to do with physical reality.

4. Single spherically symmetric impurity in the normal state

In order to elucidate the reason for the failure of the Fermi surface restricted approach, we investigate scattering off a single spherically symmetric impurity in the normal state. To make contact with the scattering phase shift analysis known from elementary quantum mechanics we introduce the scattering wave functions $\psi_k(r)$. From these one can construct the Green function

$$G(r, r', \omega_+) = \int \frac{d^2k}{(2\pi)^2} \frac{\psi_k(r)\psi_k^*(r')}{\omega_+ - \varepsilon(k)} \tag{53}$$

which fulfills the integral equation (5), suitably simplified to the normal state. Inserting this representation for G into (5) and projecting out the r'-dependence, one can neglect the inhomogeneous term when the limit $\omega \to \varepsilon(k)$ is taken. One thus obtains an equation for $\psi_k(r)$:

$$\psi_k(r) = e^{ik \cdot r} + \int d^2\rho\, G^0(r - \rho; \varepsilon(k))\, V(\rho)\psi_k(\rho). \tag{54}$$

This equation could, of course, also have been derived directly from the time-independent Schrödinger equation.

Inserting the plane-wave representation of $G^0(r - \rho; \varepsilon(k))$ and defining a T-matrix through

$$T(p, k) = \int d^2r\, e^{ip \cdot r}\, V(r)\psi_k(r), \tag{55}$$

we arrive at

$$\psi_k(r) = e^{ik \cdot r} + \int \frac{d^2p}{(2\pi)^2}\, e^{ip \cdot r}\, \hat{G}^0(p, \varepsilon(k))\, T(p, k) \tag{56}$$

The Lippman-Schwinger equation is obtained by multiplying (56) by $e^{-ik \cdot r} V(r)$ and then integrating with respect to d^2r:

$$T(k, k') = v(k - k') + \int \frac{d^2p}{(2\pi)^2} v(k - p)\, G^0(p, \varepsilon(k'))\, T(p, k') \tag{57}$$

This is identical with (15) when the energy variable ω in the generalized T-matrix is replaced by $\varepsilon(k)$ (Mahan, 1981). Solving (57) can be simplified if only a single defect, described by a spherically symmetric potential $v(r)$ is considered. Then the T-matrix depends on the moduli of p and k and the angle between them, even when these vectors belong to different energies. Eq. (57) can thus be rewritten as

$$T(k', k, \cos\phi) = v(k', k, \cos\phi)$$
$$+ \int_0^\infty \frac{dp\, p}{2\pi} \int_0^{2\pi} \frac{d\theta}{2\pi} v(k', p, \cos(\phi - \theta))\, G^0(p, \varepsilon(k))\, T(p, k, \cos\theta)$$

$$(58)$$

For an isotropic system the expansion of T into a Fourier series analogous to (35), leads to a set of decoupled 1D integral equations

$$T_m(k', k) = v_m(k', k) + \int_0^\infty \frac{dp\, p}{2\pi} v_m(k', p)\, G^0(p, \varepsilon(k))\, T_m(p, k). \qquad (59)$$

When the local density of states near an impurity is the quantity of interest, it is a reasonable approximation to use the Green function for the clean system $G^0(p, \varepsilon(k)) = \left[\frac{k^2}{2\mu} - \frac{p^2}{2\mu} + i\delta\right]^{-1}$ so that we can rewrite (59) as

$$T_m(k', k) = v_m(k', k)$$
$$+ \frac{\mu}{\pi} \mathcal{P} \int_0^\infty dp \frac{p}{k^2 - p^2} v_m(k', p) T_m(p, k) - i\frac{\mu}{2} v_m(k', k) T_m(k, k) \quad (60)$$

Assuming cylindrically symmetric impurities, the T-matrix equation in the alloy model (17) can be reduced to an equation identical to (59), except that the Green function is to be replaced by

$$G(p, \omega_+) = \left[\omega - \frac{p^2}{2\mu} - \Sigma(p, \omega_+)\right]^{-1}. \qquad (61)$$

The pole of $G^0(p, \omega_+)$ has moved into the complex plane so that the separation into a principal value part and a δ-function is not possible. Furthermore, the T-matrix has to be calculated as function of the parameter ω in order to obtain the selfenergy $\Sigma(p, \omega_+)$ (19), which could then be compared with the high frequency limit in Fig. 1. This work is in progress.

We are interested in strong (repulsive) scattering potentials because they can create states inside energy gaps. The strength of the scattering potential can be arbitrary, even a hard sphere should pose no problem! It seems that one could simplify equation (59) and (60) in the case of very strong potentials, because then the T-matrix on the left hand sides could be neglected. One thus arrives at a Fredholm integral equation of the first kind which, however, represents an ill-posed problem. The only numerical method for solving this equation that we have found feasible was a discretization, keeping the potential finite.

In order to calculate the wave function from (60), we expand the T-matrix (58) and both exponentials in (56) using

$$e^{ikr\cos\varphi} = J_0(kr) + 2\sum_{m=1}^{\infty} i^m J_m(kr)\cos m\varphi \tag{62}$$

and obtain

$$\psi_k(r) = \sum_{m=-\infty}^{\infty} i^m \left\{ J_m(kr) + \int_0^{\infty} \frac{dp\,p}{2\pi} J_m(pr)\, G_k^0(p)\, T_m(p,k) \right\} e^{im\varphi}. \tag{63}$$

4.1. SCATTERING PHASE SHIFT ANALYSIS

We want to establish the relations between the Fourier coefficients of the T-matrix and the partial scattering phase shifts. These relations put some constraints on the T-matrix on the energy shell which serve as a useful check for the numerical calculations. The desired relations are derived by considering the asymptotic behavior of the scattered wave function (54).

The normal state Green function is obtained from $G^{00}(r, \omega_+) + G^{03}(r, \omega_+)$ in the limit $\Delta \to 0$ (Eqs. (11,13):

$$G^0(r, \omega_+) = -\frac{\mu}{2}(iJ_0(kr) - Y_0(kr)) \quad \text{with} \quad k^2 = k_F^2 + 2\mu\omega. \tag{64}$$

According to (73) we immediately obtain the density of states per spin (per area)

$$N(E(k)) = \frac{\mu}{2\pi}. \tag{65}$$

Since this is independent of energy, it is identical with the parameter N_F introduced in Section 3. Using the asymptotic expansions of the Bessel functions we find:

$$G^0(r - \rho, \omega_+) \asymp -i\,e^{-\frac{i}{4}\pi} \frac{\mu}{\sqrt{2\pi}} \frac{1}{\sqrt{k|r-\rho|}} e^{ik|r-\rho|}. \tag{66}$$

When this is inserted into (54) we can use the approximation $k|r - \rho| \approx kr$ in the denominator and $k|r - \rho| \approx kr - k' \cdot \rho$ with $k' = ke_r$ in the exponent, assuming that the potential $v(r)$ is short ranged. Thus, the behavior of the scattered wave function at large distances from the scattering center is given by:

$$\psi_k(r) \asymp e^{ik\cdot r} - i\,e^{-\frac{i}{4}\pi} \frac{\mu}{\sqrt{2\pi}} \frac{e^{ikr}}{\sqrt{kr}} T(k', k). \tag{67}$$

This involves the T-matrix on the energy shell $|k'| = |k| = \sqrt{2\mu E}$ only. Because we started from the retarded Green function, there is only an outgoing

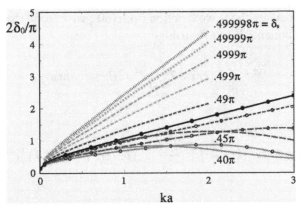

Figure 3. s-wave scattering phase shifts as functions of *ka*. Curves marked with circles are the results for a disk, the unmarked ones are for a Gaussian potential. The parameters are $\tan\delta_s = \pi N \bar{v}$, characterizing the strengths of the scattering potential by its spatial average (33). For $a \to 0$, δ_0 and hence the T-matrix vanish logarithmically!

cylindrical wave. The prefactor $1/\sqrt{r}$ is required for normalization in two dimensions. Note that this wave function decays even more slowly than in three dimensions which renders the assumption usually made with regard to the independence of scattering events at different defects rather questionable.

Using (62), with the asymptotic expansions of the Bessel functions inserted, we expand (67) in a cosine series, which we again write in complex form to save space:

$$\psi_k(r) \asymp \frac{1}{\sqrt{2\pi kr}} e^{-\frac{i}{4}\pi} \sum_{m=-\infty}^{\infty} \left((-1)^m \, i e^{-ikr} + (1 - i\mu T_m(k,k)) \, e^{+ikr} \right) e^{im\phi} \quad (68)$$

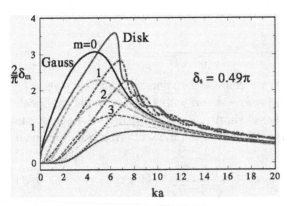

Figure 4. s-, p-, d-, and f-wave scattering phase shifts for a Gaussian and a disk-shaped potential (33) as function of *ka*. The range of *ka* is much larger than in Fig. 3.

Particle conservation implies that the currents represented by the incoming and the outgoing waves should cancel. Hence, the factor multiplying e^{+ikr}

in (68) can only be a phase factor which is usually written in the form

$$1 - i\mu T_m(k, k) = e^{+2i\delta_m} \qquad \text{or} \qquad \frac{\mu}{2} T_m(k, k) = \frac{1}{\cot \delta_m + i}. \qquad (69)$$

This defines the scattering phase shift δ_m for the m'th angular momentum channel. The complex quantity $T_m(k, k)$, which is obtained by numerically solving Eq. (59), can thus be expressed in terms of one real quantity. This provides us with an excellent check for our numerical calculations! Note that no such relation exists for the generalized T-matrix.

Results for the s-wave scattering phase shift for varying potential strengths $\tan \delta_s = \pi \mathcal{N} \bar{v}$ are shown in Fig. 3. For disk-shape potentials the change in δ_0 when the potential strength $\bar{v}$ is increased from some large value to infinity is small. For the Gaussian potential the results are very similar as long as the potential is fairly weak. However, δ_0 increases indefinitely with $\bar{v}$. This is due to the fact that any potential with smoothly decreasing tails will be infinite everywhere in space when the maximum goes to infinity. We note that $\delta_0 \to 0$ when the potential range a goes to zero, no matter how strong the potential. To get resonant scattering ($2\delta_0/\pi = 1$) one does require a rather large potential, but it will occur only for $ka \approx 1$, i.e. when the wavelength of the scattered particle is comparable with the range of the potential.

In Fig. 4 scattering phase shifts for various angular momentum channels are shown for a fairly strong potential and a wide range of ka. As expected, δ_m goes to zero for $a \to 0$ more and more rapidly as m increases. Nonetheless, resonance occurs in all m-channels considered, albeit at very different values of ka. For $ka \gg 1$, the scattering phase shifts become almost independent of m!

Results for the disk-shaped potential shown in these two figures have been calculated from (60) which is very time consuming because integrals have to be evaluated to obtain the $v_m(k', k)$'s. Our numerical results agree perfectly with

$$\tan \delta_m(k) = -\frac{qaI'_m(qa)J_m(ka) - kaI_m(qa)J'_m(ka)}{qaI'_m(qa)Y_m(ka) - kaI_m(qa)Y'_m(ka)} \qquad \text{for} \qquad v_{\max} \geq \frac{k^2}{2\mu}, \qquad (70)$$

$$\text{with} \quad qa = \sqrt{\frac{4}{\pi} \tan \delta_s - (ka)^2} \geq 0, \qquad (71)$$

which is derived by solving the Schrödinger equation for $r < a$ and for $r > a$ and then matching the logarithmic derivatives of the wave functions at $r = a$.

This comparison shows that it is absolutely essential to keep the principle value integral in (60). When this is neglected, as is the case in the Fermi surface restricted approach, one immediately finds

$$\tan \delta_m(k) = \pi \mathcal{N} v_m(k, k) \leq \tan \delta_s. \qquad (72)$$

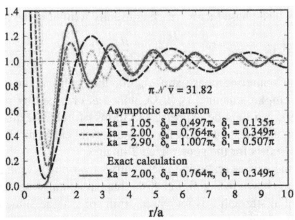

Figure 5. Local density of states as function of distance for a disk-shaped potential (33) with radius a centered at the origin. A very high value $\pi N \bar{v} = \tan \delta_s = 31.82 \Leftrightarrow \delta_s = 0.49\pi$ has been assumed for the potential. Some curves have been calculated using the asymptotic expansion (68) rather than (63) The values of ka have been chosen such that either s-wave ($\delta_0 = 0.5\pi$) or p-wave ($\delta_1 = 0.5\pi$) scattering or neither of them are resonant. For $ka = 2.00$, exact results can be compared with results obtained using the asymptotic form of the wave function.

Hence, resonant scattering occurs only for $\bar{v} \to \infty$. Now, however, all angular momentum channels become resonant simultaneously. This is not only un-physical, it also causes convergence problems with the Fourier expansion of (58)!

4.2.   RESULTS FOR THE LOCAL DENSITY OF STATES

From (60) and (63) we calulated the Green function (53) and, in particular, the Local Density of States (LDOS), which is experimentally accessible through Scanning Tunneling Microscopy (Crommie et al., 1993):

$$N(\boldsymbol{r}, E) = -\frac{1}{\pi} Im\, G_E(\boldsymbol{r}, \boldsymbol{r}) = \int \frac{d^2k}{(2\pi)^2} |\psi_k(\boldsymbol{r})|^2 \delta\left(E - \frac{k^2}{2\mu}\right). \qquad (73)$$

The result is rotationally invariant, as expected. Note that, in order to calculate $N(\boldsymbol{r}, E)$ at arbitrary distances from the scattering center, we need $T_m(p, k)$ off the energy shell. In the asymtotic regime we can use (68) to calculate the LDOS. For the disk-shaped potential this is almost trivial because we have the analytic results (70) for $T_m(k, k)$. In Fig. (5) we included results for the LDOS, normalized to the bulk DOS N, which are valid only in the asymptotic regime and compare them with one exact result. The most remarkable conclusion, in view of the discussions in the literature centered around δ-function scatterers, to be drawn from this figure is that nothing exceptional happens when the

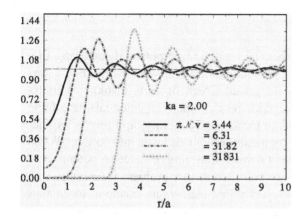

Figure 6. Local density of states as function of distance for a Gaussian potential (33) with range a centered at the origin.

scattering becomes resonant. This can be attributed to an overdamping of resonant states due to the bulk density of states being finite at all energies. This can be seen more formally from Eq (69) because this gives

$$\pi \mathcal{N} |T_m(k,k)| \leq 1. \tag{74}$$

Well-defined resonant states can only be expected inside energy gaps of the bulk DOS. It seems to be rather difficult, though, to devise a consistent toy model to study such effects. (Joynt, 1997) The most appropriate applications that come to mind are unconventional superconductors, which we are working on.

Since we have varied the energy of the scattered particle rather than the potential, the wavelength of these Friedel oscillations changes. Also remarkable is the amplitude of these oscillations which is much greater than in three dimensions (Fetter, 1965). The slow decay of the oscillation is related to the two-dimensionality of the system considered.

In Fig. 6 corresponding exact results are shown for a Gaussian potential. For the same potential strength $\pi \mathcal{N} \bar{v} = 31.82$, the results for the two model potentials are very similar. For a very high potential, $\delta_s = 0.49999\pi$, the LDOS vanishes at distances much larger than the potential range. This is related to the peculiar behavior of the scattering phase shift discussed in connection with Fig. 3.

The Green function, and with it $\mathcal{N}(r, E)$, could also be obtained from Eqs. (14) and (15), where $E = \omega + \epsilon_F$. Then one has to replace $\varepsilon(k)$ in (59) by ω and solve this set of equations for a range of values of this new variable. Introducing Fourier expansions, integrals with respect to angles in (14) can be done, rendering $G(r, r, \omega)$ isotropic. We are then left with a double integral, while the calculation of the LDOS from (73) only involves a set of one-dimensional integrals. Assuming δ-function scatterers, these equations immediately lead back to Eq. (9).

5. Summary

We have shown that an approximate treatment of potential scattering in a metal (superconductor), which puts momentum variables on the Fermi surface and evaluates the energy-integrated Green function invoking particle-hole symmetry, leads to unacceptable results when the potentials have finite ranges. The key problem is that in the strong scattering limit results will depend on the number of angular momentum channels taken into account. When only a single scattering channel (s-wave scattering, δ-function potential) or very few scattering channels are considered, this problem is not apparent. Except, of course, for the discrepancy that elementary quantum mechanics tells us that a δ-function potential does not scatter in two dimensions.

The only way to obtain acceptable results and to reproduce standard results within a continuum description is to take all momentum and energy dependencies accurately into account. This involves the evaluation of principle value integrals and, in the most general cases, the numerical solution of two-dimensional Fredholm integral equations of the second kind. In the strong scattering limit these go over into those of the first kind. However, these cannot be solved numerically without some regularization. One such regularization consists of reverting back to an equation of the second kind. For a disk-shape potential we have shown that the results for a strong potential and an infinite potential do not significantly differ, as one would expect. For potentials that go to zero continuously, like Gaussians or Lorentzians, it has turned out to be unphysical to let the height of the potential to go to infinity.

Comparison with alternative approaches based on a localized description of the material is an important project for the future.

Acknowledgements

The authors gratefully acknowledge useful discussions with A. V. Balatsky, J. C. Davis, A. Lichtenstein, N. Schopohl, S. Trugman, and J.-X. Zhu. One of the authors (K.S.) is very grateful for the hospitality extended to him at the Los Alamos National Laboratory, where part of this work was done.

References

Abrikosov, A. A. and Gor'kov, L. P. (1961) Contribution to the theory of superconducting alloys with paramagnetic impurities, *Sov. Phys. JETP* **12**, 1243.

Atkinson, W. A., Hirschfeld, P. J., MacDonald, A. H., and Ziegler, K. (2000) Details of Disorder Matter in 2D d-Wave Superconductors, *Phys. Rev. Lett.* **85**, 3926.

Atkinson, W. A., Hirschfeld, P. J., and Zhu, L. (2003) Quantum interference in nested d-wave superconductors: A real-space perspective, *Phys. Rev. B* **68**, 054501.

Balatsky, A. V., Vekhter, I., and Zhu, J.-X. (2006) Impurity-induced states in conventional and unconventional superconductors, *Rev. Mod. Phys.* **78**, 373.

Balian, R. and Werthamer, N. R. (1963) Superconductivity with Pairs in a Relative *p* Wave, *Phys. Rev.* **131**, 1553.

Crommie, M. F., Lutz, C. P., and Eigler, D. M. (1993) Imaging standing waves in a two-dimensional electron gas, *nature* **363**, 524.

Dahm, T., Erdmenger, J., Scharnberg, K., and Rieck, C. T. (1993) Superconducting states from spin-fluctuation mechanisms, *Phys. Rev. B* **48**, 3896.

Eilenberger, G. (1968) Transformation of Gorkov's equation for type II superconductors into transport-like equations, *Z. Physik* **214**, 195.

Fetter, A. L. (1965) Sperical impurity in an infinite superconductor, *Phys. Rev.* **140**, A1921.

Flatté, M. E. and Byers, J. M. (1999) Local Electronic Structure of Defects in Superconductors, In H. Ehrenreich and F. Spaepen (eds.), *Solid State Physics*, Vol. 52, New York, , Academic Press, p. 137.

Foulkes, I. F. and Gyorffy, B. L. (1977) *p*-wave pairing in metals, *Phys. Rev. B* **15**, 1395.

Hensen, S., Müller, G., Rieck, C. T., and Scharnberg, K. (1997) In-plane Surface Impedance of Epitaxial YBa$_2$Cu$_3$O$_{7-\delta}$ Films: Comparison of Experimental Data Taken at 87 GHz with d- and s-wave Models of Superconductivity, *Phys. Rev. B* **56**, 6237.

Hirschfeld, P. J. and Atkinson, W. A. (2002) Π à la Node: Disordered *d*-wave superconductors in two dimensions for the random masses, *J. Low Temp. Phys.* **126**, 881.

Joynt, R. (1997) Bound States and Impurity Averaging in Unconventional Superconductors, *J. Low Temp. Phys.* **109**, 811.

Klemm, R. A., Rieck, C. T., and Scharnberg, K. (2000) Order-parameter symmetries in high-temperature superconductors, *Phys. Rev. B* **61**, 5913.

Kulić, M. L. and Dolgov, O. V. (1999) Anisotropic impurities in anisotropic superconductors, *Phys. Rev. B* **60**, 13062.

Larkin, A. I. and Ovchinnikov, Y. N. (1969) Quasiclassical Method in the Theory of Superconductivity, *Sov. Phys.-JETP* **28**, 1200.

Mahan, G. D. (1981) *Many-Particle Physics*, New York, Plenum Press.

Millis, A. J., Sachdev, S., and Varma, C. M. (1988) Inelastic scattering and pair breaking in anisotropic and isotropic superconductors, *Phys. Rev. B* **37**, 4975.

Rieck, C. T., Scharnberg, K., and Scheffler, S. (2005) Effects of disorder with finite range on the properties of *d*-wave superconductors, In J. Ashkenazi (ed.), *NATO Advanced Research Workshop on New Challenges in Superconductivity: Experimental Advances and Emerging Theories.*, p. 151, Springer.

Salkola, M. I., Balatsky, A. V., and Scalapino, D. J. (1996) Theory of Scanning Tunneling Microscopy Probe of Impurity States in a *D*-Wave Superconductor, *Phys. Rev. Lett.* **77**, 1841.

Scheffler, S. (2004) Störstellen beliebiger Reichweite und Stärke in *d*-Wellen Supraleitern, Diplomarbeit, Universität Hamburg.

Serene, J. W. and Rainer, D. (1983) The quasiclassical approach to superfluid ^{3}He, *Phys. Reports* **101**, 221.

Taylor, J. R. (1972) *Scattering Theory*, New York, John Wiley & Sons.

Watson, G. N. (1952) *Theory of Bessel Functions*, p. 424, Cambridge University Press, 2. edition.

FIRST PRINCIPAL CALCULATIONS OF EFFECTIVE EXCHANGE INTEGRALS FOR COPPER OXIDES AND ISOELECTRONIC SPECIES

K. Yamaguchi[*], Y. Kitagawa, S. Yamanaka, D. Yamaki, T. Kawakami,
M. Okumura
Department of Chemistry, Graduate School of Science, Osaka University, Toyonaka, Osaka 560-0043, Japan.

H. Nagao
Department of Computational Science, Faculty of Science, Kanazawa University, Kakuma, Kanazawa, Ishikawa 920-1192, Japan.

S. P. Kruchinin
Bogolyubov Institute for Theoretical Physics, The Ukranian National Academy of Science, Kiev 252143, Ukraine.

Abstract. Our theoretical efforts for strongly correlated electron systems such as transition metal oxides have been reviewed in relation to electonic structures of these species. The effective exchange integrals (J) of several transition-metal oxides have been calculated by hybrid DFT methods. The ab initio results for the species are also mapped to the N-band Hubbard model. The two band model for copper oxides has been extended to possible isoelectronic π-d, π-R and σ-R systems, which often exhibit magnetic conductivity and superconductivity. For example, magnetic modifications of conducting polymers, CT complexes and TTF derivatives are examined on the theoretical grounds. Triangular and cubane-type clusters are investigated by general spin orbital (GSO) DFT in relation to spin frustration. The spin orbit interaction is also included to calculate the Dzyaloshinskii-Moriya term.

Key words: Spin-mediated mechanism, J model, Spin fluctuation model, Cooperative mechanism, π-d, π-R and σ-R systems, Material design.

K. Scharnberg and S. Kruchinin (eds.),
Electron Correlation in New Materials and Nanosystems, 223–233.
© 2007 Springer.

1. Introduction

The instability in chemical bonds[1], leading to electron localization via electron repulsion, is now one of the important and crucial concepts even in material science as well as in chemical reactions. It is closely related to electronic, magnetic and optical properties of strongly correlated electron systems such as p-d, π-d, π-R and σ-R conjugated systems. Theoretical investigations of these systems are indeed essential for elucidation of interrelationship between magnetism and high-Tc superconductivity in general, since both characteristics have been commonly observed experimentally. About 20 years ago, our ab initio calculations indicated that the CuOCu unit exhibits very strong effective exchange interaction ($| J |$ $>> 0$)[2]. But we could not imagine that such finding might be related to the high-Tc superconductivity, because our main interest is to elucidate electronic structure and reactivity of transition metal oxides. After the discovery of high-Tc cuprates by Bednorz and Müller[3], we immediately proposed a spin-mediated J-model ($T_c = cJ(k_B)$)[4]. It is noteworthy that our J model has been presented on the basis of ab initio calculations before the Zhang-Rice t-J model[5]. On the other hand, the spin fluctuation (SF) model has been employed to rationalize metallic and superconducting behaviors of overdoped cuprates[7]. The large $|J|$ value is replaced with strong susceptibility $\chi(q)$ or large SF frequency (ω_{SF}) for overdoped cuprates in spin-mediated model[8]. In this review, we summarize theoretical efforts toward ab initio calculations of J-values and other interaction parameters (t, U, so on) of transition-metal oxides and related species.

2. Theoretical background

2.1. GROUP-THEORETICAL INTERRELATIONSHIP BETWEEN THEORETICAL MODELS

There are several theoretical models for strongly correlated electron systems, which arise from the instability in chemical bonds from the view point of quantum chemistry[1]. Our group have investigated several model Hamiltonians for these species as shown in Fig. 1, and have elucidated group-theoretical interrelationships[1] among them on the basis of five different symmetry operations; spatial symmetry (P_n), spin rotation (S), time-reversal (T), gauge transformation (ϕ) and permutation group (S_N).

The classical Heisenberg (spin vector) model is characterized by the magnetic group[9]

$$M = H_n + T(P_n - H_n)$$ (1)

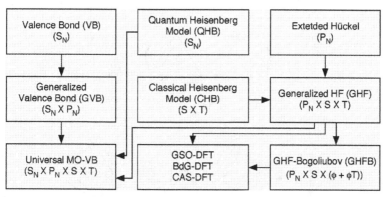

Figure 1. Group-theoretical interrelationships between several theoretical models for strong correlated electron systems (see text).

where H_n denotes the subgroup of P_n. The spin vector model has been applied to triangular and tetrahedral magnetic clusters as shown in Fig. 2. The corresponding generalized Hartree-Fock (GHF) solutions for the clusters are constructed by using the group-theoretical operation: $P_n \times S \times T$.

However, the GHF solutions are not the eigen functions of $\hat{S}^2$ and $\hat{S}_z$ operators, leading to the extended Hartree-Fock (EHF) and spin optimized (SO) SCF general spin orbitals (GSO) methods, which are characterized by $S_N \times P_n \times S \times T$ [1,10]. The GSO SO-SCF wavefunctions for the D_{3h} and T_d clusters are equivalent to the corresponding full CI wave functions in the case of the Hubbard model. The spin structures of multi-center radicals in Fig. 2 can be regarded as pictorial expression of spin correlation function defined by $\langle S(1) \cdot S(2) \rho_2(1,2) \rangle$, where $\rho_2(1,2)$ is the second-order density matrix of GHF, EHF and SO SCF solutions[1,11]. This implies that the spin structures by GSO HF express spin correlations instead of spin populations[11]. The spin correlation function is related to the dynamical magnetic susceptibility χ in the k-space[7]. The quantum effects for multi-center radicals can be expressed by the spin-symmetry recovery from broken-symmetry (BS) GHF, which entails the multi-configuration description by EHF and SO SCF in conformity with the quantum resonance effects, particularly for S=1/2 spins[10]. The number density projection is

number density projection is

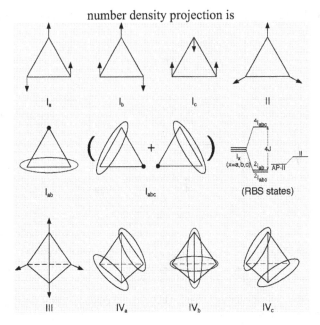

Figure 2. Spin vector (classical) models (II, III) for triangular and tetrahedral radical species. Spin fluctuation models for D_{3h} clusters (Ia-c). The resonating BS (RBS) states are expected by I_{ab}, I_{abc} and IV_{a-c}.

required for GHF-Bogoliubov solution of superconducting state of finite clusters[1].

The GSO HF and GSO density functional theory (DFT) have been applied to elucidate electronic and magnetic properties of triangular and cubane-type clusters of manganese oxides and iron-sulfur compounds[12-14]. Some of these species indeed exhibit noncollinear spin structures (see Fig. 2) which entail antiferromagnetic (singlet-type) spin couplings between spins of Mn (or Fe) ions. The cluster models of triangular lattice of CoO_2[15] and κ-(BEDF-TTF)(Cu$_2$(CN)$_3$)[16] have also been investigated in relation to the spin fluctuation effect of the 1/2 spin and the unconventional superconductivity via spin fluctuation.

The valence-bond (VB), generalized VB (GVB)[17] and resonating VB (RVB)[18] approaches have been applied to multi-center polyradicals in Fig. 2. The RVB approach is inevitable for the spin-frustrated systems (I_{a-c} in Fig. 2). Thus the multi-configuration pictures are common in both GSO SO SCF and RVB approaches for electron and spin correlated polyradicals. If the spin-orbit (SO) and related magnetic interactions play important roles, GSO-approaches involving SO terms are rather practical for ab initio computations of those terms. Ab initio calculations of the Dzyaloshinskii-

Moriya (DM) term for spin-frustrated systems by SO GHF, SO GDFT and SO-GHF CASCI is such an example[19]. The GSO-HF Bogoliubov (GHFB) solutions are required for superconductors with electronic origin as shown in Fig. 1[1,20].

The resonance of BS solutions often takes place because of quantum effect. The resonating BS (RBS) solutions for triangular [3,3] system can be easily constructed by the superposition of the BS solutions in Fig. 2[21]

$$^2I_{ab} = \left(I_a - I_b\right) \big/ \sqrt{N} \tag{2a}$$

$$^2I_{abc} = \left(I_{ac} + I_{bc}\right)\big/\sqrt{N'} = \left(I_a + I_b - 2I_c\right)\big/\sqrt{N'} \tag{2b}$$

where N and N' denote the normalization constants: note that the BS solutions are non-orthogonal. The RBS solutions become equivalent to GSO SOSCF (Full CI) in the case of simple [3,3] system. The RBS solutions are, however, applicable to more larger systems if the DFT parametrizations are utilized[21], giving rise to an alternative approach to the RVB[18] and GVB[17] methods. The RBS methods are indeed utilized for π-d, π-R and σ-R conjugated systems.

2.2. APPROXIMATE AND EXACT SOLUTIONS OF HUBBARD MODELS FOR POLYRADICAL SPECIES

About 30 years ago, our group initiated theoretical studies on electron localizations via electron correlations in molecular systems[22,23]. The Hubbard model has often been used for polyradical species to elucidate important roles of electron and spin correlation effects[1]

$$H = \sum_{i,j,\sigma} t_{ij} a_{i\sigma}^+ a_{j\sigma} + U_{eff} \sum n_{i\uparrow} n_{j\downarrow} \tag{3}$$

where t_{ij} and U_{eff} denote the transfer and on-site repulsion integrals, respectively. The so-called Mott transition via electron repulsion (U_{eff}) has occurred even in small clusters for the mean-field GSO HF and related theories. The GSO EHF and GSO SO-SCF equations of the multi-center polyradicals in Fig. 2 have also been solved on the basis of the Hubbard model[10], where the binding parameter is defined by

$$x = -t\big/U_{eff} = \beta\big/U_{eff}. \tag{4}$$

Electronic and spin correlation effects become very important in the weak bonding region : $0 \leq x \leq 1/2$, where $x=1/2$ is a quasi Mott transition point for the finite system[1].

In the localized electron region ($x<1/2$), the Hubbard model can be reduced to the Heisenberg spin Hamiltonian model

$$H = - \sum_{a,b} J_{ab} S_a \cdot S_b \tag{5}$$

where J_{ab} is the effective exchange integral. For example, the J_{ab} values for the triangular spin systems by the symmetry-projected GHF method are given by

$$J_{ab} = \frac{1}{2 + \cos^2 \omega} \left[-3x\cos\omega - \frac{1}{3}\left(\sin\omega + \sin^2\omega\right) + \frac{2}{3} \right] \tag{6}$$

where ω is the orbital mixing (order) parameter ($0 \le \cos\omega \le 1$). The J_{ab} value by the symmetry-projected GHF[10] becomes almost identical to those of GSO EHF and GSO SO-SCF in the strong electron correlation region : $0 \le x \le 1/2$. The situation is also recognized for four-center four-electron [4, 4] systems[24] in Fig.2, showing the utility of the broken-symmetry approach.

The J_{ab} value for the triangle (or tetrahedral) multi-center radicals is calculated by first principle (FP) methods in Fig. 1.

$$J_{ab}(FP) = \frac{{}^{LS}E_Y - {}^{HS}E_Y}{{}^{HS}\langle \hat{S}^2 \rangle_Y - {}^{LS}\langle \hat{S}^2 \rangle_Y} \tag{7}$$

where ${}^X E_Y$ and ${}^X \langle \hat{S}^2 \rangle_Y$ denote, respectively, the total energy and total spin angular momentum of the spin state X (X = the lowest spin (LS) and/or the highest spin (HS) state) by a computational method Y[2,25-27]. In our scheme, both $\hat{S}^2$ and $\hat{S}_z$-symmetry adapted (SA) GSO EHF and GSO SO-SCF are used to calculate isotropic J_{ab} values in the Heisenberg model. On the other hand, the $\hat{S}^2$ and $\hat{S}_z$-symmetry-projection of broken-symmetry (BS) GHF solution has been carried out in combination with the energy splittings of the Heisenberg model, leading to the same equation in eq. (6), where ${}^X E_Y$ and by GHF (together with GHF MP and GHF CCSD) and GDFT can be used[25-27].

Recent developments of ab initio hybrid density function theory (HDFT) enable us to determine t and U_{eff} parameters in eq. (3) and J_{ab} values in eq. (6) for p-d, π-d, π-R and σ-R conjugated systems[28,29]. UB3LYP calculations of finite clusters have been carried out to determine t, U_{eff} and J parameters for organic superconductors such as κ-(BEDT-TTF)$_2$(X) and κ-(BETS)$_2$X. For the purpose, the singlet-triplet (ST) energy gap by FP calculation is mapped to that of Hubbard model as

$$\Delta(ST)/2 = J_{ab}(FP) = J_{ab}(Hubbard) = \frac{1}{2}\left(U_{eff} + \sqrt{4t^2 + U_{eff}^2} \right) \quad (8a)$$

$$= -t\left(2t/U_{eff} \right) = J_{ab}(Heisenberg) \quad (U \gg t) \quad \quad 8b)$$

$$= -t\left(1 - \left(U_{eff}/2t \right) + \left(U_{eff}/2t \right)^2 \right) \quad \text{(band model)} \ (U < t) \quad (7c)$$

It is noteworthy that $J_{ab}(FP)$ and $J_{ab}(Hubbard)$ are parameters responsible for magnetic excitations in the whole interaction region: (7a)–(7c).

3. Molecular design of isoelectronic p-d, π-d, π-R and σ-R systems

3.1. N-BAND HUBBARD MODELS VIA AB INITIO CALCULATIONS

The ab initio broken-symmetry (BS) molecular orbital (MO)[4,31] and MCSCF[32] calculations of copper oxides have shown that cuprates exist in the intermediate electron correlation regime with finite on-site Coulomb repulsion ($U_{eff} = 5\sim6$ eV). The ab initio computational results have been mapped into N-band Hubbard models[31]. For example, two-site (two-band) model considering both copper and oxygen sites is such an example.

This d-p (p-d) model Hamiltonians for metal-oxide clusters M_nO_m (M = Fe, Co, $\cdots$, Cu) have been diagonalized to elucidate charge and spin populations, and effective exchange integrals (J) in these systems. It was found that holes induced in CuO clusters are populated mainly on oxygen sites[31]. Interestingly, the 1/N expansion of this d-p model by Nagoya group[33] has provided reasonable explanations of the phase diagrams and pseudo gaps in cuprates.

On the other hand, Zhang and Rice[5] have derived the so-called t-J Hamiltonian model from the p-d model, assuming that hole (electron) doped on the oxygen site forms the singlet pair with the unpaired electron on the Cu(II) site. The exact diagonalization of the t-J model has indicated that the *const.* in the J-model ($T_c = cJ$) is about 0.1[34]. The slave-boson approximation to the t-J model has revealed possible phase diagrams for cuprates[35], which have been modefied to obtain one of reasonable explanations of our J model[1,8].

3.2. EQUIVALENCE TRANSFORMATION OF CUO (CUO$_2$) TO ISOELECTRONIC P-D, π-R AND σ-R SYSTEMS

From the view point of material science, molecular design of new materials is one of important and interesting problems. Concerning with the high-Tc superconductivity, we have proposed new p-d, π-R and σ-R conjugated systems which are considered to be isoelectronic to the CuO bond of cuprates on the basis of the two band (p-d) model[8,31]. The doubly occupied p-orbital of O^{2-} in CuO can be replaced by π-orbital of organic donors such as TTF derivatives. While the open-shell dx^2-y^2 orbital of Cu(II) may be substituted with the open-shell d-orbital of transition metal complexes. These chemical modifications enable us to propose various π-d conjugated systems. Similarly, we may have π-R conjugated systems by replacing the open-shell d-orbital of π-d systems with that of organic radical species (R) such as nitroxide. On the other hand, the π-orbital of π-R conjugated systems can be regarded as a σ-orbital of σ-R conjugates systems. All these p-d, π-R, π-R and σ-R conjugated systems in Fig. 3 are strongly correlated electron systems because of existence of unpaired electrons on d or R sites before carrier doping.

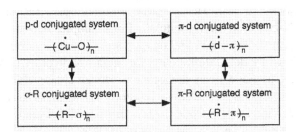

Figure 3. Molecular design of p-d, π-R and σ-R systems which are isoelectronic to CuO bonds on the basis of the two-band models.

Ab initio calculations have been performed to elucidate theoretical possibilities of magnetic modifications of molecular-based materials by introduction of spin sources. Fig. 4 illustrates such theoretical proposals, where the conducting parts are constructed with (1) conducting polymers[24], (2) charge-transfer (CT) complexes[36], (3) organic metals such as TTF-derivatives (BEDT-TTF, BETS, etc)[31,37], (4) nonmagnetic metal clusters[38] or DNA wires[39,40]. On the other hand, stable organic radicals such as nitroxides and transition metal complexes (MX$_4$) are employed as spin sources[31].

First principle computational results of cluster models of these species are mapped into the Anderson model used in heavy fermion systems as

$$H = \sum_{i,\sigma} \varepsilon_i c_{i\sigma}^\dagger c_{i\sigma} + \sum_{j,\sigma} E_j d_{j\sigma}^\dagger d_{j\sigma} + U_{eff} \sum_j n_{j\uparrow} n_{j\downarrow}$$

$$+V \sum_{k\sigma} \left(d_{j\sigma}^\dagger c_{i\sigma} + c_{i\sigma}^\dagger d_{j\sigma} \right) \tag{9}$$

where V denotes the coupling constant between delocalized electron i and localized spin j. The natural orbital analysis of the first-order density matrix by the computations has elucidated the spin polarization path of conduction electron via localized spins[36-40].

Past 15 years, many experiments have been carried out to realize π-d, π-R and related systems[40-44]. Now, many interesting molecular-based materials have been discovered experimentally: (a) high-spin ion-radicals, (b) antiferromagnetic metal (AFM), (c) d-wave superconductor (SC), (d) coexistence of AFM and d-wave SC, so on. These exotic materials have already been extensively examined on the theoretical grounds in relation to the theoretical proposals[36-40] in Fig. 4. Exact diagonalization of the derived model Hamiltonian of the species provides a guarantee of future achievements.

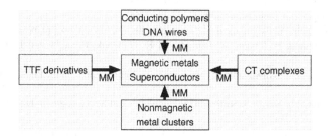

Figure 4. Theoretical proposals of magnetic modification (MM) of molecule-based materials.

References

1. K. Yamaguchi, in Self-consistent Field Theory and Applications (R. Carbo and M. Klobukowski Eds, Elsevier, 1990) p727.
2. K. Yamaguchi, Y. Takahara and T. Fueno, in Appl. Quantum. Chem. (V. H. Smith et.al, Eds, D. Reidel, Lancaster, 1986) p155.
3. J. C. Bednorz and K. A. Müller, Z,phys, B64, 189 (1986).
4. K. Yamaguchi, Y. Takahara, T. Fueno and K. Nasu, Jpn. J. Appl. Phys. 26, L1362 (1987).

5.   F. C. Zhang and T. M. Rice, Phys. Rev. B37, 3759 (1989).
6.   S. C. Zhang, Science, 275, 1089 (1997).
7.   T. Moriya and K. Ueda, Adv. Phys. 49, 555 (2000).
8.   K. Yamaguchi, D. Yamaki, Y. Kitagawa, M. Takahata, T. Kawakami, T. Ohsaku and
     H. Nagao Int. J. Quant. Chem. 92, 47 (2003).
9.   K. Yamaguchi, Chem. Phys. Lett, 34, 434 (1975).
10. K. Yamaguchi, Y. Yoshioka. K. Takatsuka, T. Fueno, Theoret. Chim. Acta, 48, 185
     (1978).
11. K. Yamaguchi, Chem. Phys, 29, 117 (1978).
12. K. Yamaguchi, T. Fueno, M. Ozaki, N. Ueyama and A. Nakamura, Chem. Phys. Lett.
     168, 56 (1990).
13. K. Yamaguchi, S. Yamanaka, M. Nishino, Y. Takano, Y. Kitagawa, H. Nagao, and Y.
     Yoshioka, Theoret. Chem. Acc. 102, 328 (1999).
14. S. Yamanaka, R. Takeda and K. Yamaguchi, Polyhedron, 22 2013 (2003).
15. S. Yamanaka, D. Yamaki, R. Takeda, H. Nagao and K. Yamaguchi, Int. J. Quant.
     Chem., 100, 1179 (2004).
16. T. Kawakami, T. Taniguchi, M. Shoji, Y. Kitagawa, S. Yamanaka, M. Okumura, K.
     Yamaguchi, Mol. Cryst. Liq. Cryst., in press.
17. W. R. Wadt and W. A. Goddard III, J. Am. Chem. Soc., 96, 1689 (1974).
18. P. W. Anderson, Mat. Res. Bull., 8, 153 (1973).
19. R. Takeda, S. Yamanaka and K. Yamaguchi, Int. J. Quant. Chem. 102, 80 (2005).
20. D. Yamaki, T. Ohsaku, H. Nagao and K. Yamaguchi, Int. J. Quant. Chem, 96, 10
     (2003).
21. S. Yamanaka, R. Takeda, T. Kawakami, S. Nakano, D. Yamaki, S. Yamada, K.
     Nakata, T. Sakuma, T. Takada and K. Yamaguchi, Int. J. Quant Chem., 95, 512
     (2003).
22. K. Yamaguchi, Chem. Phys. Lett, 33, 330 (1975).
23. K. Yamaguchi, Chem. Phys .Lett, 35, 230 (1975).
24. K. Yamaguchi, Y. Toyoda and T. Fueno, Synthetic Metals, 19, 81 (1987).
25. K. Yamaguchi, F. Jensen, A. Dorigo and K. N. Houk, Chem. Phys. Lett, 149, 537
     (1988).
26. S. Yamanaka, T. Kawakami, H. Nagao and K. Yamaguchi, Chem. Phys. Lett, 231, 25
     (1994).
27. Y. Yoshioka, S. Kubo, S. Kiribayashi, Y. Takano and K. Yamaguchi, Bull. Chem. Soc.
     Jpn., 71, 573 (1998).
28. T. Kawakami, T. Taniguchi, S. Nakano, Y. Kitagawa and K. Yamaguchi, Polyhedron,
     22, 2051 (2003).
29. T. Kawakami, T. Taniguchi, Y. Kitagawa, Y. Takano, H. Nagao and K. Yamaguchi,
     Mol. Phys., 100, 2641 (2002).
30. L. Noodleman and D. A. Case, Adv. Inorg. Chem. 38, 423 (1994).
31. K. Yamaguchi, Int. J. Quant. Chem, 37, 167 (1990).
32. S. Yamamoto, K. Yamaguchi and K. Nasu, Phys. Rev. B, 42, 266 (1990).
33. A. Kobayashi, A. Tsuruta, T. Matsuura and Y. Kuroda, J. Phys. Soc. Jpn, 71, 1640
     (2002).
34. H. Yokoyama and M. Ogata, J. Phys. Soc. Jpn, 65, 3615 (1996).
35. N. Nagaosa and P. A. Lee, Phys. Rev. B 4, 966 (1992).
36. K. Yamaguchi, H. Namimoto, T. Fueno, T. Nogami and Y. Shirota, Chem. Phys. Lett.,
     166, 408 (1990).

37. K. Yamaguchi, M. Okumura, T. Fueno and K. Nakasuji, Synthetic Metals 41-43, 3631 (1991).

38. M. Okumura, Y. Kitagawa, T. Kawakami, and K. Yamaguchi, Synthetic Metals, 154, 313-316 (2005).

39. T. Kawakami, Y. Kitagawa, F. Matsioka, Y. Yamashita, H. Isobe, H. Nagao and K. Yamaguchi, Int. J. Quant. Chem., 85 619 (2001).

40. K. Yamaguchi, T. Taniguchi, T. Kawakami, T. Hamamoto and M. Okumura, Polyhedron, 24, 2758 (2005).

41. P. Day, M. Kurmoo, T. Makkah, I. R. Marsden, R. H. Friend, F. L. Pratt, W. Hayes, D. Chasseau, J. Gaultier, G. Bravic, L. Ducasse, J. Am. Chem. Soc., 114, 10722 (1992).

42. M. Kurmoo et al., Inorg. Chem., 35, 4719 (1996).

43. H. Kobayashi, H. Tomita, T. Naito, A. Kobayashi, F. Sakai, T. Watanabe, and P. Cassoux, J. Am. Chem. Soc., 118, 368 (1996).

44. H. Kobayashi, A. Kobayashi and P. Cassoux, Chem. Soc. Rev., 29, 325 (2000).

MICROSCOPIC EVIDENCE OF THE FFLO STATE

IN THE STRONGLY-CORRELATED SUPERCONDUCTOR

CeCoIn$_5$ PROBED BY ^{115}In-NMR

Ken-ichi Kumagai[1] (kumagai@phys.sci.hokudai.ac.jp),
Kosuke Kakuyanagi[1], Masato Saitoh[1], Sinnya Takashima[2], Minoru Nohara[2],
Hidenori Takagi[2] and Yuji Matsuda[3]
[1]*Division of Physics, Graduate School of Science, Hokkaido University, Sapporo 060-0810, Japan*
[2]*Department of Advanced Materials Science, University of Tokyo, Kashiwa, Chiba 277-8581, Japan*
[3]*Department of Physics, Kyoto University, Kyoto 606-8502, Japan*

Abstract. ^{115}In NMR measurements of the strongly correlated superconductor CeCoIn$_5$ reveals an unusual temperature dependence of the NMR spectra in the newly-observed high field and low temperature superconducting phase. The anomalies are attributed to a spatially inhomogeneous local density of states of quasiparticles. From a simulation, the oscillation wave length of the superconducting gap is found to decrease rapidly with decreasing temperature. The present NMR study provides solid evidence from a microscopic point of view that the Fulde-Ferrell-Larkin-Ovchinnikov (FFLO) state is realized in the newly-dicovered superconducting phase of CeCoIn$_5$.

Key words: CeCoIn$_5$, NMR, Knight Shift, FFLO, Superconductivity

1. Introduction

In spin-singlet superconductors, Cooper pairs are suppressed by a magnetic field as a consequence of its coupling to the conduction electron spins (Pauli paramagnetism) or to the orbital angular momentum (vortices). In strongly-correlated systems, such as cuprates and heavy fermion materials, unconventional pair states with a nodal structure of the superconducting energy gap are believed to exist, resulting in novel vortex core states. A different kind of inhomogeneous superconductivity has been proposed in the mid 1960's. Fulde and Ferrell (Fulde and Ferrell, 1964) as well Larkin and Ovchinnikov (Larkin and Ovchinnikov, 1965) (FFLO) developed theories for a novel superconducting (SC) phase with non-zero total momentum, and consequently a spa-

K. Scharnberg and S. Kruchinin (eds.),
Electron Correlation in New Materials and Nanosystems, 235–249.
© 2007 *Springer.*

tially modulated superconducting order parameter, when Pauli pair-breaking dominates over the orbital effect. Pair-breaking arising from the Pauli effect is reduced by forming a new pairing state $(k \uparrow, -k + q \downarrow)$ with $|q|$ $\sim 2\mu_B H / \hbar v_F$ (v_F is the Fermi velocity) instead of $(k \uparrow, -k \downarrow)$-pairing in ordinary singlet superconductors. (Fulde and Ferrell, 1964; Larkin and Ovchinnikov, 1965; Gruenberg and Gunther, 1966).

In spite of enormous efforts to find the FFLO state, it has never been observed in conventional superconductors. In the last decade, heavy fermion superconductors $CeRu_2$ (Huxley et al., 1993), UPt_3 (Tenya et al., 1995), and UPd_2Al_3 (Gloos et al., 1993) have been proposed as candidates for the observation of the FFLO state, since the anomalous peak effect was observed in the magnetization just below the upper critical field (Tachiki et al., 1996). However, subsequent research has called the interpretation of the data for these materials in terms of the FFLO state into question. (Norman, 1993)

The attempt of observing the FFLO phase in strongly correlated superconductors has only been addressed recently. $CeCoIn_5$, a new type of heavy fermion superconductor with quasi two-dimensional (2D) electronic structure, (Petrovic et al., 2001; Shishido et al., 2002) has the highest transition temperature ($T_c = 2.3$ K) among all heavy fermion superconductors discovered until now. Subsequent measurements have identified $CeCoIn_5$ as an unconventional superconductor with $d_{x^2-y^2}$-wave gap symmetry (Kohori et al., 2001; Movshovich et al., 2001; Izawa et al., 2001; Eskildsen et al., 2003). Moreover, non-fermi liquid behavior observed in many measurements indicates the importance of antiferromagnetic spin fluctuations and the possibility that $CeCoIn_5$ is located in the vicinity of a quantum critical point. (Paglione et al., 2003; Bianchi et al., 2003b)

$CeCoIn_5$ shows the extremely large $H_{c2}^{\parallel}(T = 0)$ of 11.9 T, far exceeding the conventional estimate of the Pauli field, which is also favorable for the occurrence of the FFLO state, because the Pauli effect may overcome the orbital effect. The Pauli limited superconducitivity is in fact supported by the fact the phase transition from SC to normal metal at the field parallel to the ab–plane, $H_{c2}^{\parallel}$, is found to be of first order below approximately 1.3 K, in contrast to the second order transition as expected for the BCS superconductors. (Izawa et al., 2001; Tayama et al., 2002; Bianchi et al., 2002)

Recent reports of heat capacity measurements (Radovan et al., 2003; Bianchi et al., 2003a) (Radovan et al., 2003; Bianchi et al., 2003a) in a field parallel to the ab–plane of $CeCoIn_5$ have revealed that a second order phase transition line, which branches from the first order $H_{c2}^{\parallel}$-line and decreases with decreasing T, was observed within the SC state below 0.35 K. This new superconducting phase was confirmed by thermal condcutivity (Capan et al., 2004), ultrasound velocity (Watanabe et al., 2004), NMR (Kakuyanagi et al., 2005) measurements, and the possibility of a high-field FFLO state has been evoked. The schematic phase diagram in the high field and low temperature regime is shown in Fig. 1. As this new phase appears in the Pauli limited region, the experimental results make the FFLO scenario a very appealing one for CeCoIn$_5$. Furthermore, CeCoIn$_5$ satisfies the requirements for the formation of the FFLO state, because the mean free path is large compared with the coherence length ξ and the Ginzburg-Landau parameter $\kappa = \lambda/\xi \gg 10$,together with the penetration length λ are also very large. (Maki, 1966)

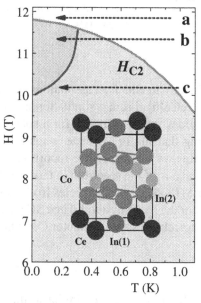

Figure 1. Superconducting phase diagram in the high field and low temperature region of CeCoIn$_5$ where the first order transition at H_{c2} appears. A new *high field SC phase* is shown by a red line. The arrows indicate the magnetic fields used for the NMR measurements. The inset shows the crystal structure of CeCoIn$_5$.

One of the most fascinating aspects of the FFLO state is that the condensate develops a spatially modulated order parameter and spin polarization with a wave length of the order of $2\pi/|\mathbf{Q}|=\Lambda$. The SC order parameter has planar nodes aligned perpendicularly to the applied field. However, there still is no corroborative and direct evidence for the modulated order parameter of the FFLO state due to the lack of microscopic information for probing spatial SC gap structure. Therefore, a more detailed experimental investigation of the quasiparticle state from a microscopic point of view is required in order to shed light on this subject.

In order to establish the existence of a FFLO state, the NMR method, one of local probes sensitive to spatial imaging, (Kakuyanagi et al., 2002; Kakuyanagi et al., 2003) is particularly well suited to elucidate the spatially-modulated gap structure of the vortex state. In a previous letter (Kakuyanagi

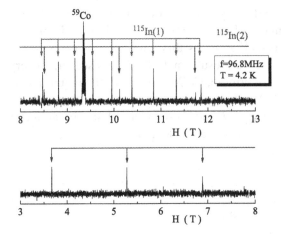

Figure 2. NMR spectrum of the single crsytal of CeCoIn obtained by a box-car integration of the spin echo for a wide range sweeping of applied magnetic field at $T = 4.2$ K. The calculated resonance positions for quadrupole split lines for ^{115}In(1), ^{115}In(2) and ^{59}Co are shown by arrows. The magnetic field applied along the a-axis of the single crystal of CeCoIn$_5$.

et al., 2005), we reported a ^{115}In-NMR study in which an anomalous distribution of the Knight shift was found in CeCoIn$_5$. From a simulation analysis we concluded that the NMR result provides strong evidence for a spatially-modulated SC order parameter, showing that the FFLO state is realized in the newly observed SC phase of CeCoIn$_5$. In this paper, we report detailed results of ^{115}In-NMR and present the analysis of the spectrum in terms of the spatially-oscillating local density of state (DOS) of quasiparticles from which we are able to provide the confirmation for the spatially-modulated SC energy gap in the FFLO phase of the strongly-correlated superconductor CeCoIn$_5$.

2. Experimental

Single crystals of CeCoIn$_5$ were grown by a flux method similar to the one reported previously. (Petrovic et al., 2001). No detectable secondary phases were found in powder X-ray diffraction and scanning electron microscopy. Specific heat measurements revealed a sharp SC transition at $T_c = 2.3$ K with a transition width of less than 0.1 K, indicating the high quality of the sample. The tetragonal crystal structure of CeCoIn$_5$ consists of alternating layers of CeIn$_3$ and CoIn$_2$ and hence has two inequivalent In sites per unit cell as shown in the inset of Fig.1. Spin echo NMR measurements were performed using a conventional phase-coherent pulse NMR spectrometer on a single crystal of ~10 mg along with $H \parallel (100)$.

For preliminary information on the sample we observed the ^{59}Co, ^{115}In(1) and ^{115}In(2)-NMR by sweeping the magnetic field over a wide range. All the observed multiply-split signals are shown in Fig. 2. We properly assigned the observed resonance lines to the electric quadrupole split NMR ones from each of the nuclei with reasonable parameters. Table I shows NMR parameters which agree with the results obtained previously. (Curro et al., 2001) Additional signals in the spectrum were not observed, indicating the absence of any secondary phases.

For low temperature measurements, we used a ^{3}He-^{4}He dilution refrigerator. Temperature was measured by a calibrated RuO$_2$ thermometer. NMR measurements were carried out in the magnetic field H parallel to the [100]-direction under the field-cooled condition. Rf excitation (H_1) power was reduced at least by 10^{-2} times compared to that in ordinary measurements in order to prevent heating the single crystal sample. The spectrum of ^{115}In(1) was obtained from a convolution of fourier transform (FT) signals of spin echo which were measured at each 20 kHz interval. The Knight shift of ^{115}In was obtained from the central ^{115}In-line ($\pm 1/2 \leftrightarrow \mp 1/2$ transition) using a gyromagnetic ratio of $^{115}\gamma_N = 9.3295$ MHz/T.

TABLE I. Parameters of nuclear moment and electric quadrupole interactions at each nuclear site. I is nuclear spin, γ_N is the gyromagnetic ratio, ν_Q is quadrupole frequency and η is asymmetric parameter of the electric field gradient.

site	I	γ_N(MHz/T)	ν_Q (MHz)	η
In(1)	9/2	9.3295	8.173	0.01
In(2)	9/2	9.3295	15.489	0.386
Co	7/2	10.03	0.234	0.01

3. Results

We measured ^{115}In-NMR spectra at three magnetic fields: namely, at $H=11.8$ T (in the normal state for whole temperature range, shown by arrow (a) in Fig. 1), at $H=11.3$ T (across the normal to the SC phase at $T\sim700$ mK shown by arrow (b)), and also to the second phase transition at $T\sim300$ mK), and at $H=10.1$ T (in the SC state down to $T=140$ mK shown by arrow (c)). The spectra obtained at various temperatures are shown in Fig. 3. The line width in the normal state at the largest field ($H=11.8$ T) is ~15 kHz and is nearly temperature-independent. The peak position shifts towards higher frequencies which indicates that the Knight shift of ^{115}In(1) increases smoothly with decreasing temperature in the normal state.

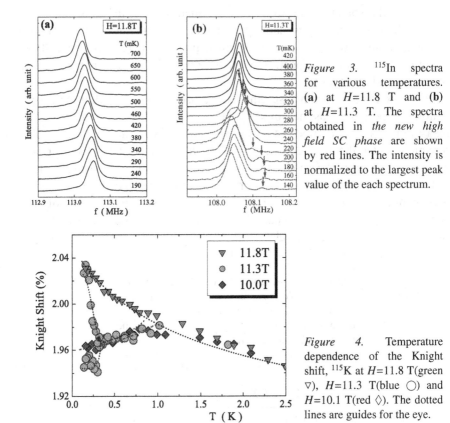

Figure 3. ^{115}In spectra for various temperatures. (a) at $H=11.8$ T and (b) at $H=11.3$ T. The spectra obtained in *the new high field SC phase* are shown by red lines. The intensity is normalized to the largest peak value of the each spectrum.

Figure 4. Temperature dependence of the Knight shift, ^{115}K at $H=11.8$ T(green ▽), $H=11.3$ T(blue ○) and $H=10.1$ T(red ◇). The dotted lines are guides for the eye.

For the NMR spectra obtained in the SC state at $H=10.1$ T (not shown in Fig. 3) the line widths are quite similar to the ones in the normal state. One may imagine the local field being distributed in the SC vortex lattice state, showing so-called the Redfield pattern. (Kakuyanagi et al., 2002; Kakuyanagi et al., 2003) However, under applying high fields very near to $H_{c2}(0)$, as in this case, the spatial distribution of the local magnetic field due to the vortices for large $\xi=20$ Å and $\lambda=2600$ Å (Oremeno et al., 2002; Chia et al., 2003) is smaller than the inhomogeneous broadening of 15 kHz as observed in the normal state. The intensity strongly decreases with decreasing temperature in the SC state in contrast to the case in the normal state. This is due to the reduction of the volume fraction in which NMR signals can be detected, as *rf* radiation finds it hard to penetrate into the single crystal sample due to skin effects at the SC surface as discussed later. With these experimental situations, we did not observe Redfield patterns in the superconducting state.

Compared with the NMR results at H=11.8 T and H=10.1 T, the NMR spectra at the intermediate field of H=11.3 T are anomalous as shown in Fig. 3(b). The spectra with single peaks are observed down to T = 320 mK and the peak positions slightly decrease with temperature below T_c. And then, the spectra show complex features just across the phase boundary at T = 300 mK. The resonance line at lower frequency grows rapidly when T decreases below $T^*(H)$. A double peak structure appears at T = 240 mK, followed by a shoulder structure at T = 260~300 mK. The separation of the upper and lower peak positions increases with decreasing temperature. The relative ratio of the intensities at each peak changes drastically with decreasing temeprature. The most remarkable feature in the NMR spectrum is the appearance of a new resoance peak with small but finite intensity at higher frequency as can be seen in Fig. 3(b) in the low temperature region. This peak at the high frequency side remains nearly T-independent at low temperature below ~180mK (Kakuyanagi et al., 2005).

The T-dependences of the Knight shift, ^{115}K, as deduced from the peak position is plotted in Fig. 4. The Knight shift in the normal state increases gradually with decreasing temperature. The values of the Knight shifts of each site are close to the extrapolated one from the data at high temperatures (Curro et al., 2001). In the SC state at H=10.1 T, the Knight shift decreases slightly below T_c(~0.7 K) down to the lowest temperature. The Knight shift at the intermediate field of H=11.3 T shows quite an unusual temperature dependence. K decreases below T_c. However, just across the second transition to *the new high field SC phase* at T=300 mK, the Knight shift deduced from the upper peak changes strongly with decreasing temperature. The Knight shifts of the the upper and the lower peak coincide with the values in the normal and the superconducting state, respectively, at the lowest temperature.

Thus, this evolution of the spectrum is an important clue to the nature of *the high field SC phase*. We stress that the occurrence of magnetic ordering is highly unlike as the source for the higher resonance in view of the large difference in the intensity of the two lines. Should antiferromagnetic order set in, with the direction of the induced moments parallel to the applied magnetic field, the alternating hyperfine fields would produce two unequivalent ^{115}In(1) sites, which would give rise to two resonance lines having equal intensities. If, on the other hand, the magnetic moments are oriented perpendicular to the applied magnetic field, additional hyperfine fields should appear for all of the In(1) nuclei, which should result in a shift of whole the spectrum to the high frequency side.

A noteworthy feature on the spectrum inside *the high field SC phase* is that the position of the higher resonance line coincides well with that of the resonance line above $H_{c2}^{\parallel}$, while the lower resonance line is located close to that of the SC state above $T^*(H)$. Therefore, it is natural to deduce that the higher resonance line originates from a normal quasiparticle regime, which is newly formed below $T^*(H)$, while the lower resonance line corresponds to the SC regime, which appears to have a similar quasiparticle structure above $T^*(H)$. These results lead us to conclude that the appearance of the new resonance line at the higher frequency is a manifestation of a novel normal quasiparticle structure in *the high field SC phase* in CeCoIn$_5$.

4. Discussion

In the previous section, we have stressed that *the high field SC phase* is characterized by the formation of normal regions. This brings us to the next question as to whether the NMR spectrum below $T^*(H)$ is an indication of the FFLO phase. It has been predicted for s-wave superconductors without vortices that in the FFLO phase the SC order parameter exhibits one-dimensional spatial modulations along the magnetic field, forming planar nodes that are periodically aligned perpendicularly to the flux lines. Therefore, the formation of normal regions is consistent with a phase expected in a FFLO state. We will show that the NMR spectra just below $T^*(H)$ in Fig. 3(b) can be accounted for by considering such planar structures.

In the FFLO state in which the SC order parameter is modulated spatially, the NMR rf field is expected to penetrate more deeply into the sample at the nodal planes. Since the skin depth of the normal metal is larger than the superconducting penetration depth, the rf field penetrates into the nodal plane as the TEM mode. The TEM mode transmits to the space which is surrounded by two conductors. In the NMR experiments, the rf magnetic field H_1 is applied perpendicular to the dc magnetic field *i.e.*, $H_0 \parallel a$ and $H_1 \parallel b$. The shielding supercurrent flows crossing the planar nodes. The distribution of the rf fields in the SC region is calculated with the help of the London equations.

$$\Delta H_1 = \frac{1}{\lambda^2} H_1 \,,$$

with the boundary conditions at the nodal planes

$$H_1(0) = H_1\left(\frac{\Lambda}{2}\right) = 1$$

Here, H_1 is nomalized to its value in the normal state. The H_1 distribution is described as follows.

$$H_1 = \frac{\sinh\left(\frac{z}{\Lambda} \cdot \frac{\Lambda}{\lambda}\right) + \sinh\left(\frac{\Lambda}{\lambda}\left(\frac{1}{2} - \frac{z}{\Lambda}\right)\right)}{\sinh\left(\frac{1}{2} \cdot \frac{\Lambda}{\lambda}\right)} \tag{1}$$

The spatial distribution of H_1 depends on both $\frac{z}{\Lambda}$ and $\frac{\Lambda}{\lambda}$. The width of the distribution of H_1 becomes larger with increasing Λ.

Because of the second order transition at $T^*(H)$, (Radovan et al., 2003; Bianchi et al., 2003a) the modulation length of the SC order parameter parallel to H_0 or the thickness of the SC layers, $\Lambda(= 2\pi/|\mathbf{q}|)$, is expected to diverge as, $\Lambda \propto (T^* - T)^{-\alpha}$ with $\alpha > 0$, upon approaching $T^*(H)$. Therefore, Λ will exceed the in-plane penetration length λ in the vicinity of $T^*(H)$. In such a situation, the rf field penetrates into the normal sheets much deeper than into the SC sheets, which results in a restoration of the NMR intensity to that of the normal sheets. At low temperatures, where Λ becomes comparable to $\xi(\ll \lambda)$, the penetration of the rf field into the normal sheets is the same as that into the SC sheets.

Next, we consider the effect of H_1 on the NMR intensity. The spin echo signal is described as follows (Hahn, 1956),

$$I_{\text{echo}} \propto \sin(\gamma H_1 t_{w_1}) \sin^2\left(\frac{\gamma H_1 t_{w_2}}{2}\right)$$

where t_{w1} and t_{w2} is a time of first and second pulse width, respectively, and $t_{w2}=2t_{w1}$ for the spin echo measurement. As we used reduced-power for the NMR pulse to avoid heating of the single crystal sample in the present measurement, the condition $\gamma H_1 t_w << \pi/2$ is satisfied. Under this condition, the NMR echo intensity is proportional to the cube of H_1.

The NMR Knight shift is proportional to the spin susceptibility through hyperfine interaction, and is given by $K = A\chi_s$. Here, A is a hyperfine coupling constant. Since the electron spin susceptibility depends on the density of states at the Fermi surface, the Knight shift in the superconducting state is given by (Yoshida, 1958)

$$K\left(\frac{\Delta}{T}\right) \propto -\int_{-\infty}^{\infty} N_S\left(\frac{\epsilon}{\Delta}\right) \frac{\partial}{\partial \epsilon} f\left(\frac{\epsilon}{T}\right) d\epsilon$$

where f is the Fermi distribution function, and N_S is the density of states in the SC state. For an anisotropic superconductor with d-wave symmetry we obtain after integration with respect to solid angle

$$N_S\left(\frac{\epsilon}{\Delta}\right) = \frac{1}{4\pi} \int_{|\epsilon|>|\Delta|} N_0 \frac{\epsilon}{\sqrt{\epsilon^2 - |\Delta^2(\theta, \phi)|}} d\Omega$$

In the FFLO state, the SC order parameter is spatially modulated. Although we need a realistic functional form of the superconducting order parameter based on a microscopic model, here we simply use a sinusoidal functional form (Tachiki et al., 1996),

$$\Delta(z) = \Delta_0 \sin \frac{2\pi z}{\Lambda}$$

By a convolution of the H_1 dependence of the signal intensity, the NMR intensity in the FFLO state is described as follows.

$$I(\omega) \propto \frac{1}{\Lambda} \int_0^\Lambda \delta\left(K\left(\frac{\Delta(z)}{T}\right) - \omega\right) g\left(H_1(z)\right) dz$$

In our experimental condition, $g(H_1(z))$ is proportional to $H_1^3(z)$. The inevitable inhomogeneity of H_1 leads to a distribution of spin echo intensity across the sample. Nevertheless, since there is a cutoff in H_1, the assumed form probably is a crude phenomenological model of the actual intensity.

By integrating the δ-function over the sample space,

$$\int \delta(f(x)) \, dx = \int \delta(y) \frac{dy}{\frac{\partial f}{\partial x}} = \sum_{f(x)=0} \frac{1}{\left|\frac{\partial f}{\partial x}\right|}$$

the spectrum is given by

$$I(\omega) \propto \frac{1}{\Lambda} \sum_{K=k} \frac{g(H_1(z))}{\left|\frac{\partial}{\partial z} K\left(\frac{\Delta(z)}{T}\right)\right|}$$

As K is a simple decreasing function z, this equation can be simplified to,

$$I(\omega) \propto \frac{T}{\Delta_0} \frac{\left[H_1\left(\frac{1}{2\pi} \arcsin\left(\frac{T}{\Delta_0} K^{-1}(\omega)\right), \frac{\Lambda}{\lambda}\right)\right]^3}{\left|K'\left(K^{-1}(\omega)\right)\right| \sqrt{1 - \left(\frac{T}{\Delta_0} K^{-1}(\omega)\right)^2}}$$

In this model, the NMR spectrum depends on only two parameters $\frac{T}{\Delta_0}$ and $\frac{\Lambda}{\lambda}$. From numerical calculation, we obtain the NMR spectrum by using the relations $\Delta_0 = 2.13 \, k_B T_c$ and $T_c = 2.3$ K ($\Delta_0 = 4.9$ K). In this case, as $\frac{T}{\Delta_0}$ just corresponds to the measuring temperature, $\frac{\Lambda}{\lambda}$ is the only free parameter of the fitting procedure.

We remind the reader that the most notable features of the [115]In(1) spectrum of $H=11.3$ T shown in Fig. 3(b) are the double peak structure in the temperature range between $T = 260$ mK and $T = 240$ mK, and the existence of the higher frequency peak below $T \sim 220$ mK. Figure 5 shows the fitting results of the [115]In(1) spectrum for various temperatures. The closed circles with solid line (black) represent the observed spectrum. The dotted line (blue) is a numerically calculated spectrum. The solid line (red) is a calculated spectrum with convoluting inhomogenous broadening of the the Lorenz function, given by,

$$I(\omega) = \frac{2}{\pi} \int \frac{\delta}{4(\omega - \omega')^2 + \delta^2} I(\omega') d\omega'$$

The fits reproduce well the spectrum with the shoulder structure at the low frequency side just below $T_c^*(H)$ and the observed double peak structure at $T=260$ mK~240 mK by using the parameters of $\frac{T}{\Delta_0} = 0.053$, $\frac{\Lambda}{\lambda}=19\sim10$ and $\delta=15$ kHz. For $T < 220$ mK, a high frequency divergent component is suppressed with convoluting inhomogenous broadening. However, although the present model well explains the tendencies of the observed spectra, some discrepancies remain between the simulated and the observed ones, especially at low temperatures. This might be ascribed to the simplicity of the present phenomenological model. For further extensions one may need to consider a more realistic function of the periodic superconducting order parameter and the effects of vortices. Nevertheless, the good fits to the observed spectra obtained in a wide tempeature range suggests strongly that the spatially-modulated superconducting gap is reasonably well confirmed, which provides us with the evidence for the appearance of the FFLO state in which the SC order parameter is periodically modulated.

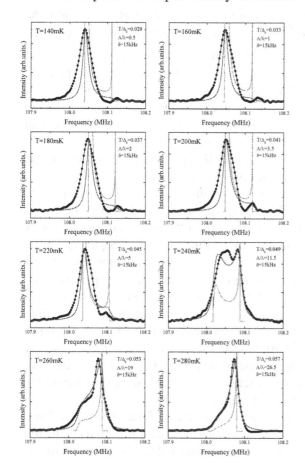

Figure 5. ^{115}In-NMR spectra (•) for various temperatures. The solid red line and blue dotted line represent the simulation spectra with and without a convolution of an inhomogeneous broadening with a Lorentian function ($\delta=15$ kHz), respectively. For details, see the text.

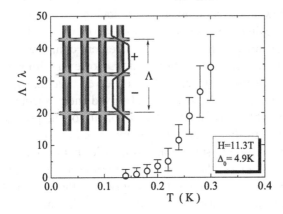

Figure 6. Temperature dependence of $\frac{\Lambda}{\lambda}$ for $H=11.3$ T and $\Delta_0=4.9$ K. The inset illustrates the quasiparticle structure in the FFLO state. The planar nodes are periodically aligned perpendicular to the vortices. The SC order parameter exhibits one dimensional spatial modulation along the vortices.

We obtain the value of $\frac{\Lambda}{\lambda}$ from fitting the NMR spectra and show the temperature dependence of $\frac{\Lambda}{\lambda}$ in Fig. 6. With the present FFLO model, the evolution of the spectrum with $\frac{\Lambda}{\lambda}$ is strong in the range of $\Lambda \sim 10\lambda$. For $\Lambda \ll \lambda$ and $\lambda \ll \Lambda$, $\frac{\Lambda}{\lambda}$, the dependence of the spectrum is not apparent, resulting the accuacry of this parameter become poor for these limiting cases. $\frac{\Lambda}{\lambda}$ decreases rapidly with decreasing temperature. This result shows that the period of the modulation of the SC order parameter becomes shorter with decreasing temperature in the FFLO state.

We finally mention about the nature of the quasiparticle structure of vortex cores in CeCoIn$_5$ inferred from the present results. The well separated double resonance lines provide important information for the quasiparticle excitation in the FFLO state. The quasiparticles excited around the FFLO planar nodes are spatially well separated from those excited around vortex cores. According to a recent theory (Mizushima et al., 2005), the quasiparticles are not excited in the region where the vortex lines intersect with the planar node because of the bound states due to the π shift of the pair potential associated with the planar node. This indicates that the quasiparticle regions within the vortex line are not spatially overlapped with those in the FFLO planar nodes, and also that the Knight shift within the vortex core deviates from that in the normal quasiparticle sheets. This implies that the vortex core is to be distinguished from the normal state above $H_{c2}^{\parallel}$, the feature in sharp contrast to conventional superconductors, where the Knight shift within the core coincides with the Knight shift in the normal state.

What is the reason behind this unusual structure of the vortex core? Even just below $H_{c2}^{\parallel}$, the vortex cores are associated with a large spatial oscillation of the SC order parameter. We recall that a strong reduction of the quasiparticle density of states within vortex cores has been reported in high-T_c cuprates (Kakuyanagi et al., 2002), and discussed in terms of the strong enhancement of the antiferromagnetic correlation within cores (Kakuyanagi et al., 2003). A similar situation may expect to present in CeCoIn$_5$, as strongly-enhanced antiferromagnetic correlations are inferred from the T-shift of Knight shift in the normal state above $H_{c2}^{\parallel}$, which increases with decreasing T, as is evident from Fig. 4. This behavior is notably different from that expected in the Fermi liquid model, which predicts the T-independent Knight shift. The non-Fermi liquid behavior in CeCoIn$_5$has been discussed in the light of the incipient antiferromagnetism with the field indued quantum critical point in the vicinity of the upper critical field (Paglione et al., 2003; Bianchi et al., 2003b). These results call for further investigations of the vortex core structure in the presnce of strong antiferromagnetic correlation.

5. Conclusion

In summary, NMR measurements of CeCoIn$_5$ in the vicinity of the upper critical field H_{c2} reveal an unusual change of the [115]In(1)-NMR spectrum across the newly-discovered phase transition within the SC phase. The anomalous spectra are well characterized by taking into account the modulation of local density of state of quasiparticles and also the spatial modulation of NMR rf filed (H_1). The length of the spatial modulation of the SC order parameter is estimated from the simulation. The present NMR results provide strong evidence that the new superconducting phase at high fields and low temperatures in CeCoIn$_5$ is the FFLO phase in which the SC gap is spatially-modulated.

Acknowledgements

We thank K. Machida, M. Ichioka, R. Ikeda, and T. Kita for stimulating discussions. The present work is supported by a grant-in-aid for Scientific Research from the Ministry of Education, Culture, Sports, Science, and Technology of Japan.

References

Bianchi, A., Movshovich, R., Capan, C., Pagliuso, P. G., and Sarrao, J. L. (2003a) Possible Fulde-Ferrell-Larkin-Ovchinnikov Superconducting State in CeCoIn$_5$, *Phys. Rev. Lett.* **91**, 187004.

Bianchi, A., Movshovich, R., Oeschler, N., Gegenwart, P., Steglich, F., Thompson, J. D., Pagliuso, P. G., and Sarrao, J. L. (2002) First-Order Superconducting Phase Transition in CeCoIn$_5$, *Phys. Rev. Lett.* **89**, 137002.

Bianchi, A., Movshovich, R., Vekhter, I., Pagliuso, P. G., and Sarrao, J. L. (2003b) Avoided Antiferromagnetic Order and Quantum Critical Point in CeCoIn$_5$, *Phys. Rev. Lett.* **91**, 257001.

Capan, C., Bianchi, A., Movshovich, R., Christianson, A. D., Malinowski, A., Hundley, M. F., Lacerda, A., Pagliuso, P. G., and Sarrao, J. L. (2004) Anisotropy of thermal conductivity and possible signature of the Fulde-Ferrell-Larkin-Ovchinnikov state in CeCoIn$_5$, *Phys. Rev. B* **70**, 134513.

Chia, E., Harlingen, D. J. V., Salamon, M. B., Yanoff, B. D., Bonalde, I., and Sarrao, J. L. (2003) Nonlocality and strong coupling in the heavy fermion superconductor CeCoIn$_5$: A penetration depth study, *Phys. Rev. B* **67**, 014527.

Curro, N. J., Simovic, B., Hammel, P. C., Pagliuso, P. G., Sarrao, J. L., Thompson, J. D., and Martins, G. B. (2001) Anomalous NMR magnetic shifts in CeCoIn$_5$, *Phys. Rev. B* **64**, 18051.

Eskildsen, M., Dewhuyst, C., Hoogenboom, B., Petrovic, C., and Canfield, P. (2003) Hexagonal and Square Flux Line Lattices in CeCoIn$_5$, *Phys. Rev. Lett.* **90**, 187001.

Fulde, P. and Ferrell, R. (1964) Superconductivity in a Strong Spin-Exchange Field, *Phys. Rev. A* **135**, 550.

Gloos, K., Modler, R., Schimanski, H., Bredl, C. D., Geibel, C., Steglich, F., Buzdin, A. I., Sato, N., and Komatsubara, T. (1993) Possible formation of a nonuniform superconducting state in the heavy-fermion compound UPd$_2$Al$_3$, *Phys. Rev. Lett.* **70**, 501.

Gruenberg, L. and Gunther, L. (1966) Fulde-Ferrell Effect in Type-II Superconductors, *Phys. Rev. Lett.* **16**, 996.

Hahn, E. L. (1956) Spin Echoes, *Phys. Rev.* **80**, 580.

Huxley, A. D., Paulson, C., Laborde, O., Tholence, J. L., Sanchez, D., Junod, A., and Calemczuk, R. (1993) Flux pinning, specific heat and magnetic properties of the laves phase superconductor CeRu$_2$, *J. Phys. Condens. Matter* **5**, 7709.

Izawa, K., Yamaguchi, H., Matsuda, Y., Shishido, H., Settai, R., and Onuki, Y. (2001) Angular Position of Nodes in the Superconducting Gap of Quasi-2D Heavy-Fermion Superconductor CeCoIn$_5$, *Phys. Rev. Lett.* **87**, 057002.

Kakuyanagi, K., Kumagai, K., and Matsuda, Y. (2002) Quasiparticle excitation in and around the vortex core of underdoped YBa$_2$Cu$_4$O$_8$ studied by site-selective NMR, *Phys. Rev. B* **65**, 060503.

Kakuyanagi, K., Kumagai, K., Matsuda, Y., and Hasegawa, M. (2003) Antiferromagnetic Vortex Core in Tl$_2$Ba$_2$CuO$_{6+\delta}$ Studied by Nuclear Magnetic Resonance, *Phys. Rev. Lett.* **90**, 197003.

Kakuyanagi, K., Saitoh, M., Kumagai, K., Takashima, S., Nohara, M., Takagi, H., and Matsuda, Y. (2005) Texture in the Superconducting Order Parameter of CeCoIn$_5$ Revealed by Nuclear Magnetic Resonance, *Phys. Rev. Lett.* **94**, 047602.

Kohori, Y., Yamato, Y., Iwamoto, Y., Kohara, T., Bauer, E., Maple, M., and Sarrao, J. (2001) NMR and NQR studies of the heavy fermion superconductors CeTIn$_5$ (T=Co and Ir), *Phys. Rev. B* **64**, 134526.

Larkin, A. and Ovchinnikov, Y. (1965) Inhomogeneous state of superconductors, *Sov. Phys. JETP* **20**, 762.

Maki, K. (1966) Effect of Pauli Paramagnetism on Magnetic Properties of High-Field Superconductors, *Phys. Rev.* **148**, 362.

Mizushima, T., Machida, K., and Ichioka, M. (2005) Topological Structure of a Vortex in the Fulde-Ferrell-Larkin-Ovchinnikov State, *Phys. Rev. Lett.* **95**, 117003.

Movshovich, R., Jaime, M., Thompson, J. D., Petrovic, C., Fisk, Z., Pagliuso, P. G., and Sarrao, J. L. (2001) Unconventional Superconductivity in CeIrIn$_5$ and CeCoIn$_5$: Specific Heat and Thermal Conductivity Studies, *Phys. Rev. Lett.* **86**, 5152.

Norman, M. (1993) Existence of the FFLO state in superconducting UPd$_2$Al$_3$, *Phys. Rev. Lett.* **71**, 3391.

Oremeno, R., Sibley, A., Gough, C. E., Sebastian, S., and Fisher, I. R. (2002) Microwave Conductivity and Penetration Depth in the Heavy Fermion Superconductor CeCoIn$_5$, *Phys. Rev. Lett* **88**, 047005.

Paglione, J., Tanatar, M. A., Hawthorn, D. G., Boaknin, E., Hill, R. W., Ronning, F., Sutherland, M., Taillefer, L., Petrovic, C., and Canfield, P. C. (2003) Field-Induced Quantum Critical Point in CeCoIn$_5$, *Phys. Rev. Lett.* **91**, 246405.

Petrovic, C., Pagliuso, P. G., Hundley, M. F., Movshovich, R., Sarrao, J. L., Thompson, J. D., Fisk, Z., and Monthoux, P. (2001) Heavy-fermion superconductivity in CeCoIn$_5$ at 2.3 K, *J. Phys.: Condensed Matter* **13**, L337 – L343.

Radovan, H. A., Fortune, N. A., Murphy, T. P., Hannahs, S. T., Palm, E. C., and Hall, S. W. T. D. (2003) Magnetic enhancement of superconductivity from electron spin domains, *Nature* **425**, 51.

Shishido, H., Settai, R., D. Aoki, S. I., Nakawaki, H., Nakamura, N., Iizuka, T., Inada, Y., Sugiyama, K., Takeuchi, T., Kindo, K., Kobayashi, T., Haga, Y., Harima, H. Aoki, Y., Namiki, T., Sato, H., and Onuki, Y. (2002) Fermi Surface, Magnetic and Superconducting Properties of LaRhIn5 and CeTIn5 (T: Co, Rh and Ir), *J. Phys. Soc. Jpn.* **71**, 162.

Tachiki, M., Takahashi, S., Gegenwart, P., Weiden, M., Lang, M., Geibel, C., Steglich, F., Modler, R., Paulsen, C., and Onuki, Y. (1996) Generalized Fulde-Ferrell-Larkin-Ovchinnikov state in heavy-fermion and intermediate-valence systems, *Z. Phys. B* **100**, 369.

Tayama, T., Harita, A., Shishido, T. S. Y. H., Settai, R., and Onuki, Y. (2002) Unconventional heavy-fermion superconductor CeCoIn$_5$: DC magnetization study at temperatures down to 50 mK, *Phys. Rev. B* **65**, 180504.

Tenya, K., Ikeda, M., Tayama, T., Mitamura, H., Amitsuka, H., Sakakibara, T., Maezawa, K., Kimura, N., Settai, R., and Onuki, Y. (1995) Anomaly of Magnetization in the Superconducting Mixed State of UPt$_3$, *J. Phys. Soc. Jpn.* **64**, 1063.

Watanabe, T., Kasahara, Y., Izawa, K., Sakakibara, T., Matsuda, Y., van der Beek, C. J., Hanaguri, T., Shishido, H., Settai, R., and Onuki, Y. (2004) High-field state of the flux-line lattice in the unconventional superconductor CeCoIn$_5$, *Phys. Rev. B* **70**, 020506.

Yoshida, K. (1958) Paramagnetic Susceptibility in Superconductors, *Phys. Rev.* **110**, 769.

MODELS OF SUPERCONDUCTIVITY IN Sr_2RuO_4

Thomas Dahm (dahm@uni-tuebingen.de)
Institut für Theoretische Physik, Universität Tübingen, Auf der Morgenstelle 14, D-72076 Tübingen, Germany

Hyekyung Won
Department of Physics, Hallym University, Chunchon 200-702, South Korea

Kazumi Maki
Department of Physics and Astronomy, University of Southern California, Los Angeles, CA 90089-0484, U.S.A.

Abstract. Experimental data on purest Sr_2RuO_4 single crystals clearly indicate the presence of nodes in the superconducting order parameter. Here, we consider one special p-wave order parameter symmetry and two two-dimensional f-wave order parameter symmetries having nodes within the RuO_2 plane. These states reasonably describe both specific heat and penetration depth data. We calculate the thermal conductivity tensor for these three states and compare the results with thermal conductivity data. The state most consistent with both thermodynamic and thermal conductivity data turns out to be the chiral f-wave state with horizontal nodes.

Key words: Ruthenates, Unconventional Superconductivity, Thermodynamic Properties, Transport Properties

1. Introduction

The superconductivity in Sr_2RuO_4 has been interpreted in terms of a p-wave triplet superconducting state having a full energy gap (Maeno et al., 1994; Maeno, 1997; Rice and Sigrist, 1995; Sigrist et al., 1999). For example, the spontaneous spin polarization seen by muon spin rotation experiments (Luke et al., 1998) and the flat Knight-shift seen by nuclear magnetic resonance (NMR) (Ishida et al., 1998) are consistent with spin triplet pairing. However, specific heat data(Nishizaki et al., 2000; Nishizaki et al., 1999), NMR data (Ishida et al., 2000) and the superfluid density (Bonalde et al., 2000) of purest single crystals of Sr_2RuO_4 with $T_c \lesssim 1.5K$ clearly show low temperature behavior consistent with nodes in the order parameter very similar to observations of d-wave superconductivity in the high-T_c cuprate superconductors (Hardy et al., 1993; Won and Maki, 1994).

K. Scharnberg and S. Kruchinin (eds.),
Electron Correlation in New Materials and Nanosystems, 251–261.
© 2007 *Springer.*

Here we shall study three examples of two-dimensional (2D) superconducting order parameters with spin triplet pairing having nodes within the RuO_2 a-b plane. The first one is the anisotropic p-wave state proposed by Miyake and Narikiyo (Miyake and Narikiyo, 1999) with $\Delta(\vec{k}) \propto \sin(k_x a) \pm i\sin(k_y a)$. Here, a denotes the lattice constant of the RuO_2 square lattice. In order to have a node with this state, however, we have to stretch the Fermi wavevector k_F towards the particular value of $k_F a = \pi$, while a more realistic value would be $k_F a = 2.7$ as judged from bandstructure calculations (Mazin and Singh, 1997). In the following we will denote this particular p-wave state as the *nodal* p-wave state. As the second and third example we consider the planar f-wave states proposed by Hasegawa et al. (Hasegawa et al., 2000). Here, the angular ϕ dependence of the order parameter is given by $\Delta(\vec{k}) \propto \cos(2\phi)e^{\pm i\phi}$ and $\Delta(\vec{k}) \propto \sin(2\phi)e^{\pm i\phi}$, respectively.

In principle, these states could couple with each other, since they belong to the same representation of the point group of the square lattice. This coupling, in principle, could remove the nodes of these states. Since at present not much is known about this coupling, however, we shall confine ourselves to the states having nodes, since these provide the best description of the experimental data mentioned above.

Within circular symmetric weak-coupling BCS theory one immediately realizes that the thermodynamics of the planar f-wave states is identical to the one of d-wave superconductors (Won and Maki, 1994). We have worked out the thermodynamics of the anisotropic, nodal p-wave state here as well. In Figs. 1 and 2 we show our results for the temperature dependence of the specific heat $C_s/\gamma T$ and the superfluid density $\rho_s(T)$ for the nodal p-wave and the 2D f-wave states together with the experimental data. For comparison, we also show the results of a 3D f-wave state, considered by two of us (Maki and Yang, 1999; Won and Maki, 2000a). As is readily seen, the 2D f-wave states give a better description of the experimental data than the 3D f-wave or the nodal p-wave states, though the differences between the 2D f-wave and the 3D f-wave states are rather small.

Recently the thermal conductivity of Sr_2RuO_4 in a planar magnetic field has been studied (Tanatar et al., 2000; Tanatar et al., 2001; Tanatar and Maeno, 2000; Izawa et al., 2001; Izawa and Matsuda, 2000). Both groups studied the thermal conductivity parallel to the a-axis in a magnetic field within the a-b plane in a direction tilted by an angle θ from the heat current. Both groups found no appreciable angular dependence. This experimental result is already inconsistent with the isotropic p-wave state having a full energy gap and the 3D f-wave state (Maki and Yang, 1999).

Indeed, we shall show that the thermal conductivity data is inconsistent with two of the nodal states considered here, because they exhibit rather large angular dependence.

In the next section we briefly summarize the thermodynamic properties of the nodal p-wave superconductor with $\Delta(\vec{k}) \propto \sin(k_x a) \pm i \sin(k_y a)$. In many respects the results are very similar to the ones for d-wave superconductors (Won and Maki, 1994) and 3D f-wave superconductors (Maki and Yang, 1999). Then we proceed to consider the thermal conductivity in a planar magnetic field. The result for the 2D f-wave state with angular dependence $\cos(2\phi)e^{\pm i\phi}$ is very similar to the one in d-wave superconductors discussed in Ref. (Won and Maki, 2000b).

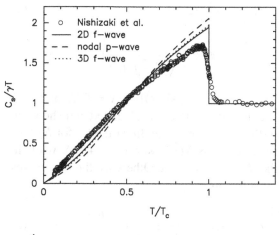

Figure 1. The specific heat $C_s/\gamma T$ as a function of T/T_c for the 2D f-wave states (solid line) and the nodal p-wave state (dashed line) considered in this work. Also shown are the experimental data by Nishizaki et al. (Nishizaki et al., 2000) (circles) and the 3D f-wave state considered in Ref. (Maki and Yang, 1999) (dotted line).

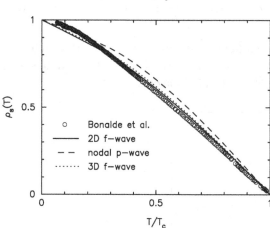

Figure 2. The superfluid density $\rho_s(T)$ as a function of T/T_c for the 2D f-wave states (solid line) and the nodal p-wave state (dashed line) together with the experimental data by Bonalde et al. (Bonalde et al., 2000) (circles) and the 3D f-wave state considered in Ref. (Maki and Yang, 1999) (dotted line).

2.   Thermodynamics of the nodal p-wave superconductor

We consider the superconducting order parameter given by $\vec{\Delta}(\vec{k}) = \hat{d}\frac{\Delta}{s_M}$ [$\sin(k_x a) \pm i \sin(k_y a)$] with $k_x a = \pi \cos(\phi)$ and $k_y a = \pi \sin(\phi)$ and the normalization $s_M = \sqrt{2}\sin\left(\frac{\pi}{\sqrt{2}}\right) = 1.125$. This is the model proposed by Miyake and

Narikiyo (1999) except that we have chosen the Fermi wavevector $k_F a = \pi$ in order to have a node in $\vec{\Delta}(\vec{k})$. The quasi-particle Green function in Nambu representation is given by

$$G(k, \omega) = (i\omega - \xi_k \rho_3 - \Delta(k)\rho_1 \sigma_1)^{-1} \tag{1}$$

where $\Delta(k) = \frac{\Delta}{s_M}[\sin(k_x a) \pm i \sin(k_y a)]$.

Then the quasi-particle density of states is given by

$$N(E)/N_0 = \mathrm{Re}\left\langle \frac{E}{\sqrt{E^2 - \Delta^2(k)}} \right\rangle \tag{2}$$

$$= \frac{4}{\pi} y \int_0^{\pi/4} d\phi \, \mathrm{Re}\left(\frac{1}{\sqrt{y^2 - f^2(\phi)}} \right)$$

where $f^2(\phi) = s_M^{-2}(1 - [\cos(2\pi \cos \phi) + \cos(2\pi \sin \phi)]/2)$, $y = E/\Delta$, and $\langle \cdots \rangle$ denotes an angular average. The density of states is calculated and shown in Fig. 3 together with the one for the 2D f-wave case. In particular for $E/\Delta \ll 1$, the density of states increases linearly as $N(E)/N_0 \simeq 0.7164 E/\Delta$, while in the 2D f-wave case it varies like $N(E)/N_0 \simeq E/\Delta$. Otherwise the two curves look very similar. Here, the gap equation

$$\lambda^{-1} = \langle f^2 \rangle^{-1} \int_0^{E_c} dE \left\langle \frac{f^2}{\sqrt{E^2 - \Delta^2 f^2(\phi)}} \right\rangle \tanh\left(\frac{E}{2T} \right) \tag{3}$$

has been solved numerically. In particular we find $\Delta(0)/T_c = 2.00$, which has to be compared with 2.14 in the 2D f-wave case. Both of these values are consistent with estimates from NMR data (Ishida et al., 2000).

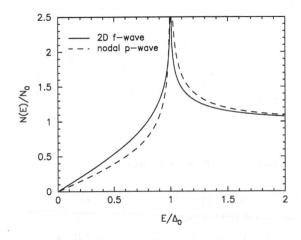

Figure 3. The density of states for the 2D f-wave state (solid line) and the nodal p-wave state (dashed line).

The entropy S is obtained from

$$S = -4 \int_0^\infty dE N(E) [f \ln f + (1 - f) \ln(1 - f)] \tag{4}$$

with f being the Fermi function and $\beta = 1/T$. Then the specific heat $C_s(T)$ is given by

$$C_s(T) = T \frac{dS(T)}{dT} \tag{5}$$

$C_s(T)/\gamma T$ has been shown in Fig. 1.

Finally, the superfluid density $\rho_s(T)$ is given by

$$\rho_s(T) = 1 - \frac{\beta}{2} \int_0^\infty dE \frac{N(E)}{N_0} \text{sech}^2 \left(\frac{\beta E}{2} \right) \tag{6}$$

which behaves almost linearly in T and is shown in Fig. 2. We note that at low temperatures an expansion of $\rho_s(T)$ leads to $\rho_s(T) = 1 - 2 \ln 2 \times 0.7164 \frac{T}{\Delta} + \cdots$.

3. Thermal conductivity tensor in the a-b plane

As shown in earlier experiments on YBCO, the thermal conductivity tensor in a planar magnetic field is very sensitive to the nodal directions (Salamon et al., 1995; Yu et al., 1995; Aubin et al., 1997) and thus may be used to further discriminate between the states studied above. We shall consider the thermal conductivity tensor in the vortex state of the nodal p-wave and the 2D f-wave separately.

3.1. NODAL P-WAVE STATE

The necessary theoretical scheme, neglecting vortex core scattering, has been worked out in the past (Won and Maki, 2000b; Kübert and Hirschfeld, 1998b; Kübert and Hirschfeld, 1998a; Vekhter and Hirschfeld, 2000; Hirschfeld, 1998; Barash and Svidzinskii, 1998). We just apply this method to the present case. In particular for $\frac{H}{H_{c2}}, \frac{T^2}{\Delta^2} \ll 1$ in the superclean limit we obtain

$$\begin{aligned} \kappa_{xx}/\kappa_n &= \frac{2}{\pi} \left(\frac{2 s_M}{\pi} \right)^2 \langle (1 + \cos(2\phi)) x \rangle \langle x \rangle \\ &= \frac{2}{\pi} \left(\frac{2 s_M}{\pi} \right)^2 \frac{vv' e H}{\Delta^2} F(\theta) \end{aligned} \tag{7}$$

where v and v' are the Fermi velocities within the a-b plane and perpendicular to it, respectively, and $x = |\mathbf{v} \cdot \mathbf{q}|/\Delta$ denotes the Doppler shift due to the superflow around the vortex (see Won and Maki (2000b)). $\kappa_n = \frac{\pi^2 T n}{6\Gamma m}$ is the normal state thermal conductivity. The function $F(\theta)$ is given by

$$
\begin{aligned}
F(\theta) = \frac{2}{\pi^2} \sqrt{1 + \sin^2 \theta} E\left(\frac{1}{\sqrt{1 + \sin^2 \theta}}\right) \\
\times \left(\sqrt{1 + \sin^2 \theta} E\left(\frac{1}{\sqrt{1 + \sin^2 \theta}}\right) \right. \\
\left. + \sqrt{1 + \cos^2 \theta} E\left(\frac{1}{\sqrt{1 + \cos^2 \theta}}\right) \right)
\end{aligned}
\tag{8}
$$

with E being the complete elliptic integral of the second kind and

$$
\kappa_{xy}/\kappa_n = 0 \tag{9}
$$

In the present situation there will be no off-diagonal component, because the heat current is parallel to the nodal direction. The angular dependence of κ_{xx} is given by the function $F(\theta)$, which is shown in Fig. 4. Surprisingly, κ_{xx} has a broad maximum for $\theta = \pi/2$. Also, the anisotropy $\kappa_{xx}(\pi/2)/\kappa_{xx}(0) = 1.910$ is quite strong. Therefore, in view of the thermal conductivity experimental data (Tanatar et al., 2000; Tanatar et al., 2001; Tanatar and Maeno, 2000; Izawa et al., 2001; Izawa and Matsuda, 2000), we have to reject this possibility.

3.2. 2D F-WAVE STATE

As already mentioned, we consider the two states $\sin(2\phi)e^{\pm i\phi}$ and $\cos(2\phi)e^{\pm i\phi}$. The order parameter $\propto \sin(2\phi)e^{\pm i\phi}$ has the same nodal structure as the nodal p-wave state studied in the last subsection and has been studied by Graf and Balatsky (Graf and Balatsky, 2000). Following the same procedure as above we find for the state $\sin(2\phi)e^{\pm i\phi}$

$$
\kappa_{xx}/\kappa_n = \frac{2}{\pi}\langle (1 + \cos(2\phi)) x\rangle\langle x\rangle = \frac{2}{\pi}\frac{vv'eH}{\Delta^2}F(\theta) \tag{10}
$$

and

$$
\kappa_{xy} = 0 \tag{11}
$$

where $F(\theta)$ has been shown in Fig. 4. This is the same angular dependence as for the nodal p-wave state only with a larger amplitude.

Therefore, also the state $\sin(2\phi)e^{\pm i\phi}$ gives a rather large θ dependence, which is inconsistent with the existent experiments (Tanatar et al., 2000; Tanatar et al., 2001; Tanatar and Maeno, 2000; Izawa et al., 2001; Izawa and Matsuda, 2000). Finally, let us consider the state $\cos(2\phi)e^{\pm i\phi}$, which has its

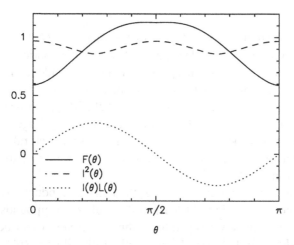

Figure 4. Angular variation of the functions $F(\theta)$, $I^2(\theta)$, and $I(\theta)L(\theta)$ as defined in Eqs. (8), (14), and (15). The angular variation of the thermal conductivity κ_{xx} for the nodal p-wave state, as given by the function $F(\theta)$, is much stronger than for the $\cos(2\phi)e^{\pm i\phi}$ f-wave state ($I^2(\theta)$) due to the different position of the nodes in the gap function.

nodes along the zone diagonal. As already noted, this state has the same thermodynamics as a d-wave superconductor. Further, the thermal conductivity tensor is now given by (Won and Maki, 2000b)

$$\kappa_{xx}/\kappa_n = \frac{2}{\pi}\frac{vv'eH}{\Delta^2}I^2(\theta) \tag{12}$$

and

$$\kappa_{xy}/\kappa_n = -\frac{2}{\pi}\frac{vv'eH}{\Delta^2}I(\theta)L(\theta) \tag{13}$$

where

$$I(\theta) = \frac{1}{\pi}\left(\sqrt{\frac{3+s}{2}}E\left(\sqrt{\frac{2}{3+s}}\right)\right.$$
$$\left. + \sqrt{\frac{3-s}{2}}E\left(\sqrt{\frac{2}{3-s}}\right)\right) \tag{14}$$

and

$$L(\theta) = \frac{1}{\pi}\left(\sqrt{\frac{3+s}{2}}E\left(\sqrt{\frac{2}{3+s}}\right)\right.$$
$$\left. - \sqrt{\frac{3-s}{2}}E\left(\sqrt{\frac{2}{3-s}}\right)\right) \tag{15}$$

with $s = \sin(2\theta)$.

In Fig. 4 the angular dependences of the functions $(I(\theta))^2$ and $I(\theta)L(\theta)$ are shown together with $F(\theta)$. Thus, as in d-wave superconductors, this state exhibits a fourfold symmetry in κ_{xx}. The angular dependence is about 10% and may be compatible with the experiments(Tanatar et al., 2000; Tanatar et al., 2001; Tanatar and Maeno, 2000; Izawa et al., 2001; Izawa and Matsuda, 2000). However, the small fourfold term and a relatively large twofold term observed in the experiments are more consistent with the chiral f-wave state with horizontal line nodes as has been discussed by two of us (Maki and Yang, 1999; Won and Maki, 2000a).

As mentioned above, our analysis of the thermal conductivity tensor neglects vortex core scattering. At least in the high-T_c compounds in a small magnetic field and at low temperatures this contribution can be neglected (Kübert and Hirschfeld, 1998a; Chiao et al., 1999). In Sr_2RuO_4 the vortex core size is larger and at present it is unclear to what extent this contribution plays a rôle. It has been shown by Vekhter and Houghton (Vekhter and Houghton, 1999), however, that vortex core scattering plays a minor role at low temperatures in d-wave superconductors, because most of the excited quasiparticles are located near the gap nodes and we expect the same to happen here.

As an additional check on the position of the nodes of the order parameter we propose a measurement of the transverse thermal conductivity κ_{xy}. As Eqs. (9) and (11) show, κ_{xy} vanishes if the nodes lie along the a or b directions. However, we expect a finite transverse thermal conductivity κ_{xy} showing a $\sin(2\theta)$ variation for the $\cos(2\phi)e^{\pm i\phi}$ f-wave state having its nodes along the zone diagonal, as Eq. (13) shows.

Tewordt and Fay have studied the influence of impurity scattering on the thermal conductivity in the vortex state and found that it tends to reduce the angular dependence (Tewordt and Fay, 2001). Since the Fermi surface of Sr_2RuO_4 consists of three bands it has been suggested that the gap structure may be different in the three bands (Zhitomirsky and Rice, 2001; Annett et al., 2002; Deguchi et al., 2004a; Deguchi et al., 2004b). A criticism of these multigap models and a recent review can be found in the work by Won et al. (2005) and Maki et al. (2005).

4. Conclusions

We compared one nodal p-wave and two 2D nodal f-wave superconducting states with experimental data from purest crystals of Sr_2RuO_4. We find that within weak-coupling theory the two 2D f-wave states give the closest description of the thermodynamic data. Unfortunately, none of them is satisfactory with respect to thermal conductivity data. A relatively large twofold term in the thermal conductivity κ_{xx} seen by Izawa et al. is only consistent with the chiral f-wave state with horizontal nodes discussed elsewhere.

Acknowledgements

We thank Y. Maeno and I. Bonalde for providing us with the digital form of their experimental data, which were used in Fig. 1 and Fig. 2.

We also thank K. Izawa, Y. Maeno, Y. Matsuda, and M. A. Tanatar for fruitful discussions on their ongoing experiments. One of us (KM) thanks for the hospitality of N. Schopohl and the University of Tübingen where part of this work has been done. HW acknowledges support from the Korea Research Foundation under the Professor Dispatching Scheme.

References

Annett, J. F., Litak, G., Györffy, B. L., and Wysokinski, K. I. (2002) Interlayer coupling and *p*-wave pairing in strontium ruthenate, *Phys. Rev. B* **66**, 134514.

Aubin, H., Behnia, K., Ribault, M., Gagnon, R., and Taillefer, L. (1997) Angular Position of Nodes in the Superconducting Gap of YBCO, *Phys. Rev. Lett.* **78**, 2624.

Barash, Y. S. and Svidzinskii, A. A. (1998) Nonmonotonic magnetic-field dependence and scaling of the thermal conductivity for superconductors with nodes of the order parameter, *Phys. Rev. B* **58**, 6476.

Bonalde, I., Yanoff, B. D., Salamon, M. B., Harlingen, D. J. V., Chia, E. M. E., Mao, Z. Q., and Maeno, Y. (2000) Temperature Dependence of the Penetration Depth in Sr$_2$RuO$_4$: Evidence for Nodes in the Gap Function, *Phys. Rev. Lett.* **85**, 4775.

Chiao, M., Hill, R. W., Lupien, C., Popić, B., Gagnon, R., and Taillefer, L. (1999) Quasiparticle Transport in the Vortex State of YBa$_2$Cu$_3$O$_{6.9}$, *Phys. Rev. Lett.* **82**, 2943.

Deguchi, K., Mao, Z. Q., and Maeno, Y. (2004a) Determination of the Superconducting Gap Structure in All Bands of the Spin-Triplet Superconductor Sr$_2$RuO$_4$, *J. Phys. Soc. Jpn.* **73**, 1313.

Deguchi, K., Mao, Z. Q., and Maeno, Y. (2004b) Gap Structure of the Spin-Triplet Superconductor Sr$_2$RuO$_4$ Determined from the Field-Orientation Dependence of the Specific Heat, *Phys. Rev. Lett.* **92**, 047002.

Graf, M. J. and Balatsky, A. V. (2000) Identifying the pairing symmetry in the Sr$_2$RuO$_4$ superconductor, *Phys. Rev. B* **62**, 9697.

Hardy, W. N., Bonn, D. A., Morgan, D. C., Liang, R., and Zhang, K. (1993) Precision measurements of the temperature dependence of λ in YBa$_2$Cu$_3$O$_{6.95}$: Strong evidence for nodes in the gap function, *Phys. Rev. Lett.* **70**, 3999.

Hasegawa, Y., Machida, K., and Ozaki, M. (2000) Spin-Triplet Superconductivity with Line Nodes in Sr$_2$RuO$_4$, *J. Phys. Soc. Jpn.* **69**, 336.

Hirschfeld, P. J. (1998) Quasiparticle transport in the vortex state of unconventional superconductors, *J. Kor. Phys. Soc.* **33**, 485.

Ishida, K., Mukuda, H., Kitaoka, Y., Asayama, K., Ma, Z. Q., Mori, Y., and Maeno, Y. (1998) Spin-triplet superconductivity in Sr$_2$RuO$_4$ identified by [17]O Knight shift, *Nature* **396**, 658.

Ishida, K., Mukuda, H., Kitaoka, Y., Mao, Z. Q., Mori, Y., and Maeno, Y. (2000) Anisotropic Superconducting Gap in the Spin-Triplet Superconductor Sr_2RuO_4: Evidence from a Ru-NQR Study, *Phys. Rev. Lett.* **84**, 5387.

Izawa, K. and Matsuda, Y. (2000), private communication.

Izawa, K., Takahashi, H., Yamaguchi, H., Matsuda, Y., Suzuki, M., Sasaki, T., Fukase, T., Yoshida, Y., Settai, R., and Onuki, Y. (2001) Superconducting Gap Structure of Spin-Triplet Superconductor Sr_2RuO_4 Studied by Thermal Conductivity, *Phys. Rev. Lett.* **86**, 2653.

Kübert, C. and Hirschfeld, P. J. (1998a) Quasiparticle Transport Properties of d-Wave Superconductors in the Vortex State, *Phys. Rev. Lett.* **80**, 4963.

Kübert, C. and Hirschfeld, P. J. (1998b) Vortex contribution to specific heat of dirty d-wave superconductors: breakdown of scaling, *Solid State Comm.* **105**, 459.

Luke, G., Fudamoto, Y., Kojima, K. M., Larkin, M. I., Merrin, J., Nachumi, B., Uemura, Y. J., Maeno, Y., Mao, Z. Q., Mori, Y., Nakamura, H., and Sigrist, M. (1998) Time-reversal symmetry-breaking superconductivity in Sr_2RuO_4, *Nature* **394**, 558.

Maeno, Y. (1997) Electronic states of the superconductor Sr_2RuO_4, *Physica C* **282**, 206.

Maeno, Y., Hashimoto, H., Yoshida, K., Nishizaki, S., Fujita, T., Bednorz, J. G., and Lichtenberg, F. (1994) Superconductivity in a layered perovskite without copper, *Nature* **372**, 532.

Maki, K., Haas, S., Parker, D., and Won, H. (2005) Perspectives on Nodal Superconductors, *Chinese J. Phys.* **43**, 532.

Maki, K. and Yang, G. (1999) Introduction to unconventional superconductivity, *Fizika A* **8**, 345.

Mazin, I. I. and Singh, D. J. (1997) Ferromagnetic Spin Fluctuation Induced Superconductivity in Sr_2RuO_4, *Phys. Rev. Lett.* **79**, 733, Note, however that the Fermi surface observed in angular-resolved photoemission does not fully agree with the bandstructure calculations. See Fig. 10 in (Yokoya et al., 1996).

Miyake, K. and Narikiyo, O. (1999) Model for Unconventional Superconductivity of Sr_2RuO_4: Effect of Impurity Scattering on Time-Reversal Breaking Triplet Pairing with a Tiny Gap, *Phys. Rev. Lett.* **83**, 1423.

Nishizaki, S., Maeno, Y., and Mao, Z. (2000) Changes in the Superconducting State of Sr_2RuO_4 under Magnetic Fields Probed by Specific Heat, *J. Phys. Soc. Jpn.* **69**, 572.

Nishizaki, S., Maeno, Y., and Mao, Z. (1999) Effect of impurities on the specific heat of the spin-triplet superconductor Sr_2RuO_4, *J. Low Temp. Phys.* **117**, 1581.

Rice, T. M. and Sigrist, M. (1995) Sr_2RuO_4: An electronic analogue of ^{3}He?, *J. Phys. Cond. Mat.* **7**, L643.

Salamon, M. B., Yu, F., and Kopylov, V. N. (1995) The field dependence of the thermal conductivity: evidence for nodes in the gap, *J. Supercond.* **8**, 449.

Sigrist, M., Agterberg, D., Furusaki, A., Honerkamp, C., Ng, K. K., Rice, T. M., and Zhitomirsky, M. E. (1999) Phenomenology of the superconducting state in Sr_2RuO_4, *Physica C* **317**, 134.

Tanatar, M. A. and Maeno, Y. (2000), private communication.

Tanatar, M. A., Nagai, S., Mao, Z. Q., Maeno, Y., and Ishiguro, T. (2000) Thermal conductivity study of Sr_2RuO_4 in oriented magnetic field, *Physica C* **341**, 1841.

Tanatar, M. A., Nagai, S., Mao, Z. Q., Maeno, Y., and Ishiguro, T. (2001) Anisotropy of Magnetothermal Conductivity in Sr_2RuO_4, *Phys. Rev. Lett.* **86**, 2649.

Tewordt, L. and Fay, D. (2001) Thermal conductivity near H_{c2} for spin-triplet superconducting states with line nodes in Sr_2RuO_4, *Phys. Rev. B* **64**, 024528.

Vekhter, I. and Hirschfeld, P. J. (2000) Angle-dependent magnetothermal conductivity in d-wave superconductors, *Physica C* **341**, 1947.

Vekhter, I. and Houghton, A. (1999) Quasiparticle Thermal Conductivity in the Vortex State of High-T_c Cuprates, *Phys. Rev. Lett.* **83**, 4626.

Won, H., Haas, S., Parker, D., Telang, S., Vanyolos, A., and Maki, K. (2005) BCS theory of nodal superconductors, In A. Avella and F. Mancini (eds.), *Lectures on Physics of highly Correlated Electron Systems IX*, pp. 3–43, AIP proceedings **789**, Melville.

Won, H. and Maki, K. (1994) d-wave superconductor as a model of high-T_c superconductors, *Phys. Rev. B* **49**, 1397.

Won, H. and Maki, K. (2000a) Possible f-wave superconductivity in Sr$_2$RuO$_4$, *Europhys. Lett.* **52**, 427.

Won, H. and Maki, K. (2000b) Quasiparticle spectrum in the vortex state of d-wave superconductors, cond-mat/0004105.

Yokoya, T., Chainani, A., Takahashi, T., Ding, H., Campuzano, J. C., Katayama-Yoshida, H., Kasai, M., and Tokura, Y. (1996) Angle-resolved photoemission study of Sr$_2$RuO$_4$, *Phys. Rev. B* **54**, 13311.

Yu, F., Salamon, M. B., Leggett, A. J., Lee, W. C., and Ginsberg, D. M. (1995) Tensor Magnetothermal Resistance in YBa$_2$Cu$_3$O$_{7-x}$ via Andreev Scattering of Quasiparticles, *Phys. Rev. Lett.* **74**, 5136.

Zhitomirsky, M. E. and Rice, T. M. (2001) Interband Proximity Effect and Nodes of Superconducting Gap in Sr$_2$RuO$_4$, *Phys. Rev. Lett.* **87**, 057001.

HIGH-T$_C$ SUPERCONDUCTIVITY OF CUPRATES
AND RUTHENATES

John D. Dow (catsc@cox.net)
Department of Physics, Arizona State University, Tempe, Arizona 85287-1504 U.S.A.*

Dale R. Harshman
Physikon Research Corporation, P. O. Box 1014, Lynden, Washington 98264 U.S.A.
Department of Physics, Arizona State University, Tempe, Arizona 85287-1504 U.S.A.
Department of Physics, University of Notre Dame, Notre Dame, Indiana 46556 U.S.A.

Anthony T. Fiory
Departent of Physics, New Jersey Institute of Technology, Newark, New Jersey 07102 U.S.A.

*To whom correspondence should be addressed.

Abstract. Recent muon spin rotation (μ^+SR) spectroscopy of YBa$_2$Cu$_3$O$_7$ finds that the superconductivity (i) is consistent with nodeless pairing (e.g., s-wave or extended s-wave), after fluxon-pinning is taken into account, and (ii) is well-described by a two-fluid model. These are the same results as found over a decade ago in samples with strongly pinned vortices. As compared with the two-fluid fit, the probability of the same quality fit to the recent data, assuming a d-wave model, was found to be nearly 4x10^{-6}. Clearly YBa$_2$Cu$_3$O$_7$ is a nodeless, strong-coupling superconductor.

However YBa$_2$Cu$_3$O$_7$, when measured with scanning tunneling microscopy (STM) or photoemission, apparently shows clear evidence of d-wave behavior, which comes from the Cu atoms and the cuprate-planes. The difference is that μ^+SR spectroscopy senses only the superconducting condensate, while STM and photoemission sense the layers closest to the surface (CuO, BaO, CuO$_2$) which need not all superconduct. The presence of only s-wave character in the μ^+SR superconductivity data indicates that the cuprate-planes (which contain Cu d-waves) do not superconduct, and implies that the BaO layers do. (If the cuprate-planes superconducted, significant d-wave character would be apparent in μ^+SR data.) The observed nodeless character must come from different planes than the d-like CuO$_2$-planes (or CuO chain layers), namely from the BaO layers, which produce nodeless superconductivity.

This picture, which places the superconducting hole condensate in the BaO layers of the cuprates, is lent further support by the following ruthenates (which superconduct, in their SrO layers): Cu-doped Sr$_2$YRuO$_6$ (which has no CuO$_2$ planes), and GdSr$_2$Cu$_2$RuO$_8$ and Gd$_{2-z}$Ce$_z$Sr$_2$Cu$_2$RuO$_{10}$ (whose CuO$_2$ planes are magnetic). All three of these compounds superconduct near 45 K in their SrO layers. The sister compound Ba$_2$GdRuO$_6$ does not superconduct at all (whether Cu doped or not) because it is a two-layer compound containing

K. Scharnberg and S. Kruchinin (eds.),
Electron Correlation in New Materials and Nanosystems, 263–274.
© 2007 *Springer.*

pair-breaking Gd (with $J \neq 0$) in the layer next to its BaO layer, thereby suppressing the BaO layer's potential superconductivity. Replacing Gd with non-pair-breaking Y leads to superconductivity beginning at 93 K in the BaO layers of doped Ba_2YRuO_6. $GdSr_2Cu_2RuO_8$ and $Gd_{2-z}Ce_zSr_2Cu_2RuO_{10}$ superconduct in their SrO layers because the potentially pair-breaking Gd ion is more than one layer distant from the nearest SrO plane, which contains the superconducting hole condensate. (If the superconducting holes were in the cuprate-planes, the Gd ions would destroy the hole-pairing, since the cuprate planes are adjacent to the Gd sites.)

The physics of the ruthenates and the cuprates are similar: $Gd_{2-z}Ce_zCuO_4$ does not superconduct for similar reasons to those for doped Ba_2GdRuO_6. It is the BaO, SrO, or interstitial oxygen layers that carry the supercurrent in both classes of material, not the CuO_2 layers.

Key words: Cuprate and Ruthenate Superconductors, Thermal Conductivity.

1. Introduction

Most properties of high-temperature superconductivity remain unresolved after many years, with many workers advocating models in which (1) all high-temperature superconductors contain cuprate-planes; (2) the cuprate-planes are the loci of the superconductivity [1][2][3]; (3) the hole-pairing which leads to superconductivity has d-wave symmetry [4]; (4) all layers in the unit cell superconduct more or less the same; and (5) some materials, most notably $Nd_{2-x}Ce_xCuO_4$ and Ce- or Am-doped $Pb_2Sr_2YCu_3O_8$, are n-type superconductors, while most high-temperature superconductors are p-type. All of these concepts are invalid.

2. Not all high-temperature superconductors have cuprate-planes

Not all high-temperature superconductors contain cuprate-planes (CuO_2). The most notable case is Ba_2YRuO_6 doped with about 1% Cu, which has a superconducting onset temperature of 93 K [5]. This material contains a few Cu atoms as isolated dopants; at 1% concentration, the Cu is too dilute to form cuprate-planes. Another, more studied example, is doped Sr_2YRuO_6, with a $T_{c,onset}$ of about 49 K [6] and the same structure as Ba_2YRuO_6 (Fig. 1). Cuprate-planes are clearly not needed for high-temperature superconductivity.

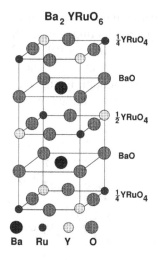

$Ba_2 YRuO_6$

Figure 1. Crystal structure of Ba_2YRuO_6.

3. The cuprate superconductor $YBa_2Cu_3O_7$

3.1. THE CUPRATE-PLANES ARE NOT THE LOCI OF SUPERCONDUCTIVITY

Neutron diffraction studies of $YBa_2Cu_3O_x$ (and $PrBa_2Cu_3O_x$) have shown that the charges on the various planes in these two compounds are actually almost the same [7]. (See the crystal structure of $YBa_2Cu_3O_7$ in Fig. 2.)

The CuO chain layers have positive charges, which decrease with composition from x=6 to x=7. Therefore the CuO chain layers cannot be the primary superconducting layers because their positive charges decrease with x as the superconducting critical temperature T_c increases (the superconductivity turns on with x increasing above x=6.4).

The Y ion is basically trivalent Y^{+3}, and does not superconduct by itself. The BaO and CuO_2 layers are the only remaining candidates for carrying the superconductivity. (See Fig. 3.) The BaO layers each have a slightly positive net charge at the onset of superconductivity (between x=6.4 and 6.5) and the positive charge increases for larger x, while the two CuO_2 planes each have a negative charge of about –2, which increases with x.

The cuprate-planes (which many researchers regard as superconducting [8]) are the only planes that have a net negative charge, and those planes definitely have a significant Cu character. When studied by μ^+SR, which is sensitive only to the superconducting layers, those layers do not show a measureable Cu (*i.e.*, *d*-wave) character [9][10]. These facts suggest that the cuprate-planes do *not* superconduct significantly, and are not responsible for the observed *p*-type strong-coupling superconductivity. Hence the BaO layers have to be the primary superconducting layers in $YBa_2Cu_3O_7$ [9][10].

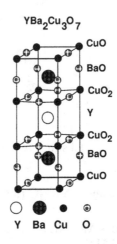

Figure 2. Crystal structure of $YBa_2Cu_3O_7$.

3.2.   THE INADEQUACY OF *d*-WAVE PAIRING THEORY FOR $YBa_2Cu_3O_7$

Figure 1 of [10] is reproduced as our Fig. 4, and shows rather clearly that the *d*-wave theory *alone* for $\lambda(T,H)$, the effective magnetic penetration depth, does not fit the data extracted from muon spectroscopy. (Neither does the *s*-wave theory *alone*.) Moreover, the *d*-wave theory does not fit even the earlier data of Sonier *et al.*, as Amin *et al.* remarked in [11]: Amin *et al.* found poor agreement with non-local *d*-wave theory and actually commented about the inadequacy of *d*-wave theory.

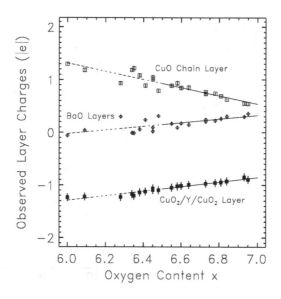

Figure 3. Layer charges of YBa$_2$Cu$_3$O$_x$ (which are virtually the same as for PrBa$_2$Cu$_3$O$_x$) against oxygen content x, after [23]. Note that the only layers that can possibly superconduct are the BaO and CuO$_2$ layers which have the required positive slope with x.

Sonier *et al.* are currently the main μ^+SR experimenters advocating *d*-wave pairing; they have ignored both flux-pinning and *s*-wave theory as they have claimed that *d*-wave theory alone (without any flux-pinning) fits their data. They have also ignored the many problems with *d*-wave theory and continue (to this day) to advocate a *d*-wave explanation of the muon data, without achieving agreement between the theory and the data, or having adequate data to support their claim. The fact is that neither *d*-wave pairing theory *alone* (as Sonier claims) nor *s*-wave pairing theory *alone* can fit the μ^+SR data [10].

3.3. ACCOUNTING FOR FLUXON PINNING AND FLUXON RE-ORDERING

In this section, we show that by first taking fluxon-pinning and fluxon re-ordering into account, the residual μ^+SR spectra are well described by a nodeless pairing model, thereby contradicting the possibility of *d*-wave pairing. (Sonier *et al.* completely ignore fluxon-pinning and fluxon re-ordering, and instead mis-interpret the data in terms of *d*-wave pairing, which is why they were unable to obtain a satisfactory fit to the μ^+SR data.)

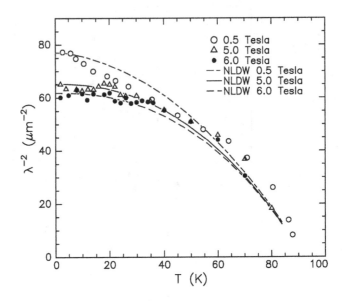

Figure 4. Comparison of the *d*-wave theory for the effective magnetic penetration depth λ (T,H) of [11] with the fitted data of Figure 3 of [11]. "NLDW" stands for non-linear *d*-wave.

After accounting for fluxon-pinning and fluxon re-ordering in the μ^+SR data one can extrapolate the temperature-dependence of the zero-field penetration depth. As shown in Fig. 5, the data for various magnetic fields collapse onto each other [10], with the curve through the data corresponding to the two-fluid model. Compared with the two-fluid model [10], a comparable fit with the *d*-wave pairing model (again, after accounting for fluxon-pinning) was found to have a probability of nearly 4×10^{-6}.

Clearly, fluxon-pinning and re-ordering must be taken into account first, and then the data show that the superconductivity is node-less, consistent with *s*-wave or extended *s*-wave pairing.

This result is certainly true for YBa$_2$Cu$_3$O$_7$, and appears to be true for Bi$_2$Sr$_2$CaCu$_2$O$_8$, although the conclusion for Bi$_2$Sr$_2$CaCu$_2$O$_8$ may not be as strong as for YBa$_2$Cu$_3$O$_7$ (due to the larger uncertainty in the measurements). Consequently, it is reasonable to posit that the nodeless pairing (even with flux-pinning) accounts for the muon data of high-temperature superconductors.

The superconductivity in YBa$_2$Cu$_3$O$_7$ comes from the BaO layers, not from the cuprate-planes, and the *d*-wave behavior measured both in scanning tunneling microscopy (STM) and in photo-emission originates in the *non-superconducting* cuprate-planes. This raises the question of whether we can explain measurements such as specific heat and thermal conductivity, with our nodeless superconductivity model.

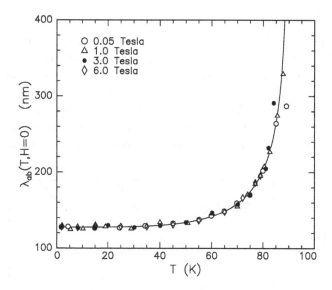

Figure 5. Muon penetration depth λ $_{ab}$ versus temperature T, with the fitted effects of pinning removed from the data, in the *s*-wave pairing model λ $_{ab}$(T,H=0)=λ $_{ab}$(T=0,H=0) [1-(T/T$_c$)^4]$^{-1/2}$. After [10].

3.4. SPECIFIC HEAT

Moler *et al.* have measured the specific heat of YBa$_2$Cu$_3$O$_7$ [12], and find an excess specific heat which can be explained if at least one type of plane is not superconducting, *e.g.*, if the electrical carriers in the CuO$_2$ planes are normal rather than superconducting. They acknowledge that their data are consistent with a gap function having no nodes (although they endorse a model with nodes that does not completely describe their data [12]). In particular they have observed a "zero-field linear-T term" in their specific heat that is not entirely consistent with a superconductor having a gap function with lines of nodes (their preferred model). A zero-field linear-T term in the specific heat data would occur if the holes in the CuO$_2$ planes are normal, while the holes in other layers (*e.g.*, the BaO layers) are superconducting. Hence the experimental results of Moler *et al.* support the idea that we propose: the superconductivity is *s*-wave or extended *s*-wave in character and in the BaO planes, not *d*-wave and not in the cuprate-planes.

3.5.   THERMAL CONDUCTIVITY

The thermal conductivity of $YBa_2Cu_3O_y$ has been neasured by Taillefer *et al.* [13] for both fully insulating material (y=6.0) and fully super-conducting material (y=6.9). Their results for the thermal conductivity divided by temperature, κ/T *versus* T^2, are displayed in Fig. 6. The important feature of this figure is the fact that κ/T is *non-zero* for $T^2 \square 0$ when the material is superconducting (y=6.9), but is zero for insulating material (y=6.0). The central question is how to interpret these data.

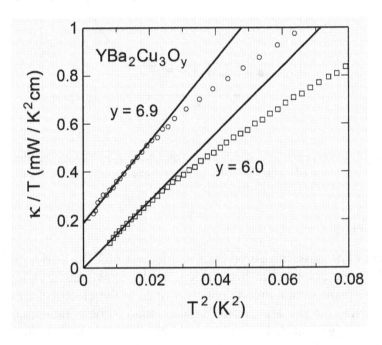

Figure 6. Thermal conductivity divided by temperature T of $YBa_2Cu_3O_y$ *versus* T^2 for insulating y=6.0 and for superconducting y=6.9, after [13].

If some electrical carriers are normal and others are superconducting, then the normal carriers are responsible for the fact that The thermal conductivity of $YBa_2Cu_3O_y$ has been measured by κ/T=0.2 (mW/K^{2}cm) in the otherwise superconducting material.

But if all holes superconduct, then the finite value of κ/T at T=0, for y=6.9, is evidence that the order parameter must have nodes [13]. Therefore the observation that we have κ/T=0.2 (mW/K^{2}cm) at T=0 in superconducting material means either (i) that there are normal carriers in some layers of YBa$_2$Cu$_3$O$_{6.9}$, or (ii) that all layers superconduct the same and feature d-wave superconductivity. The only interpretation that is consistent with the fact that the μ^+SR data do not agree with the cuprate-plane STM and photoemission results, and with the fact that the specific heat has a "zero-field linear-T term" is the one that favors a nodeless interpretation of superconductivity in the BaO layers plus non-superconducting CuO$_2$ planes [14].

4. Ruthenates

4.1 Cu-DOPED Sr$_2$YRuO$_6$ AND Ba$_2$YRuO$_6$

The ruthenates Sr$_2$YRuO$_6$ and Ba$_2$YRuO$_6$ do not have any cuprate-planes [15]. Nevertheless, when they are Cu-doped on Ru sites at the rather low doping levels of u=0.01, 0.05, 0.1, and 0.15, they superconduct, with superconducting onset temperatures of around 49 K and 93 K, respectively [5][6]. The Sr$_2$YRuO$_6$ compound has been studied more thoroughly than the Ba$_2$YRuO$_6$ compound with its higher superconductivity onset temperature, so we shall assume that the two compounds behave the same and have the same structure, and discuss the more thoroughly studied Sr$_2$YRuO$_6$ compound here, while referring to the figure for Ba$_2$YRuO$_6$ (Fig. 1).

Muon spin rotation (μ^+SR) measurements on Sr$_2$YRuO$_6$ doped with 10% Cu on Ru sites (u=0.1) reveal two distinct muon stopping sites: one in the YRuO$_4$ layer which we call $\mu_{O(1,2)}$ and another in the SrO layer termed $\mu_{O(3)}$. Actually $\mu_{O(1,2)}$ is *two* sites that are very close together, but we cannot resolve the differences of the locations of the two sites. Both O(1,2) sites are near the center of the YRuO$_4$ planes and differ only slightly (so little we cannot resolve the difference) depending on the proximity of the sites to the Ru or to the Y sites on the corners of the nearly square YRuO$_4$ plane. The $\mu_{O(3)}$ site is in the SrO layer, midway between two O atoms on the edge of a cell (the Sr ion is at the center of the SrO plane). Each YRuO$_4$ layer is ferromagnetic in the a-b plane, with its magnetization vector lying in the YRuO$_4$ plane, pointing from Ru to Ru, and reversing direction with each different YRuO$_4$ plane along the c axis, to point first from Ru to Ru in one plane and then in the opposite direction from Ru to Ru in the next plane-making the entire structure anti-ferromagnetic. The Y(Ru$_{1-u}$Cu$_u$)O$_4$ layers contain Ru moments which fluctuate at high-temperatures; at 29.25 K they experience (over a narrow temperature range) a spin-glass state, and at a Néel temperature of T$_N \approx$ 23 K, they order ferromagnetically in each plane [16]. Strictly speaking, bulk (full Meissner fraction) superconductivity does not occur until T$_c$ = T$_N \approx$ 23 K, even though the onset of superconductivity (with fluctuating Ru moments) occurs at temperatures as high as

$T_{c,onset} \approx 49$ K. Therefore the fluctuating Ru moments *break* pairs and suppress both T_c and the bulk Meissner fraction.

4.2 $GdSr_2Cu_2RuO_8$ AND $Gd_{2-z}Ce_zSr_2Cu_2RuO_{10}$

A central question is whether $GdSr_2Cu_2RuO_8$ and Gd_{2-z} $Ce_zSr_2Cu_2RuO_{10}$, which have almost the same temperatures for the onset of superconductivity as doped Sr_2YRuO_6 [17][18], superconduct in their cuprate-planes or in their SrO layers. But the cuprate-planes in these materials (and in the compounds obtained by replacing Ru with Nb) are either weakly ferromagnetic or anti-ferromagnetic, while the materials superconduct, which strongly indicates that the power dissipation spectra have no unique feature associated with Ru, and that the superconductivity is in the SrO layers, not in the cuprate planes [19].

4.3   ARE THERE ANY *n*-TYPE HIGH-TEMPERATURE UPERCONDUCTORS?

Many authors believe that $Nd_{2-x}Ce_xCuO_4$ and its homologues superconduct *n*-type, not *p*-type [9]. This is a topic of current controversy, and we advocate the viewpoint that (i) there are no known *n*-type high-temperature superconductors; (ii) in the material $Nd_{2-x}Ce_xCuO_4$ the doping by Ce is not isolated-Ce doping (which is *n*-type), but doping by Ce-(interstitial O) pairs (which is *p*-type). This issue is still being resolved experimentally.

But there is a system, $Pb_2Sr_2YCu_3O_8$ (PSYCO) which can be doped with most ordinary rare-earths (in place of Y) that is *p*-type in most cases, but *n*-type when doped with Ce or Am. When doped with Ce or Am, the material does not superconduct [20][21]. This leads to the conclusion that *n*-type high-temperature materials that superconduct probably do not exist.

5.   Summary

Not all high-temperature superconductors have cuprate-planes. Indeed some ruthenates superconduct without cuprate-planes, indicating that the superconductivity occurs not in the cuprate-planes, as some authors claim [22], but in the SrO, BaO, or interstitial-oxygen regions.

Consequently, the cuprate-planes are not the loci of high-temperature superconductivity in the cuprate superconductors or in materials of the Cu-doped Sr_2YRuO_6 class; the BaO, SrO, or interstitial-oxygen regions are. The hole-pairing is nodeless in character (consistent with strong-coupled *s*-wave pairing) which becomes easily recognizable after flux-pinning is accounted for.

By recognizing that different layers in the unit cell can behave differently, and that some layers can superconduct, while other layers do not, one can explain the specific heat data. The thermal conductivity data also are easily understood in this picture.

The ruthenates, most notably Cu-doped Sr_2YRuO_6 and Ba_2YRuO_6 (which have no cuprate-planes) superconduct at rather high temp-eratures, with their superconducting hole condensates in their SrO or BaO layers (with onset temperatures for superconductivity of ~49 K and ~93 K). The rutheno-cuprates $GdSr_2Cu_2RuO_8$ and $Gd_{2-z}Ce_zSr_2Cu_2RuO_{10}$ also have their superconducting holes in the SrO layers, not in their cuprate-planes (which are weakly ferromagnetic or antiferro-magnetic at low temperatures). All high-temperature superconductors are p-type [9]. Perhaps, with these revised ideas, the problem of high-temperature superconductivity will become solvable.

Acknowledgments

We are grateful for the support of the U.S. Army Research Office, under contract W911NF-05-1-0346.

References

[1] K. McElroy, R. W. Simmonds, J. E. Hoffman, D.-H. Lee, J. Orenstein, H. Eisaki, S. Uchida, and J. C. Davis, Nature 422, 592 (2003).
[2] A. Damascelli, Z.-X. Shen, and Z. Hussain, Rev. Mod. Phys. 75, 473 (2003).
[3] J. C. Campuzano, M. R. Norman, and M. Randeria, in Physics of Superconductors, Vol. II, "Physics of Conventional and Unconventional Superconductors," ed. by K. H. Benneman and J. B. Ketterson (Springer, Berlin, 2004), pp. 167-273.
[4] D. J. Scalapino, Phys. Rpts. 250, 1 (1995).
[5] S. M. Rao, J. K. Srivastava, H. Y. Tang, D. C. Ling, C. C. Chung, J. L. Yang, S. R. Sheen, and M. K. Wu, J. Crystal Growth 235, 271 (2002).
[6] M. K. Wu, D. Y. Chen, D. C. Ling, and F. Z. Chien, Physica B 284-288, 477 (2000).
[7] J. D. Jorgensen, B. W. Veal, A. P. Paulikas, L. J. Nowicki, G. W. Crabtree, H. Claus, and W. K. Kwok, Phys. Rev. B 41, 1863 (1990).
[8] K. A. Müller, Phil. Mag. Lett. 82, 279 (2002).
[9] J. D. Dow and M. Lehmann, Phil. Mag. 83, 527 (2003).
[10] D. R. Harshman, W. J. Kossler, X. Wan, A. T. Fiory, A. J. Greer, D. R. Noakes, C. E. Stronach, E. Koster, and J. D. Dow, Phys. Rev. B 69, 174505 (2004) and Phys. Rev. B 72, 146502 (2005). This latter paper demonstrates why the d-wave pairing of the Canadian group [J. E. Sonier, D. A. Bonn, J. H. Brewer, W. N. Hardy, R. F. Kiefl, and R. Liang, Phys. Rev. B 72, 146501 (2005)] lacks credibility.
[11] H. S. Amin, M. Franz, and I. Affleck, Phys. Rev. Lett. 84, 5864 (2000); 82, 3232 (1999).
[12] K. A. Moler, D. L. Sisson, J. S. Urbach, M. R. Beasley, A. Kapitulnik, D. J. Baar, R. Liang, and W. N. Hardy, Phys. Rev. B 55, 3954 (1997).

[13] L. Taillefer, B. Lussier, R. Gagnon, K. Behnia, and H. Aubin, Phys. Rev. Lett. 79, 483 (1997).

[14] D. R. Harshman and J. D. Dow. Int. J. Mod. Phys. B 19, 147-151 (2005).

[15] The neutron data show that $Sr_2Y(Ru_{1-u}Cu_u)O_6$ has less than ~1% impurity of any kind.

[16] D. R. Harshman, W. J. Kossler, A. J. Greer, D. R. Noakes, C. E. Stronach, E. Koster, M. K. Wu, F. Z. Chien, J. P. Franck, I. Isaac, and J. D. Dow, Phys. Rev. B 67, 054509 (2003).

[17] J. D. Dow and D. R. Harshman, Physica C 388-389, 447 (2003).

[18] J. D. Dow, "High-temperature superconductivity without cuprate planes." J. Supercond., in press (2005). (Special issue: Stripes04 — Nanoscale Heterogeneity and Quantum Phenomena in Complex Matter).

[19] J. D. Dow, H. A. Blackstead, Z. F. Ren, and D. Z. Wang, Pis'ma v. Zh. Exp. Teor. Fiz. 80, 216 (2004). Engl. transl.: JETP Lett. 80, 190 (2004).

[20] L. Soderholm, C. Williams, S. Skanthakumar, M. R. Antonio, and S. Conradson, Z. Physik B 101, 539 (1996).

[21] M. Lehmann, J. D. Dow, and H. A. Blackstead, Physica C 341-348, 309 (2000).

[22] Numerous experiments have been published which claim that the oxygen isotope effect is intrinsic, supposedly proving that the cuprate-planes superconduct. [H. Keller, "Unconventional Isotope Effects in Cuprate Superconductors", Struct. Bond, pp. 114-169, Springer-Verlag, Berlin, Heidelberg, New York, 2005; R. Khasanov, A. Shengalaya, E. Morenzoni, M. Angst, K. Conder, I. M. Savic, H. Keller, J. Phys. Condensed Matter 15, L17 (2003); R. Khasanov, A. Shengalaya, E. Morenzoni, M. Angst, K. Conder, I. M. Savic, D. Lampalis, E. Liarokapis, A. Tatsi, H. Keller, Phys. Rev. B 68, 220506 (2003).] A paper contradicting this work is D. R. Harshman, J. D. Dow, and A. T. Fiory, "Vanishing Isotope Effect in 'Ideal' High-T_C Superconductors" [to be published]. This paper shows that the Khasanov-Keller work is executed on non-optimized samples and ignores the degradation in the quality of the superconducting hole condensate. In particular, Keller *et al.* treat $Y_{1-x}Pr_xBa_2Cu_3O_7$ samples, which superconduct at 90 K when pure. [Z. Zou, J.Ye, K. Oka, and Y. Nishihara, Phys. Rev. Lett. 80, 1074-1077 (1998); A. Shukla, . Barbiellini, A. Erb, A. Manuel, T. Buslaps, V. Honkimäki, and P. Suortti, Phys. Rev. 59, 12127 (1999); F. M. Araujo-Moreira, P.N. Lisboa-Filho, A. J. C. Lanfretti, S. M. Zanetti, E. R. Leite, A. W. Morbru, L. Ghivelder, Y. G. Zhao, W. A. Ortiz, and V. T. Venkatesan, Physica C 341-348, 413 (2000)]. But these materials have significant numbers of Pr-on-Ba-site defects when they are impure (Keller's case), and so with increasing defect concentration, the defects degrade the superconductivity and eventually destroy it. [For a review of materials with Pr-on-Ba-site defects, see H. B. Radousky, J. Mater. Res. 7, 1917-1955 (1992).]

[23] J. D. Dow, H. A. Blackstead, and D. R. Harshman, Physica C 364-365, 74 (2001) show that the BaO, SrO, or other similar layers are the loci of superconductivity.

DOPING DEPENDENCE OF CUPRATE COHERENCE LENGTH, SUPERCARRIER EFFECTIVE MASS AND PENETRATION DEPTH IN A TWO-COMPONENT SCENARIO

N. Kristoffel[1, 2], T. Örd[1] and P. Rubin[2]
[1] Institute of Theoretical Physics, University of Tartu,
Tähe 4, 51010 Tartu, Estonia
[2] Institute of Physics, University of Tartu,
Riia 142, 51014 Tartu, Estonia

Abstract. A model with pair transfer between defect and itinerant subsystems with bare gaps quenched by doping is used to calculate the cuprate coherence length, the paired carrier effective mass and the penetration depth on the whole doping scale. Observed dependencies are qualitatively reproduced. Effective mass diminishing with extended doping is predicted as a general trend. The supercarrier density and critical fields dependencies are bell-like. The pairing strength and the coherence develop simultaneously.

Key words: Cuprates, Doping-dependent Spectrum, Interband Pairing, Coherence Length, Effective Mass, Penetration Depth, Critical fields

A huge amount of experimental data has been collected for cuprates being stimulated by the high-temperature superconductivity. One activates these materials by doping. The two-component scenario (Müller, 2000) stresses accordingly the functioning of a doping-created defect electronic subsystem besides the itinerant one in cuprate superconductivity. The knowledge of the dependence of various superconductivity characteristics on the whole doping scale becomes extremely important. These dependencies reflect the influence of doping on the basic ingredients of the material essential for the superconductivity. However, there is not consensus present on the cuprate superconductivity mechanism itself.

The data on cuprate superconductivity energetic characteristics and inter-relations between them on the doping scale have been ordered quite recently (Tallon and Loram, 2001; Timusk, 2003). The data characterizing the co-herence and various thermodynamic properties are far from being complete, being fragmental in doping and partly controversial. A very general outlook will state the following. Cuprate coherence length (ξ_{ab}) builds up a valley-profile type curve on doping as found recently in vortex core experiments

K. Scharnberg and S. Kruchinin (eds.),
Electron Correlation in New Materials and Nanosystems, 275–282.
© 2007 *Springer.*

(Wen et al., 2003). The c-axis penetration depth (λ) as λ^{-2} shows a curve with a well expressed maximum (Tallon and Loram, 2001; Tallon et al., 2003; Schneider, 2003) after the optimal doping. The supercarrier effective mass doping dependence is practically unknown. One supposes usually a constant value $m \sim 2m_0$ where m_0 is the free electron mass. The superfluid density is usually accepted to repeat the trend of the transition temperature. However its quenching at overdoping with the parallel raising of doped carrier concentration is considered as a problem (Bernhard et al., 2001).

In the present contribution the thermodynamic characteristics for a "typical" cuprate are calculated on the whole doping scale using a simple model (Kristoffel and Rubin, 2002; Kristoffel and Rubin, 2004). Together with the phase diagram for energetic characteristics obtained in (Kristoffel and Rubin, 2002; Kristoffel and Rubin, 2004) these data form a self-consistent complex which seems to reproduce qualitatively the known behaviour of cuprate properties with doping. A trend to decrease with doping for the CuO_2 plane paired carrier effective mass m_{ab} is predicted.

The background electron spectrum of the model is essentially doping variable. Bare normal state gaps are assumed to be created between the itinerant and defect states and further quenched by progressive doping. Overlap dynamics of the band components determines distinct critical doping concentrations. According to the position of the chemical potential in, or out of a band, the minimal quasiparticle energy defines a superconducting gap or a pseudogap. Controversial statements on the interrelations and coexistence of various gaps in specific doping regions become elucidated in this model. The pseudogap(s) can be considered as a precursor of the superconducting gap on the doping scale, but not on the energy scale. In the normal state the gap features are expected to manifest at the phase diagram points, where one observes the superconducting gap at $T < T_c$. The pairing interaction is supposed to be of the pair-transfer type (Kristoffel et al., 1994) between the itinerant (γ) band and the defect system subbands and is characterized by the coupling constant $W > 0$. These "cold" (β) and "hot" (α) subbands take account of different functioning of the ($\frac{\pi}{2}, \frac{\pi}{2}$) and ($\pi, 0$)-type regions of the momentum space in the cuprate superconductivity. The essential point of the model consists in creation by doping of an electron spectrum with multiband interference near the Fermi energy. The prerequisite to reach high transition temperatures by the interband pairing mechanism becomes fulfilled by this supposition for cuprates.

The necessary details of the model are the following. The valence band states lie between the energy zero and $\xi = -D$ and are normalized to $1 - c$, where c is a measure of the doped hole concentration (p). The corresponding scaling can be made by joining special phase diagram points. We use

$p = 0.28c$. The defect subbands created near the top of the valence band (with the weight of states $c/2$) diminish the charge-transfer gap and are characterized by the energy intervals $d_1 - \alpha c$ and $d_2 - \beta c$, i.e. they expand down from the energies at $c = 0$. The overlap with the γ-band is reached at $c_\alpha = d_1 \alpha^{-1}$ and $c_\beta = d_2 \beta^{-1}$. The choice $c_\beta < c_\alpha$ means that the lowest doping-created states belong to the "cold" subsystem. The $2D$ band densities read $\rho_\alpha = (2\alpha)^{-1}$, $\rho_\beta = (2\beta)^{-1}, \rho_\alpha = (1 - c)D^{-1}$.

At very underdoping $c < c_\beta$, the chemical potential $\mu_1 = d_2 - \beta c$ is connected with the carriers of the "cold" band. For $c > c_\beta, \mu_2 = (d_2 - \beta c)[1 + 2\beta(1 - c)D^{-1}]^{-1}$ intersects both (β, γ) bands. The Fermi level is shifted into the valence band. The overlap of the narrow β-band with the wide itinerant band leads to the formation of the Fermi surface with a hole-like barrel sheet and the other of electronic "flat" band type. For $c > c_0, d_2 - \alpha c_0 = \mu_2$ the "hot" subsystem enters the game in the whole amount and $\mu_3 = [\alpha d_2 + \beta d_1 - 2\alpha\beta c][\alpha + \beta + (1 - c)2\alpha\beta D^{-1}]^{-1}$ intersects all the overlapping bands. At extended overdoping $c > c_1, \mu_3 = d_2 - \beta c_1, \mu_4 = (d_1 - \alpha c)[1 - 2\alpha(1 - c)D^{-1}]^{-1}$ falls out of the "cold" defect band.

The calculation procedure of the energetic characteristics can be followed by the papers (Kristoffel and Rubin, 2002; Kristoffel and Rubin, 2004; Kristoffel et al., 1994). The supercarrier density (n_s) repeats the trend of the transition temperature and of the superconducting gaps $(\Delta_\alpha, \Delta_\gamma)$, with an expressed maximum near the optimal doping. The characteristic ratio $2\Delta_\gamma/\Theta_c$ is non-monotonic in p $(\Theta = k_B T)$, whereas $2\Delta_\alpha/\Theta_c$ diminishes with enchanced doping over the scale. Two pseudogaps survive as normal state gaps at $T > T_c$. They do not manifest themselves in the supercarrier density because of the interband nature of the pairing. In the case of missing bare $\beta - \gamma$ gap there will be only one charge-channel pseudogap, cf. (Kristoffel and Rubin, 2002).

For the calculation of the thermodynamic characteristics the free energy expansion for the model under consideration has been derived analogously to (Örd and Kristoffel, 1999). The mixed nature of the order parameters of the two-component system is reflected in the appearance of the "soft" and "rigid" components with the fluctuations governed by the coherence scales

$$\xi_{si}^2 = \frac{\hbar^2}{4M_{si}a_s\tau}, \tag{1}$$

$$\xi_{ri}^2 = -\frac{\hbar^2}{4M_{ri}(a_0 - a_r\tau)}, \tag{2}$$

where $\tau = \frac{T - T_c}{T_c}$. The equilibrium value of the soft order parameter

$$|\bar{\psi}_s|^2 = -\frac{|W|}{2V}u\Xi\tau \tag{3}$$

determines both the superconducting gaps, whereas for the rigid one $|\bar{\psi}_r|^2 = 0$. For $T > T_c$ $\bar{\psi}_s = \bar{\psi}_r = 0$. ξ_s acts as an usual critical coherence length and diverges at $T > T_c$. The other, ξ_r, behaves noncritically (is rigid) and as an imaginary quantity characterizes a periodic spatial coherence wave. In the limit $\tau \to 0$ ξ_r remains finite.

The necessary band characteristics for the case with μ lying inside the integration limits $\Gamma_{c\sigma}$ to $\Gamma_{0\sigma}$ read ($\gamma = 1.78$, $\zeta(3) = 1.2$)

$$\eta_\sigma = W\rho_\sigma \ln |\Gamma_{0\sigma} - \mu||\Gamma_{c\sigma} - \mu| \left(\frac{2\gamma}{\pi\Theta_c}\right)^2 , \tag{4}$$

$$\kappa_\sigma = 2W\rho_\sigma , \tag{5}$$

$$\delta_\sigma = \frac{7}{2} \frac{\zeta(3)W\rho_\sigma|\mu - \Gamma_{0\sigma}|}{(\pi\Theta_c)^2} , \tag{6}$$

$$\nu_\sigma = \frac{7}{4} \frac{\zeta(3)W^3\rho_\sigma}{(\pi\Theta_c)^2} . \tag{7}$$

In the case with μ out of the band

$$\eta_\sigma = W\rho_\sigma \ln \left|\frac{\Gamma_{c\sigma} - \mu}{\Gamma_{0\sigma} - \mu}\right| , \quad \kappa_\sigma = \delta_\sigma = \nu_\sigma = 0 . \tag{8}$$

The band effective masses are $m_\sigma = 2\pi\hbar^2\rho_\sigma V^{-1}$, where $V = a^2$ is the CuO$_2$ plane plaquette area. The coefficients entering the expr.(1-3) are defined as

$$\Xi = \frac{(\eta_\alpha + \eta_\beta)\kappa_\gamma + \eta_\gamma(\kappa_\alpha + \kappa_\beta)}{(\eta_\alpha + \eta_\beta)^2\nu_\gamma + \eta_\gamma^2(\nu_\alpha + \nu_\beta)} , \tag{9}$$

$$\begin{cases} a_0 = u^{-1}(\eta_\alpha + \eta_\beta + \eta_\gamma)^2 , \quad u = \frac{1}{2}(\eta_\alpha + \eta_\beta + \eta_\gamma)(\delta_\alpha + \delta_\beta + \delta_\gamma) \\ a_s = u^{-1}[(\eta_\alpha + \eta_\beta)\kappa_\gamma + \eta_\gamma(\kappa_\alpha + \kappa_\beta)] \\ a_r = u^{-1}\{(2(\eta_\alpha + \eta_\beta) + \eta_\gamma)\kappa_\gamma + (2\eta_\gamma + \eta_\alpha + \eta_\beta)(\kappa_\alpha + \kappa_\beta)\} , \end{cases} \tag{10}$$

$$M_{si}^{-1} = u^{-1} \left\{ (\eta_\alpha + \eta_\beta)\frac{\delta_\gamma}{m_\gamma} + \eta_\gamma \left(\frac{\delta_\alpha}{m_\alpha} + \frac{\delta_\beta}{m_\beta}\right) \right\} , \tag{11}$$

$$M_{ri}^{-1} = u^{-1} \left\{ (2(\eta_\alpha + \eta_\beta) + \eta_\gamma)\frac{\delta_\gamma}{m_\gamma} + (2\eta_\alpha + \eta_\beta + \eta_\gamma) \left(\frac{\delta_\alpha}{m_\alpha} + \frac{\delta_\beta}{m_\beta}\right) \right\} . \tag{12}$$

We calculate the doping dependencies of $\xi_0^2 = \xi_s^2\tau$ (1), of the c-axis second critical field $H_{c2} = \phi_0(2\pi\xi_0^2)^{-1}$ and the CuO$_2$ plane paired carrier effective mass m_{ab} from (11). The thermodynamic critical field

$$H_{c0} = \sqrt{4\pi[(\rho_\alpha + \rho_\beta)\Delta_\alpha^2 + \rho_\gamma\Delta_\gamma^2]} \tag{13}$$

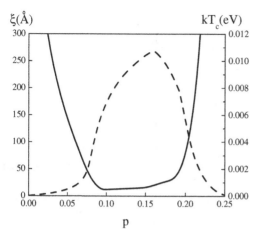

Figure 1. Transition temperature (dashed) and the CuO_2 plane coherence length *vs* hole doping.

represents the condensation energy and the penetration depth is expressed as

$$\lambda = \left[\frac{m_{ab}c^2}{4\pi e^2 N_s} \right]^{1/2} . \tag{14}$$

Here the number density of the superfluid (N_s) is given as $n_s pN_0$, where pN_0 characterizes the normal state carriers number density.

Illustrative calculations of the doping dependencies of the thermodynamic properties have been performed using the parameter set of Ref. (Kristoffel and Rubin, 2004) applied in the investigation of the energetic characteristics of a "typical" cuprate. The in-plane coherence length $\xi_0 = \xi_{ab}$ nonmonotonic dependence is shown in Fig. 1 together with the transition temperature curve. This type ξ_0 curve agrees with the recent experimental result (Wen et al., 2003) presented for the whole doping scale. At limiting dopings ξ_0 diverges. The "valley profile" type curve as projected on the T_c dependence is an expected result. Weakly coupled pairs have larger dimensions. The order of magnitude for ξ_0 about ten Å in the actual T_c-s region agrees with the estimations given in the literature [11]. This points to a self-consistency of the model. The condensation energy (H_{c0}) and the supercarrier density $n_s(0)$ given in Fig. 2 show similar behaviour. The second c-axis critical field $H_{c2}(0)$ in Fig. 3 shows also a well expressed maximum. The latter corresponds to the effective contribution of the cold subsystem after the overlap concentration c_β is passed. One concludes that in the present model the strength of the pairing and the phase coherence develop and vanish simultaneously in accordance with the conclusions of the recent experimental investigations (Bernhard et al., 2001; Tallon et al., 2003; Feng et al., 2000; Trunin et al., 2004; Schneider, 2003; He et al., 2004). The overdoped regime corresponds to higher carrier concentrations but to smaller supercarrier density. The reason lies in that the interband pairing mechanism becomes suppressed here,

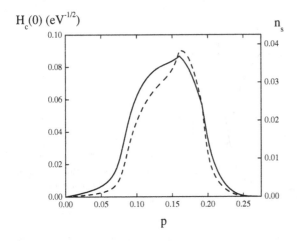

Figure 2. The superfluid density and the thermodynamic critical field (dashed) on the doping scale.

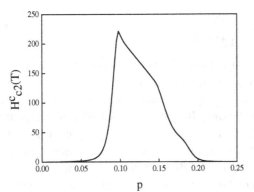

Figure 3. The c-axis second critical magnetic field *vs* doping.

cf. (Bernhard et al., 2001). The paired carrier effective mass (expressed as xm_0 through the free electron mass) doping dependence, to the authors best

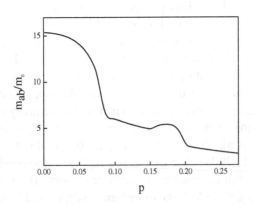

Figure 4. The supercarrier effective mass theoretical dependence on doping.

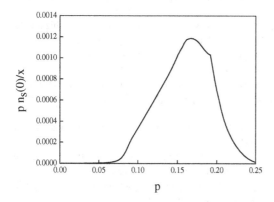

Figure 5. The ratio $pn_s(0)/x$ *vs* doping, representing the behaviour of the inverse squared penetration depth (superfluid density).

knowledge not being calculated earlier, is shown in Fig. 4. The reduction of m_{ab} with doping reproduces the well-known trend to restore the Fermi-liquid behaviour of the superconducting carriers. The large m_{ab} values at very underdoping are connected with the large m_β of the narrow defect β-band. Starting from c_β the contribution of the wide itinerant band carriers is added. After c_1 the contribution of the heaviest β-carriers vanishes. In the actual region of remarkable T_c-s m_{ab} changes relatively slowly and the values of x near 3 can be taken for estimations as often supposed (Plakida, 1995). The squared inverse penetration depth doping dependence represented by the ratio $pn_s(0)/x$ is shown in Fig. 5. Its behaviour is dictated by the supercarrier density and this leads to a λ^{-2} curve with an expressed maximum in accordance with the experimental findings (Bernhard et al., 2001; Tallon et al., 2003; Feng et al., 2000; Schneider, 2003). The enhancement of the normal state carrier concentration becomes manifested in a shift of the superfluid maximum density to higher dopings as compared with n_s.

The model under consideration despite of its simplicity seems to be able to describe qualitatively the charge-channel associated properties of cuprate superconductors on the whole hole doping scale in an unified approach.

This work was supported by Estonian Science Foundation grant No 6540.

Acknowledgements

This work was supported by Estonian Science Foundation grant No 6540.

References

Bernhard, C., Needermayer, C., Binninger, U., Hofer, A., Wenger, C., Tallon, J. L., Williams, G. V. M., Ansaldo, F. J., Budnick, J. I., Stronach, C. F., Noakes, D. R., and Blanksam-Mills, M. A. (2001) Magnetic penetration depth and condensate density of cuprate high-T_c

superconductors determined by myon-spin-rotation experiments, *Phys. Rev. B* **52**, 10488–10498.

Feng, D. L., In, D. H., Shen, K. M., Kim, C., Eisaki, H., Damascelli, A., Yashizaki, R., Shimoyama, J., Kishio, K., Gu, G., Oh, S., Andrus, A., O'Donnel, J., Eckstein, J., and Shen, Z.-H. (2000) Signature of the superfluid density in the single particle excitation spectrum of $Bi_2Sr_2CaCuO_{8+\delta}$, *Science* **289**, 277–281.

He, R., Feng, D. L., Eisaki, H., Schimoyama, J.-L., Kishio, K., and Gu, G. D. (2004) Superconducting order parameter in heavily overdoped $Bi_2Sr_2CaCu_2O_{8+\delta}$, *Phys. Rev. B* **69**, 220502–4(R).

Kristoffel, N., Konsin, P., and Örd, T. (1994) Two-band model for high-temperature superconductivity, *Riv. Nuovo Cimento* **17**, 1–41.

Kristoffel, N. and Rubin, P. (2002) Pseudogap and superconductivity gaps in a two-band model with the doping determined components, *Solid State Commun.* **122**, 265–268.

Kristoffel, N. and Rubin, P. (2004) Superconducting gaps and pseudogaps in a composite model of two-component cuprate, *Physica C* **402**, 257–262.

Müller, K. A. (2000) Recent experimental insights into HTSC materials, *Physica C* **341-348**, 11–18.

Örd, T. and Kristoffel, N. (1999) Paired carrier effective mass isotope effect in the two-band model, *Phys. Stat. Sol. B* **216**, 1049–1056.

Plakida, M. (1995) *High-temperature superconductivity*, Springer.

Schneider, T. (2003) Universal properties of cuprate superconductors, *Physica B* **326**, 283–295.

Tallon, J. L. and Loram, J. W. (2001) The doping dependence of T^* – what is the real high-T_c phase diagram?, *Physica C* **349**, 53–68.

Tallon, J. L., Loram, J. W., Cooper, R. J., Panagopoulas, C., and Bernhard, C. (2003) Superfluid density in cuprate high-T_C superconductors. A new paradigm, *Phys. Rev. B* **68**, 180501–180504 (R).

Timusk, T. (2003) The mysterious pseudogap in high-T_c superconductors, *Solid State Commun.* **127**, 337–348.

Trunin, M. R., Nefyodov, Y. A., and Schevchuk, A. F. (2004) Superfluid density in the underdoped $YBa_2Cu_3O_{7-x}$, *Phys. Rev. Lett.* **92**, 067006–4.

Wen, H. H., Yang, H. P., Li, S. L., Zeng, X. H., Sankissian, A. A., Si, W. D., and Xi, X. X. (2003) Hole doping dependence of the coherence length in $La_{2-x}Sr_xCuO_4$ thin films, *Europhys. Lett.* **64**, 790–796.

ORDER PARAMETER COLLECTIVE MODES IN
UNCONVENTIONAL SUPERCONDUCTORS

Peter Brusov (PNB1983@YAHOO.COM)
Low temperature laboratory, Physical Research Institute, 194 Stachki Ave.,
Rostov-on-Don, 344090, Russia

Pavel Brusov
Department of Physics, Case Western Reserve University, 10900 Euclid
Avenue, Cleveland, OH 44106-7079,USA

Abstract. Direct observation of collective modes in unconventional superconductors (USC) by microwave impedance technique experiments has made very important the study of the collective excitations in these systems. We consider two-dimensional and three-dimensional models of p- and d-pairing for superconductors. Within these models we calculate by path integration techniques the collective excitations in different unconventional superconductors, including high temperature superconductors (HTSC) and heavy fermion superconductors (HFSC). We considered both bulk and 2D systems. Some ideas concerning the realization in HTSC of mixtures of different states are investigated. In particular, we consider the mixture of d_{x2-y2} and d_{xy} states. The results obtained distinguish pure d-wave states from mixtures of two d-wave states in HTSC and therefore could be used for the interpretation of the sound attenuation and microwave absorption data as well as for identification of the type of pairing in these unconventional superconductors.

Key words: Collective Modes, Unconventional Superconductors, p- and d- pairing

1. Introduction

Until recently the study of the collective modes (CM) in unconventional superconductors (USC) was considered exotic for a number of reasons. First of all, while there were some evidences of nontrivial type of pairing in some superconductors (SC), including heavy fermions (HFSC) and high T_c super-conductors (HTSC), there were no superconductors in which unconventional pairing had been established unequivocally. Secondly, there was no strong experimental evidence for the existence of CM in any of these superconductors.

The situation has changed drastically within last few years. First of all, a few years ago an amplitude mode (with frequency of order 2Δ) has been observed in films of conventional superconductors. Secondly, the type of pairing has now been established for many SC. We have s-pairing in

K. Scharnberg and S. Kruchinin (eds.),
Electron Correlation in New Materials and Nanosystems, 283–292.
© 2007 *Springer.*

ordinary superconductors and electron-type HTSC; p-pairing in pure ^{3}He; ^{3}He in aerogel, Sr_2RuO_4 (HTSC), UPt$_3$ (HFSC) and d-pairing in hole-type HTSC, organic superconductors, some HFSC (UPd$_2$Al$_3$, CePd$_2$Si$_2$, CeIn$_3$, CeNi$_2$Ge$_2$ etc.). Recently the Northwestern University group[2] has presented results of a microwave surface impedance study of the HFSC UBe$_{13}$. They observed an absorption peak whose frequency- and temperature-dependence scales with the BCS gap function $\Delta(T)$. This was the first direct observation of the resonant absorption into a CM with energy approximately proportional to the super-conducting gap. This discovery opens a new page in study of the CM in USC. The significance of studying CM is connected to the fact that they exhibit themselves in ultrasound attenuation[3] and microwave absorption[2] experiments, neutron scattering, photoemission and Raman scattering[4]. The large peak in the dynamical spin susceptibility in HTSC arises from a weakly damped spin density-wave CM. This gives rise to a dip between the sharp low energy peak and the higher binding energy hump in the ARPES spectrum. Also, the CM of amplitude fluctuation of the d-wave gap yields a broad peak above the pair-breaking threshold in the B$_{1g}$ Raman spectrum[4]. The contribution of CM to ultrasound and microwave absorption may be substantial. Thus the study of CM in USC has become very important[1].

2. The path integral models of p- and d-pairing for bulk superconductors

In the method of functional integration, the initial Fermi system is described by anticommuting functions $\chi_s(\mathbf{x},\tau), \overline{\chi}_s(\mathbf{x},\tau)$ defined in the volume $V = L^3$, which are antiperiodic in time τ with a period $\beta = T^{-1}$. Here s is the spin index. These functions can be expanded into a Fourier series

$$\chi_s(x) = (\beta V)^{-1/2} \sum_p a_s(p) \exp\left(i(\omega\tau + \mathbf{k} \square \mathbf{x})\right), \tag{1}$$

where $p = (\mathbf{k}, \omega)$; $\omega = (2n+1)\pi T$ are Fermi-frequencies and $x = (\mathbf{x}, \tau)$. Let us consider the action functional for an interacting Fermi-system

$$S = \int_0^\beta d\tau d^3x \sum_s \overline{\chi}_s(\mathbf{x},\tau)\partial_\tau \chi_s(\mathbf{x},\tau) - \int_0^\beta H'(\tau)d\tau, \tag{2}$$

with

$$H'(\tau) = \int d^3x \sum_s (2m)^{-1}\nabla\overline{\chi}_s(\mathbf{x},\tau)\nabla\chi_s(\mathbf{x},\tau) - (\lambda + s\mu_0 H)\overline{\chi}_s(\mathbf{x},\tau)\chi_s(\mathbf{x},\tau)$$

$$+ \frac{1}{2}\int d^3x d^3y\, U(\mathbf{x}-\mathbf{y})\sum_{ss'} \overline{\chi}_s(\mathbf{x},\tau)\overline{\chi}_{s'}(\mathbf{y},\tau)\chi_{s'}(\mathbf{y},\tau)\chi_s(\mathbf{x},\tau) \tag{3}$$

To obtain the effective action functional, we divide Fermi fields into "fast" and "slow" fields. Fast fields χ_{1s} and $\overline{\chi}_{1s}$ are determined by components of expansion (1) either with frequencies $|\omega| > \omega_0$, or with momenta $|k - k_F| > k_0$. The remaining component $\chi_{0s} = \chi_s - \chi_{1s}$ of the Fourier expansion defines slow fields χ_{0s}. Integrating over fast fields, we obtain

$$\int \exp SD\overline{\chi}_{1s}D\chi_{1s} = \exp \widetilde{S}\left(\chi_{0s}, \widetilde{\chi}_{0s}\right). \tag{4}$$

The most general form of $\widetilde{S}$ is the sum of even-order forms of $\chi_{0s}, \overline{\chi}_{0s}$:

$$\widetilde{S} = \sum_{n=0}^{\infty} \widetilde{S}_{2n}. \tag{5}$$

Neglecting $\widetilde{S}_0$ and sixth and higher-order terms (this can be done when the layer $|k - k_F| < k_0$ is narrow), we retain only terms $\widetilde{S}_2$ and $\widetilde{S}_4$ describing noninteracting quasiparticles near the Fermi surface and their pairing interaction

$$\widetilde{S}_2 \approx \sum_{s,p} Z^{-1}\left[i\omega - c_F(k - k_F) + s\mu H\right]a_s^+(p)a_s(p). \tag{6}$$

$\widetilde{S}_4$ is different for different types of pairing.

2.1. P-PAIRING

In the case of triplet pairing $\widetilde{S}_4$ is equal to

$$\widetilde{S}_4 = -(\beta V)^{-1} \sum_{p1+p2=p3+p4} t_0(p_1, p_2, p_3, p_4)a_+^+(p_1)a_-^+(p_2)a_-(p_4)a_+(p_3)$$

$$-(2\beta V)^{-1} \sum_{p1+p2=p3+p4} t_1(p_1, p_2, p_3, p_4)(2a_+^+(p_1)a_-^+(p_2)a_-(p_4)a_+(p_3)$$
$$+ a_+^+(p_1)a_+^+(p_2)a_+(p_4)a_+(p_3) + a_-^+(p_1)a_-^+(p_2)a_-(p_4)a_-(p_3)),$$

where $t_0(p_i)$ is symmetric, $t_1(p_i)$ is antisymmetric scattering amplitudes under exchanges $p_1 \leftrightarrow p_2$; $p_3 \leftrightarrow p_4$. In the vicinity of the Fermi sphere one could put $\omega_i=0$, $k_i=k_F\,n_i$ ($i=1,2,3,4$). We could write $t_0=f((n_1, n_2), (n_1-n_2, n_3-n_4))$; $t_1=(n_1-n_2, n_3-n_4)\,g((n_1, n_2), (n_1-n_2, n_3-n_4))$, where f and g are even with respect to second argument. We consider model with $f=0$, $g=$const<0. To go over to Bose-fields of Cooper pairs, we introduce the Gaussian integral over Bose-fields c_{ia}, c_{ia}^+ in the integrand of the integral over Fermi-fields:

$$\int dc_{ia}^+ dc_{ia} \left[g^{-1} \sum_{p,i,,a} c_{ia}^+(p)c_{ia}(p) \right]. \tag{7}$$

Carrying out a shift in Bose-fields by quadratic form of slow fields annihilating the fourth-order form, we obtain a Gaussian integral over slow fields. Evaluation of this integral leads to the required effective functional of action

$$S_{\text{eff}} = g^{-1} \sum_{p,i,a} c_{ia}^+(p)c_{ia}(p) + \frac{1}{2}\ln \ \det \frac{\hat{M}\left(c_{ia}, c_{ia}^+\right)}{\hat{M}\left(c_{ia}^{(0)}, c_{ia}^{(0)+}\right)}, \tag{8}$$

where $c_{ia}^{(0)}$ is the condensate value of Bose-fields c_{ia} and $\hat{M}\left(c_{ia}, c_{ia}^+\right)$ is the 4×4 matrix depending on Bose-fields and parameters of quasi-fermions.

$$M_{11} = Z^{-1}[i\omega + \xi - \mu(\mathbf{H}\sigma)]\delta_{p_1 p_2}, \ M_{22} = Z^{-1}[-i\omega + \xi + \mu(\mathbf{H}\sigma)]\delta_{p_1 p_2},$$

$$M_{12} = \frac{(n_{1i} - n_{2i})c_{ia}(p_1 + p_2)\sigma_a}{(\beta V)^{-1/2}}, \ M_{12} = -\frac{(n_{1i} - n_{2i})c^+{}_{ia}(p_1 + p_2)\sigma_a}{(\beta V)^{-1/2}},$$

2.2. D-PAIRING

In the case of singlet pairing $\tilde{S}_4$ is equal to

$$\tilde{S}_4 = -(\beta V)^{-1} \sum_{p_1 + p_2 = p_3 + p_4} t(p_1, p_2, p_3, p_4)a_+^+(p_1)a_-^+(p_2)a_-(p_4)a_+(p_3). \tag{9}$$

The first version of the model of d-pairing in SC constructed by the path integral technique was proposed by Brusov and Brusova[10] in 1994 when the idea of d-pairing in HTSC had just been proposed. We describe below an improved self-consistent model of d-pairing in SC. In the case of d-pairing we have

$$t(p_1, p_2, p_3, p_4) = V\left(\hat{k}, \hat{k}'\right) = \sum_{m=-2}^{2} g_m Y_{2m}\left(\hat{k}\right)Y_{2m}^*\left(\hat{k}'\right), \tag{10}$$

where here $k_1 = k + q/2$, $k_2 = -k + q/2$, $k_3 = k' + q/2$, $k_4 = -k' + q/2$, and $Y_{2m}(k)$ are spherical harmonics with $l=2$. Going over to Bose-fields c_{ia}, c_{ia}^+ describing Cooper pairs (in the case of d-pairing, c_{ia} are symmetric traceless 3×3 matrices and the number of the degrees of freedom c_{ia} is $2\cdot(6-1)=10$) one gets the effective action, which has in the canonical variables($c_1 = c_{11} + c_{22}$, $c_3 = c_{12} + c_{21}, c_4 = c_{13} + c_{31}, c_5 = c_{23} + c_{32}$) the following form:

$$S_{eff} = (2g)^{-1} \sum_{p,j} c_j^+(p) c_j(p)(1 + 2\delta_{j1}) + \frac{1}{2} \ln \det \frac{\hat{M}(c_j^+, c_j)}{\hat{M}(c_j^{+(0)}, c_j^{(0)})}, \quad (11)$$

where

$$M_{11} = Z^{-1}[i\omega + \xi - \mu(\mathbf{H}\boldsymbol{\sigma})]\delta_{p_1 p_2},$$

$$M_{22} = Z^{-1}[-i\omega + \xi + \mu(\mathbf{H}\boldsymbol{\sigma})]\delta_{p_1 p_2},$$

$$M_{12} = M_{21}^* = (\beta V)^{-1/2} \left(\frac{15}{32\pi}\right)^{1/2} \left[c_1 (1 - 3\cos^2\theta) \right.$$

$$+ c_2 \sin^2\theta \cos^2\varphi + c_3 \sin^2\theta \sin 2\varphi$$

$$\left. + c_4 \sin 2\theta \cos\varphi + c_5 \sin 2\theta \sin\varphi \right]. \quad (12)$$

This functional determines all the properties of the model superconducting Fermi-system with d-pairing including collective mode spectrum (CMS).

3. Collective properties of bulk unconventional superconductors

3.1. P-PAIRING

The first results for case of p-pairing were obtained by us[5] for A-, B-, A_1-, 2D- and polar states, which relate to the superfluid phases of ^{3}He. Here we consider additional possible SC states, which could be realized in HTSC as well as in HFSC under p-pairing. Below we summarize the obtained results. Recall that the CMS in each SC state consists of 18 modes (HF – and Gd – modes).

A-phase: $E = \Delta_0(T)(1.96 - i\,0.31)$ (3modes); $E = \Delta_0(T)(1.17 - i\,0.13)(6)$; $E^2 = c_F^2 k^2/3(3)$; $E^2 = c_F^2 k_\parallel^2$ (6).

B-phase: $E^2 = 12\Delta^2/5(5)$; $E^2 = 8\Delta^2/5(5)$; $E^2 = 4\Delta^2(4)$; $E^2 = c_F^2 k^2/3(1)$; $E^2 = c_F^2 k^2/5(1)$; $E^2 = 2c_F^2 k^2/5(2)$.

2D-phase: $E = \Delta_0(T)(1.96 - i\,0.31)$ **(2)**; $E = \Delta_0(T)(1.17 - i\,0.13)$**(4)**;

$E = 2\mu H$ (2); $E_0^2 = \Delta_0^2(T)(1.96 - i\,0.31)^2 + 4\mu^2 H^2$ (2); $E = 0(6)$; $E^2 = \Delta_0^2(T)(0.518)^2 + 4\mu^2 H^2$(1); $E^2 = \Delta_0^2(T)(0.495)^2 + 4\mu^2 H^2$ (1).

A_1-phase: $E = \Delta_0(T)(1.96 - i\,0.31)(1)$; $E = \Delta_0(T)(1.17 - i\,0.13)(2)$; $E = 2\mu H$ (8); $E = 0$ (1). Six other modes have an imaginary spectrum (this fact is connected with the instability of the A_1-phase under small perturbations).

For following three states $\begin{pmatrix} 100 \\ 0-10 \\ 000 \end{pmatrix}$, $\begin{pmatrix} 010 \\ 100 \\ 000 \end{pmatrix}$ and $\begin{pmatrix} 010 \\ -100 \\ 000 \end{pmatrix}$ spectra are

identical and for high frequency modes we found

$E = \Delta_0(T)(1.83 - i\,0.06); E = \Delta_0(T)(1.58 - i\,0.04);$

$E = \Delta_0(T)(1.33 - i\,0.10);$

$E = \Delta_0(T)(1.33 - i\,0.08); E = \Delta_0(T)(1.28 - i\,0.04);$

$E = \Delta_0(T)(1.09 - i\,0.22);$

$E = \Delta_0(T)(0.71 - i\,0.05); E = \Delta_0(T)(0.33 - i\,0.34);$

$E = \Delta_0(T)(0.23 - i\,0.71).$

Two last modes have imaginary parts of the same order as real ones. They are damped very strongly and could not be considered as resonances.

For $\begin{pmatrix} 110 \\ ii0 \\ 000 \end{pmatrix}$ state we found the following high frequency modes

$E = \Delta_0(T)(1.55 - i\,0.32); E = \Delta_0(T)(1.2 - i\,0.06);$

$E = \Delta_0(T)(0.62 - i\,0.05);$

$E = \Delta_0(T)(0.4 - i\,0.55); E = \Delta_0(T)(0.3 - i\,1.0);$

Two last modes have imaginary parts of the same order as real ones. They are damped very strongly and could not be considered as resonances.

For $\begin{pmatrix} 001 \\ 000 \\ 100 \end{pmatrix}$ state we found the following HF modes $E = \Delta_0(T)(1.80 - i\,0.09);$

$E = \Delta_0(T)(0.55 - i\,0.80).$ Last mode has imaginary part of the same order as real one. It is damped very strongly and could not be considered as resonance.

3.2. D-PAIRING

3.2.1. *Bulk HTSC under d-pairing*

We shall use the effective functional of action obtained above for analyzing the CM spectrum in HTSC.[1] Let us consider the results of calculation of the

CMS for SC phases appearing in the symmetry classification of HTSC. We considered the following states: $d_{x^2-y^2}$, d_{xy}, d_{xz}, d_{yz}, $d_{3z^2-r^2}$. For each SC phase, five HF modes were determined as well as five Gd (quasi Gd) modes. The following results were obtained for the high-frequency modes:

$d_{3z^2-r^2}: E_1 = \Delta_0(2.0 - i\,1.65)$; $E_{2,3} = \Delta_0(1.85 - i\,0.69)$; $E_{4,5} = \Delta_0(1.64 - i0.50)$.

$d_{x^2-y^2}; d_{xy}: E_1 = \Delta_0(1.88 - i\,0.79)$; $E_2 = \Delta_0(1.66 - i\,0.50)$; $E_3 = \Delta_0(1.14 - i\,0.68)$;

$E_4 = \Delta_0(1.13 - i\,0.71)$; $E_5 = \Delta_0(1.10 - i\,0.65)$.

$d_{xz}, d_{yz}: E_1 = \Delta_0(1.76 - i\,1.1)$; $E_2 = \Delta_0(1.70 - i\,0.48)$; $E_3 = \Delta_0(1.14 - i\,0.68)$;

$E_4 = \Delta_0(1.13 - i\,0.73)$; $E_5 = \Delta_0(1.04 - i\,0.83)$.

The results on HF modes can be useful in determining the OP and the type of pairing in HTSC as well as for interpreting the ultrasound and microwave absorption experiments. CM are damped much more strongly in case of d-pairing than in case of p-pairing. This is connected with the nodal structure of energy gap: points of nodes under p-pairing and lines of nodes under d-pairing.

3.2.2. Bulk HFSC under d-pairing

Here we consider three states in HFSC ($d\gamma$, Y_{2-1}, $\sin^2\theta$). The conventional BCS pairing is in disagreement with the nonexponential temperature dependence of the most of thermodynamic quantities, such as the heat capacity etc. Two popular candidates are p-and d-pairing . Here we consider the d-pairing in HFSC within Brusov - Brusova model[10,11] for d-pairing in SC. Here we calculate the CMS in all three phases of HFSC – $d\gamma$, Y_{2-1} and $\sin^2\theta$.

a) $d\gamma$–state: $E_1 = \Delta_0(1.66 - i\,0.5)$ $E_4 = \Delta_0(1.21 - i\,0.60)$
$E_2 = \Delta_0(1.45 - i\,0.48)$ $E_5 = \Delta_0(1.19 - i\,0.60)$ $E_3 = \Delta_0(1.24 - i\,0.64)$

b) Y_{2-1} and $\sin^2\theta$ states:
$E_{1,2} = \Delta_0(1.93 - i\,0.41)$; $E_3 = \Delta_0(1.62 - i\,0.75)$; $E_{4,5} = \Delta_0(1.59 - i\,0.83)$.
The CMS in each state consists of five HF modes and five Gd–modes.

4. How to distinguish mixtures of two d-wave states from a pure d-wave state of HTSC

Recent experiments[6] and theoretical considerations[7,8,9] show that in HTSC the mixture of different d-wave states is realized. We have calculated for the first time the CMS in a mixed $d_{x2-y2} + id_{xy}$ state of HTSC and have shown that while spectra in both states d_{x2-y2} and d_{xy} are identical, the spectrum in the mixed $d_{x2-y2} + id_{xy}$ state turns out to be quite different from them. Thus the probe of the spectrum in ultrasound and/or microwave absorption experiments could be used to distinguish the mixture of two d-wave states from pure d-wave states.

We use the model of d-pairing, described by Eqs. (11) and (12) and considered the mixed $d_{x2-y2}+id_{xy}$ state of HTSC with gap $\Delta(T)=\Delta_0(T)$ $\sin^2\theta$. The CMS consists of five HF and five Gd-modes. The results for HF modes:

$$E_{1,2} = \Delta_0(T)(1.93 - i\ 0.41);$$

$$E_3 = \Delta_0(T)(1.62 - i\ 0.75); E_{4,5} = \Delta_0(T)(1.59 \div i\ 0.83) \qquad (13)$$

We could compare these results with spectrum of pure d_{x2-y2} and d_{xy} states:

$$E_1 = \Delta_0(T)(1.88 - i\ 0.79); E_2 = \Delta_0(T)(1.66 - i\ 0.50);$$

$$E_3 = \Delta_0(T)(1.40 - i\ 0.68); E_4 = \Delta_0(T)(1.13 - i\ 0.71);$$

$$E_5 = \Delta_0(T)(1.10 \div i\ 0.65) \qquad (14)$$

While spectra in both pure states are identical, spectrum in mixed state is quite different from that in the pure states. In pure states all modes are non-degenerate while in the mixed state two HF modes are doubly degenerate and CM have higher frequencies. Damping of CM in pure the d-states is bigger than in the mixed state (30%-65% in the pure states and 20%-50% in the mixed state). This is because in pure states the gap vanishes along chosen lines while in mixed state it vanishes just at two points (poles). The difference of spectrum of CM in pure and mixed d - wave state can be a probe of the state symmetry by ultrasound and/or microwave absorption experiments. By observations of the CM spectrum we can answer very important questions:1)does the gap disappear along chosen lines? 2) do we have a pure or mixed d-wave state in HTSC ?

5. Two dimensional P- and d-wave superconductivity

5.1. 2 D MODELS OF P- AND D-PAIRING IN USC

The existence of CuO_2 planes — the common structural factor of HTSC — suggests we consider 2D models. Brusov and Popov[5] developed a 2D p-pairing model and Brusov, Brusova and Brusov[12] developed a 2D d- pairing model.

5.1.1. p-pairing

In 2D model of p-pairing the orbital moment l $(l=1)$ should be perpendicular to the plane and has only two projections on the $\hat{z}$ - axis: ± 1. P - pairing is a triplet so under 2D p- pairing one has $3\times 2\times 2 = 12$ degrees of freedom (instead of 18 in 3D-case). Results for the CMS for 2D p-pairing are as following:

a-phase $\begin{pmatrix} 100 \\ i00 \end{pmatrix}$ $E^2 = c_F k^2/2$ (3 modes); $E^2 = 2\Delta^2 + c_F k^2/2$ (6);

$E^2 = 4\Delta^2 + (0.5 - i0.433)c_F k^2/2$ (3);

b-phase $\begin{pmatrix} 100 \\ 010 \end{pmatrix}$ $E^2 = c_F k^2/2$ (2); $E^2 = 3c_F k^2/4$ (1); $E^2 = c_F k^2/4$ (1);

$E^2 = 2\Delta^2(4)$; $E^2 = 4\Delta^2 + (0.15 - i0.22)c_F k^2/2$ (3); $E^2 = 4\Delta^2 + (0.85 - i0.22)c_F k^2/2$ (1);
$E^2 = 4\Delta^2 + (0.5 - i0.43)c_F k^2/2$ (2);

phase $\begin{pmatrix} 001 \\ 00i \end{pmatrix}$ $E^2 = 0(3)$; $E^2 = 2\Delta^2(6)$; $E^2 = 4\Delta^2(3)$;

phase $\begin{pmatrix} 0 \pm 10 \\ 100 \end{pmatrix}$ $E^2 = 0(4)$; $E^2 = 2\Delta^2(4)$; $E^2 = 4\Delta^2(4)$;

phase $\begin{pmatrix} 100 \\ 0 - 10 \end{pmatrix}$ $E^2 = 0(4)$; $E^2 = 2\Delta^2(4)$; $E^2 = 4\Delta^2(4)$.

5.2. D-PAIRING

Brusov, Brusova and Brusov[12] developed a 2D model of d - pairing in the CuO_2 planes of HTSC using a path integration technique. The orbital moment l ($l = 2$) should be perpendicular to the plane and can have only two projections on the $\hat{z}$ - axis: ± 2. d - pairing is a singlet state, so one has $1 \times 2 \times 2 = 4$ degrees of freedom (instead of 10 in case of 3D d - pairing). The pairing potential $t = v(\hat{k}, \hat{k}') = \sum_{m=-2,2} g_m Y_{2m}(\hat{k}) Y_{2m}^*(\hat{k}')$. (26)

The effective action functional S_{eff} in canonical variables $c_1 = c_{11} - c_{22}$, $c_2 = c_{12} + c_{21}$, $c_1^+ = c_{11}^+ - c_{22}^+$, $c_2^+ = c_{12}^+ + c_{21}^+$ takes the form similar to 3D one (13) with

$$M_{11} = Z^{-1}[i\omega - \xi + \mu(\mathbf{H\sigma})]\delta_{p1p2}, M_{22} = Z^{-1}[-i\omega + \xi + \mu(\mathbf{H\sigma})]\delta_{p1p2},$$
$$M_{12} = M_{21}^+ = \sigma_0 \alpha(\beta S)^{-1/2}(c_1 \cos 2\varphi + c_2 \sin 2\varphi).$$

Two SC states arise in the symmetry classification of CuO_2 planes with OP which are proportional to $\begin{pmatrix} 1 & 0 \\ 0 & -1 \end{pmatrix}$ and $\begin{pmatrix} 0 & 1 \\ 1 & 0 \end{pmatrix}$, respectively. In the former case the gap is proportional to $Y_{22} + Y_{2-2} \sim \sin^2\theta|\cos 2\varphi| \sim |\cos 2\varphi|$ while in the latter one it is proportional to $-i(Y_{22} - Y_{2-2}) \sim \sin^2\theta|\sin 2\varphi| \sim |\sin 2\varphi|$. For our 2D case we put $\theta = \pi/2$ and $\sin\theta = 1$.

The CMS turns out to be identical in both states with the following energies:

$$E_1 = \Delta_0\left(1.42 - i0.65\right), E_2 = \Delta_0\left(1.74 - i0.41\right) \tag{15}$$

The energies of both modes are complex because under d-pairing the gap vanishes in certain directions and the Bose-excitations decay into fermions which results in CM damping. Both modes can be regarded as resonances. Other two modes are Gd ones.

6. Conclusions

We have calculated the collective excitations in different USC under p- and d- pairing. We considered both bulk and 2D systems. The possibility of distinguishing a mixture of two d-wave states from a pure d-wave state of HTSC has been studied. The results on HF modes can be useful in determining the OP and the type of pairing in USC as well as for interpretation of the ultrasound and microwave absorption experiments. CM are damped much more strongly under d-pairing than under p-pairing. This is connected with the nodal structure of energy gap: points of nodes (p-pairing) and lines of nodes (d-pairing). In the light of recent experiments CM study in UCS is very important.

References

1. P.N. Brusov "*Mechanisms of High Temperature Superconductivity*", vol. 1, 2; Rostov State University Publishing, 1999.
2. J.R. Feller, C.-C. Tsai, J.B. Ketterson et al., Phys. Rev. Lett. **88**, 247005 (2002).
3. H. Matsui *et al.*, Phys. Rev. B, **63**, 060505 (2001).
4. T. Dahm, D. Manske and L. Tewordt, Phys. Rev. B, **58**, 12454 (1998).
5. P. N. Brusov, V. N. Popov "*The superfluidity and collective properties of quantum liquids,* Nauka, Moscow, 1988.
6. K. Krishana *et al.*, *Science* **277**, 83, 1997.
7. R.B. Laughlin, *Phys. Rev. Lett.* **80**, 5188, 1998.
8. J.F. Annett, *et al.*, in "*Physical Properties of High Temperature Superconductors V*", D.M. Ginsberg (Ed.), World Scientific, 1996.
9. S. P. Kruchinin , Physica C, **282-285**, p. 1397, 1997
10. P.N .Brusov, N. P. Brusova, *Physica B,* **194-196,** p. 1479, 1994.
11. P.N. Brusov, N. P.Brusova , *J. Low Temp. Phys* , **103**, p. 251, 1996.
12. P.N .Brusov et al., *Physica C* **282- 287**, p.1883, 1997.

VORTEX MATTER AND TEMPERATURE DEPENDENCE OF THE GINZBURG-LANDAU PHENOMENOLOGICAL LENGTHS IN LEAD NANOWIRES

G. Stenuit (gstenuit@tyndall.ie)
Tyndall National Institute, University College, Lee Maltings, Prospect Row, Cork, Ireland

J. Govaerts
Institut de Physique Nucléaire (FYNU), Université catholique de Louvain, Louvain-la-Neuve, Belgium

S. Michotte, L. Piraux
Unité de Physico-Chimie et de Physique des Matériaux (PCPM), Université catholique de Louvain, Louvain-la-Neuve, Belgium

Abstract. Numerical solutions of the Ginzburg–Landau (GL) equations for cylindrical configurations have been developed to study the magnetization of two superconducting lead nanowires arrays, electrodeposited under either constant or pulsed voltage conditions. By freely adjusting the GL phenomenological lengths $\lambda(T)$ and $\xi(T)$, the experimental magnetization curves, far from and close to the critical temperature T_c are reproduced to within a 10% error margin. Beyond this agreement, the temperature dependence of the adjusted phenomenological lengths are also compared to different theoretical and empirical laws. The Gorter-Casimir two-fluid model then gives the most satisfactory agreement for both samples. A distinction between them is next achieved by studying the extrapolated penetration depths at zero Kelvin. In particular, a comparison in terms of their electronic mean free paths agrees with the experimental expectation given with both experimental electrodeposition techniques.

Key words: Superconducting Nanowires, Ginzburg-Landau, Vortex, Magnetic Properties

1. Introduction

Recent developments in nanotechnologies and measurement techniques today allow the experimental investigation of the magnetic and thermodynamic superconducting properties of mesoscopic samples far away from the critical temperature T_c (Geim *et al.*, 1997; Michotte *et al.*, 2002). The present work concerns the magnetization curves (magnetization versus external magnetic field at a given temperature T) of an array of lead nanowires. The lead

K. Scharnberg and S. Kruchinin (eds.),
Electron Correlation in New Materials and Nanosystems, 293–302.
© 2007 *Springer.*

nanowire arrays were electrodeposited inside a nanoporous alumina membrane. Such a "template" contains a densely packed array of regularly shaped pores (SEM pictures of the alumina give a mean radius of 120 nm and a standard deviation of 15 nm) all straight and parallel to one another and to the applied external magnetic field, without lateral cross-over between adjacent ones. The mesoscopic superconducting nanowires were grown under either constant voltage potential (Stenuit et al., 2003) or pulsed potential (Stenuit et al., 2005) conditions, with their diameters and heights fixed by the alumina template. In order to compare the experimental magnetization curves to the theoretical predictions, some assumptions must be considered to simplify the numerical model described below.

First of all, due to the small radius of the nanowires, only the one dimensional radial Ginzburg-Landau (Ginzburg et al., 1950) equations are used (Singha Deo et al., 1999; Schweigert et al., 1999; Palacios, 1999; Pogosov, 2002; Govaerts et al., 2000; Stenuit et al., 2000). Indeed, applying the definition of the vorticity $L[C]$ associated with a path C and defined by the contour integral,

$$L[C] = -\frac{1}{2\pi} \oint_C \vec{\delta}\theta \cdot d\vec{l},$$
(1)

where θ is the phase of the Cooper pair wavefunction, one finds that for such sizes and temperatures, the phase diagram (B_{ext}, Energy) exhibits only those cylindrically symmetric states with a vorticity L which is strictly less than 3 since Giant Vortex states with higher vorticity are too large compared to the radius [1]. Nevertheless, the symmetry under rotation was verified by solving the 2D-GL equations in the (r, ϕ) plane of the cylindrical coordinates (Stenuit et al., 2003; Stenuit, 2004).

In addition, the magnetization presents a weak diamagnetic (in the $L = 0$ state) and paramagnetic (in the $L \geq 1$ state) response in the mesoscopic limit. The mutual magnetic interactions between each pairs of nanowires may thus be neglected in our model.

Therefore, considering a Gaussian distribution for the radii of nanowires making up the array, the total magnetization may be expressed as the sum of the magnetization of each type of nanowire multiplied by their statistical weight (Stenuit et al., 2002; Stenuit et al., 2003; Stenuit et al., 2005; Stenuit, 2004):

$$M_{tot}(B_{ext}, T) = A \sum_{r_i} \alpha[r_i, \mu, \sigma] \times m_{GL}[B_{ext}, r_i, \lambda(T), \xi(T)],$$
(2)

where,

---

[1] From Eq. (1), the cylindrically symmetric states whose vorticity are equal to $L = 0, 1$ and $L \geq 2$ describe the Meissner state, one Abrikosov vortex and the Giant Vortex state respectively.

- A is a scaling factor related to the number N of nanowires inside the sample: $N = A/(h\Phi_0)$ (with h being the height of the nanowires);

- $m_{GL}[B_{ext}, r_i, \lambda, \xi]$ is the theoretical magnetization of one nanowire with a radius r_i and a phenomenological parameter $\kappa(T) = \lambda(T)/\xi(T)$, fixed by the freely adjustable $\lambda(T)$ and $\xi(T)$. This quantity, m_{GL}, is obtained numerically by using a minimization procedure on the GL free energy over the field configuration space (see (Stenuit et al., 2001; Stenuit et al., 2005; Stenuit, 2004) for further details);

- $\alpha[r_i, \mu, \sigma] = \dfrac{1}{\sigma\sqrt{2\pi}}e^{-\frac{(r_i-\mu)^2}{2\sigma^2}}$ is the weight factor associated to the magnetization $m_{GL}(r_i)$, thus assuming a gaussian distribution with a mean radius μ and a variance σ^2 for the radii r_i;

- the summation extends over all radii from $\mu - 3\sigma$ to $\mu + 3\sigma$ in 2 nm steps.

The present paper focuses on the comparison between numerical predictions based on the GL equations and the experimental magnetization curves of arrays of lead nanowires electrodeposited under constant (Stenuit et al., 2003) or pulsed (Stenuit et al., 2005) voltage conditions. Our main ambition is then to compare the different GL lengths $\lambda(T)$ and $\xi(T)$ extracted from magnetization fits with the different theoretical and empirical temperature dependences, and finally to extract the electronic mean free paths of these nanowires.

2.   Analysis of the experimental results

The analysis, the details of which are available elsewhere (Stenuit et al., 2002; Stenuit et al., 2003; Stenuit et al., 2005; Stenuit, 2004), involves a comparison between experimental and theoretical magnetization curves of lead nanowires (T_c =7.2 K) electrodeposited under constant and pulsed voltage conditions. For the present section, only the sample electrodeposited under a pulsed voltage with a mean radius μ of 120 nm and variance σ of 15 nm will be considered. In the next section we will compare the results obtained in both samples (electrodeposited under a constant and pulsed voltage conditions). The values of the mean radius and its standard deviation being fixed and confirmed by analysing SEM pictures of the alumina matrix, the free parameters of the model (2) are $\lambda(T_i)$ ($= \lambda_i$, the penetration length) and $\xi(T_i)$ ($= \xi_i$, the coherence length) at a chosen temperature $T_i \leq T_c$. Finally, in order to study the hysteretic behaviour of the sample, it must also be mentioned that the experimental magnetization curves have been obtained when the external magnetic field (parallel to the z-axis of the nanowires) is swept up and down after zero field cooling. Figure 1-a presents a comparison between the

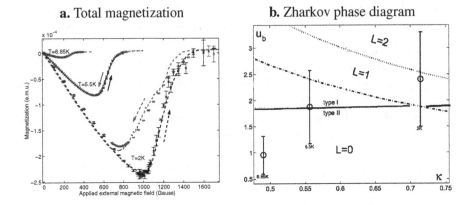

a. Total magnetization   **b.** Zharkov phase diagram

Figure 1. **a.** Comparison between the experimental (markers with error bars) and theoretical (dashed lines) total magnetization. From top to bottom, magnetizations at $T_1 = 6.85$ K, $T_2 = 5.5$ K and $T_3 = 2$ K, respectively. The adjusted values for the characteristic lengths are $(\lambda(T_i), \xi(T_i)) = (\lambda_i, \xi_i) = (125, 255), (64, 115)$ and $(50, 70)$ (nm,nm), respectively. **b.** Representation in the Zharkov phase diagram $(u_b = R/\lambda, \kappa)$ of the adjusted pairs $(\frac{\mu \pm \sigma}{\lambda_i}, \kappa_i = \frac{\lambda_i}{\xi_i})$. The solid line delimits the type-I (discontinuous) and type-II (continuous) superconducting-normal phase transition. The dashed (dotted) line is the boundary for the presence of the $L = 1$ ($L = 2$) state. These figures bring to the fore the three main characteristic magnetic behaviours observed at mesoscopic scales.

experimental and theoretical results for three different temperatures below T_c. These temperatures are chosen in order to exhibit the three main magnetic behaviours encountered at mesoscopic scales: a type-II phase transition between the normal and the superconducting states for the highest temperatures; a type-I transition to the Meissner state at intermediate temperatures and, the presence of vortex states (and therefore of the Bean-Livingston barrier (Bean *et al.*, 1964)) for the lowest temperatures. Such magnetic properties are best displayed in a Zharkov phase diagram $(R/\lambda(T), \kappa(T) = \lambda(T)/\xi(T))$, where R is the radius of an infinite height cylinder submitted to an external magnetic field parallel to its z−axis (Zharkov, 2001). On Figure 1-b, the solid line delimits the type-I and type-II superconducting-normal transitions, while the dashed (dotted) line fixes the limit of existence of the $L = 1$ ($L = 2$) state. The circles represent then the $(\mu/\lambda_i, \kappa_i)$-pairs, where the error bars are related to the dispersion $\pm \sigma/\lambda_i$ of the radii constituting the nanowires array, and the (λ_i, ξ_i)-pairs are the values used to reproduce the experimental magnetization curves (*Figure* 1-a). According to this diagram, the three main magnetic behaviours are again emphasized. In particular, for these three temperatures, the majority of the normalized radii $(\mu \pm \sigma)/\lambda(T_i)$ (error bars on Figure 1-b) belongs well to the three main magnetic properties already described.

TABLE I. Numerical values of the phenomenological lengths $\lambda(T_i)$ and $\xi(T_i)$ adjusted to fit the experimental magnetization curves measured at the temperature T_i. The superscript c (p) corresponds to a sample electrodeposited under constant (pulsed) voltage condition, respectively.

T_i (K)	6.85	6.5	6.25	6	5.75	5.5	5	4.5	3.25	2
$\lambda^c(T_i)$ (nm)	120	75	70	65	58		51	51	48	48
$\xi^c(T_i)$ (nm)	260	197	168	153	139		113	100	81	72
$\lambda^p(T_i)$ (nm)	125	92	83	72	67	64	58	55	53	50
$\xi^p(T_i)$ (nm)	255	180	154	137	124	115	102	90	75	70

Beyond the qualitative and quantitative agreement between the magnetization curves, it should finally be stressed that for the lowest temperatures, the roughness of the nanowires surface decreases the hysteretic behaviour (Geim et al., 2000). An additional free parameter whose value remains less than 20% of the sample's critical field was thus introduced in order to account for such a reduction effect (Stenuit et al., 2005; Stenuit, 2004).

To summarize this section, Table I lists the adjusted values of the characteristic lengths used to fit the experimental magnetizations at different temperature T_i. The superscripts p and c refer to the type of voltage (pulsed (Stenuit et al., 2005) or constant (Stenuit et al., 2003)) used during the electrodeposition of the samples.

3. Temperature dependence of the penetration and coherence lengths

By comparing the adjusted values of $\lambda(T_i)$ (Table I) to some theoretical (Tinkham, 1996) and empirical predictions, i.e. the London limit for pure or dirty materials, the Pippard limit, and the two-fluid model, one concludes (Stenuit et al., 2005; Stenuit, 2004) that the best fit is obtained here with the empirical two-fluid model (Gorter et al., 1934),

$$\lambda(\lambda_0, T_c, T) = \frac{\lambda_0}{\sqrt{1 - t^4}}, \qquad (3)$$

where λ_0 and $t = T/T_c$ are the penetration depth at zero Kelvin and the reduced temperature, respectively.

Once this conclusion has been established, one can now include the determined magnitudes of the Ginzburg-Landau coherence lengths (Table I),

$\xi(T_i)$, through an extended two-fluid model for $\xi(T)$,

$$\xi(\lambda_0, T_c, T) = \frac{\xi_0 \lambda_L(0)\pi}{2\sqrt{3}} \frac{1}{\lambda_0} \frac{\sqrt{1 - t^4}}{1 - t^2}, \tag{4}$$

with the Pippard coherence length, $\xi_0 = 87$ nm, and the London penetration depth, $\lambda_L(0) = 39$ nm (Yang *et al.*, 2003; Anderson, 1989). Equation Eq. 4 has been derived from the relation (4.23) in (Tinkham, 1996) and by using the two-fluid temperature dependence for $\lambda(T)$ (Eq. 3) and for the critical magnetic field $H_c(T) = H_c(0)(1 - t^2)$. The two penetration depths $\lambda_L(0)$ and λ_0 in Eq. 4 describe respectively the bulk London penetration depth at zero Kelvin, while λ_0 is the actual penetration depth of the nanowires at zero Kelvin. A chi-squared method which minimizes the following expression,

$$\chi^2(\lambda_0, T_c) = \sum_{T_i} \left\{ \left(\frac{\lambda(T_i) - \lambda_{fit}(\lambda_0, T_c, T_i)}{\Delta\lambda(T_i)} \right)^2 + \left(\frac{\xi(T_i) - \xi_{fit}(\lambda_0, T_c, T_i)}{\Delta\xi(T_i)} \right)^2 \right\}, \tag{5}$$

has been used to determine the free parameters λ_0 and T_c for both samples (pulsed (p) and constant (c) conditions) (Stenuit *et al.*, 2005; Stenuit, 2004). In this relation, $\Delta\lambda(T_i) = 0.1 \times \lambda(T_i)$ and $\Delta\xi(T_i) = 0.1 \times \xi(T_i)$ represent a 10% error margin associated to the adjusted values of $\lambda(T_i)$ and $\xi(T_i)$, respectively.

a. Constant voltage condition   **b.** Pulsed voltage condition

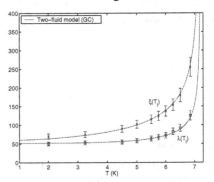

Figure 2. The two figures display the fit through the values of $\lambda(T_i)$ and $\xi(T_i)$ (Table I) determined from the magnetization model (2). The figure on the left (right) refers to the sample electrodeposited under constant (pulsed) voltage condition. The error bars represent a 10% variation around the given values of $\lambda(T_i)$ and $\xi(T_i)$.

Figures 2-a (constant voltage sample) and -b (pulsed voltage sample) show the temperature dependence (Eq. 3 and 4) of the characteristic lengths $\lambda(T)$ and $\xi(T)$ adjusted with the previous chi-squared method (Eq. 5). All lengths agree within 10% (the error bars) with the empirical two-fluid law when

$$\lambda_0^c = 47.0^{+1.1}_{-1.1} \text{ nm, and } T_c^c = 7.247^{+0.055}_{-0.046} \text{ K,} \tag{6}$$

for the sample electrodeposited under constant voltage condition (*Figure* 2-a) and,

$$\lambda_0^p = 51.9^{+1.2}_{-1.1} \text{ nm, and } T_c^p = 7.21^{+0.045}_{-0.040} \text{ K,} \tag{7}$$

for the pulsed one (*Figure* 2-b). The corresponding errors were estimated by increasing by one unit the χ^2 value around its minimal value (Barlow, 2004).

4. Comparison between both samples

4.1. THE CRITICAL TEMPERATURE

Although the critical temperatures T_c^c and T_c^p differ slightly, no conclusion between both values may be drawn, since the associated error margins largely overlap. Furthermore, it should be emphasized that inclusive of their errors, our values are consistent with the well-known experimental value for lead, $T_c = 7.22$ K (Gray, 1957).

4.2. MEAN FREE PATH

By contrast with the previous analysis on the critical temperature, the two adjusted penetration lengths at zero Kelvin, $\lambda_0^c = 47.0^{+1.1}_{-1.1}$ nm and $\lambda_0^p = 51.9^{+1.2}_{-1.1}$ nm, are significantly different and thus reveal the different electronic properties (in terms of the mean free path l) of the two samples.

According to the BCS and Pippard theories (Bardeen *et al.*, 1957; Pippard, 1953), the dependence of the penetration depth at zero Kelvin on the electronic mean free path can be expressed as,

$$\lambda_0 = \frac{2}{\pi} \int_0^\infty \frac{dq}{K_p(q) + q^2}, \tag{8}$$

where,

$$K_p(q) = \frac{1}{\lambda_L^2} \frac{\xi_p}{\xi_0} \frac{3}{2(q\xi_p)^3} \left\{ [1 + (q\xi_p)^2] \arctan(q\xi_p) - q\xi_p \right\}, \text{ and } \frac{1}{\xi_p} = \frac{1}{\xi_0'} + \frac{1}{l}.$$

At $T = 0$ Kelvin, the BCS coherence length, $\xi_0' = \xi_0/J(0, T = 0) = \xi_0$, becomes similar to the Pippard length ξ_0 (the Kernel $J(0, T) = (1 - 0.25 T/T_c)$ then reduces to unity). Therefore, the penetration depth at zero Kelvin ($\lambda_0 = \lambda_0^c$ or λ_0^p), the London penetration depth ($\lambda_L(0) = 39$ nm) and the Pippard coherence length (ξ_0 = 87 nm) being fixed, the only dependence remaining concerns the electronic mean free path l. By solving the transcendental equation (Eq. 8), one finds,

$$l^c = 360^{+159}_{-83} \text{ nm and } l^p = 142^{+21}_{-19} \text{ nm.}$$

The errors were estimated by reproducing the penetration depths within their error margins. Notice that the asymmetrical behaviour arises from the $\lambda(l) \propto 1/\sqrt{l}$ dependence of the penetration depth with the electronic mean free path.

Through this analysis, since $l^p < l^c$ and that the mean diameters of the samples are very close to one another (232 nm and 240 nm for the samples electrodeposited under constant and pulsed voltage condition respectively), one can conclude that the sample electrodeposited under a pulsed voltage condition is the "dirtiest". This comparative result is in agreement with the expectations given the two techniques involved in the electrodeposition: the pulsed technique generates smaller grains (Natter et al., 2003), which of course reduce the mean free path, as is indeed observed ($l^p < l^c$).

5. Comments and conclusions

Experimental magnetization curves of lead nanowire arrays, measured at different temperatures, were analysed. The samples were electrodeposited under constant and pulsed voltage conditions. The Ginzburg-Landau characteristic lengths were then freely adjusted for the different temperatures considered experimentally, in order to reproduce the observed magnetic properties measured with a SQUID magnetometer. The two dimensional GL approach allows us to determine the symmetry of the solutions that fits best the experimental results. In such samples, only cylindrically symmetric solutions are realized. In particular, the Meissner state ($L = 0$) and, for the lowest temperatures, the Abrikosov ($L = 1$) and Giant Vortex (GV, $L = 2, 3$) states were experimentally observed in these apparently type I superconductors. Through the Zharkov phase diagram, one clearly sees that the presence of a state with $L \neq 0$ is to be expected since the distinction between type I and II looses its significance at mesoscopic scales (Zharkov, 2001).

Beyond these agreements, our results also confirm the validity of the GL model even for temperatures significantly less than the critical temperature T_c. By adjusting both GL phenomenological lengths, we have been able to identify an extension of the temperature dependence of these two quantities far below T_c. In particular, for lead nanowires, this temperature dependence is well reproduced by the well-known two-fluid model.

Finally, our analysis involving the electronic mean free path shows, in agreement with the predictions, a significant difference between both samples electrodeposited under constant and pulsed voltage conditions.

Acknowledgements

When this work was performed, GS was supported as a research assistant of the "Institut Interuniversitaire des Sciences Nucléaires" (IISN, Belgium). GS is now supported as a post-doctoral researcher of the Science Foundation Ireland (SFI, Ireland). This work has also been partly supported by the Belgian Interuniversity Attraction Pole Program (PAI-IUAP P5/1/1) and by the "Communauté Française de Belgique" through the Program "Actions de Recherches Concertées".

References

A.K. Geim, I.V. Grigorieva, S.V. Dubonos, J.G.S. Lok, J.C. Maan, A.E. Filippov and F.M. Peeters, Nature 390 (1997) 259.

S. Michotte, L. Piraux, S. Dubois, F. Pailloux, G. Stenuit, J. Govaerts, Physica C377 (2002) 267-276.

G. Stenuit, S. Michotte, J. Govaerts and L. Piraux, Vortex Matter in Lead Nanowires, Eur. Phys. J. B 33 (2003) 103-107.

G. Stenuit, S. Michotte, J. Govaerts and L. Piraux, Supercond. Sci. Technol. 18 (2005) 174-182.

V.L. Ginzburg and L.D. Landau, Zh. Eksp. Teor. Fiz. 20, 1064 (1950) (in Russian).

P. Singha Deo, V.A. Schweigert, F.M. Peeters, Phys. Rev. B 59, 6039 (1999).

V.A. Schweigert, F.M. Peeters, Phys. Rev. Lett. 83, 2409 (1999).

J.J. Palacios, Phys. Rev. B 58, 5948 (1999).

W. V. Pogosov, Phys. Rev. B 65, 224511 (2002).

J. Govaerts, G. Stenuit, D. Bertrand and O. Van der Aa, Phys. Lett. A 267, 56 (2000).

G. Stenuit, J. Govaerts, D. Bertrand and O. Van der Aa, Physica C 332, 277 (2000).

G. Stenuit, S. Michotte, J. Govaerts, L. Piraux and D. Bertrand, Vortex Configurations in Mesoscopic Superconducting Nanowires, Mod. Phys. Lett. B17 (2003) 537-547 (Proceedings of the International Conference on Modern Problems in Superconductivity, Yalta (Ukraine), September 9 - 14, 2002).

G. Stenuit, Configurations de vortex magnétiques dans des cylindres mésoscopiques supraconducteurs, Ph.D. Thesis, Université catholique de Louvain, Louvain-la-Neuve, 2004.

G.F. Zharkov, Phys. Rev. B63 (2001) 224513.

G. Stenuit, J. Govaerts and D. Bertrand, Half-Integer Number Vortices in the Ginzburg-Landau-Higgs Model, edited by J. Annett, S. Kruchinin, NATO Science Series II, Vol.67, pp. 375-384 (Proceedings of the NATO Advanced Research Workshop "New Trends in Superconductivity", Yalta (Ukraine), September 16-20, 2001).

C.P. Bean and J.B. Livingston, Phys. Rev. Lett. 12 (1964) 14.

A.K. Geim, S.V. Dubonos, I.V. Grigorieva, K.S. Novoselov, F.M. Peeters and V.A. Schweigert, Nature 407 (2000) 55.

M. Tinkham, Introduction to Superconductivity, 2nd ed. (McGraw-Hill, New York, 1996).

C.J. Gorter and H.B.G. Casimir, Phys. Z. 35 (1934) 963; Physica 1 (1934) 306.

W.-H. Li, C.C. Yang, F.C. Tsao and K.C. Lee, Phys. Rev. B68 (2003) 184507.

A Physicist's Desk Reference, 2nd ed., edited by H.L. Anderson (AIP, New York, 1989), p. 117, cited in the Ref. (Yang et al., 2003).

R. Barlow, Asymmetric errors, Preprint physics/0401042 (2004).

American Institute of Physics Handbook, edited by D.E. Gray, (Mc-Graw-Hill, New York, 1957), pp. 4-49.

J. Bardeen, L.N. Cooper and J.R. Schrieffer, Phys. Rev. 108, 1175 (1957).

A.B. Pippard, Proc. R. Soc. A 216, 547 (1953).

H. Natter and R. Hempelmann, Electrochimica Acta 49 (2003) 51-61.

ANGULAR DIMENSIONAL CROSSOVER
IN SUPERCONDUCTOR – NORMAL METAL MULTILAYERS

Serghej L. Prischepa (prischepa@bsuir.unibel.by),
Belarus State University of Informatics and Radioelectronics, P.Brovka str.6, Minsk 220013, Belarus

Carmine Attanasio, Carla Cirillo
Dipartimento di Fisica "E.R. Caianiello" and Laboratorio Regionale SuperMat CNR/INFM-Salerno, Università degli Studi di Salerno, Baronissi (Sa), I-84081, Italy

Abstract. The angular dependences of the upper critical magnetic fields, $B_c(\Theta)$, for proximity coupled Nb/Pd multilayers are presented. Dimensional properties of the superconducting phase with respect to the orientation of the external magnetic field are determined. The crossover angle Θ^*, at which the dimensionality of the multilayer system varies, is function of temperature. It is shown that the proximity coupling strongly affects the two dimensional behaviour.

Key words: Dimensional Crossover; Superconducting Multilayers, Upper Critical Magnetic Field, Proximity Effect

1. Introduction

Layered superconductors attract great attention due to the possibility of studying the effect of correlation and the nature of the coupling between the superconducting layers on the magnetic phase diagram, vortex and pinning properties, superconductivity nucleation, etc[1]. From this point of view the periodic superconductor/normal metal (S/N) multilayers preparing by layer-by-layer vacuum deposition are still very interesting system. The critical parameters of S/N periodic structures differ from the corresponding quantities for thin films and anisotropic superconductors. Varying the layer thickness and the S and N materials it is possible to study both the proximity effect between the different metals in coupled multilayers and the behavior of the lattice of uncoupled superconducting films of different dimensionality.

K. Scharnberg and S. Kruchinin (eds.),
Electron Correlation in New Materials and Nanosystems, 303–315.
© 2007 *Springer.*

One of the fundamental quantities characterizing a superconducting state is the upper critical magnetic field B_c. A lot of efforts has been applied to study the temperature and the angular dependencies of B_c in different S/N systems[2-12]. In particular, rather comprehensive understanding of the nature of the dimensional crossover, namely, transition from the three-dimensional (3D) to two-dimensional (2D) mode upon a variation of the temperature T, the real finite value of the parallel critical field $B_{c\parallel}$ in the 2D case, leads to deeper insight of the interaction between the superconducting layers, effects of Pauli paramagnetism and spin-orbit coupling and of the superconducting phase nucleation[13-16] (see also the review[17] and references therein).

The crossover to the 2D state can be highlighted by studying the angular dependence of B_c. Initially the interest to the $B_c(\Theta)$ study (Θ being the angle between an applied magnetic field and a plane of a sample) was pushed by the possibilities to be an alternative test of the dimensionality and of studying the influence of the interface boundaries on the properties of the superconducting phase. The fundamental results concerning the $B_c(\Theta)$ dependences in homogeneous superconductors were obtained in the earlier works[18-21] on the base of the Ginzburg-Landau (GL) theory. For infinite anisotropic superconductors the dependence $B_c(\Theta)$ is a smooth, differentiable curve in the whole Θ range (including $\Theta = 0$). This curve is easily reproduced within the anisotropic effective mass GL theory[21]:

$$B_c^{3D}(\Theta) = B_c(0)\Big/\big(\cos^2\Theta + \gamma^2\sin^2\Theta\big)^{1/2} \tag{1}$$

where $\gamma = B_c(0)/B_c(\pi/2)$ is the anisotropy parameter. Later it was shown, that the same behavior for $B_c(\Theta)$ is also valid for stack of Josephson coupled superconducting sheets[22].

On the contrary, when one takes into account the presence of boundaries in the sample, the analytical properties of $B_c(\Theta)$ change drastically. First of all, for semi infinite sample, one should consider the angular dependence of the surface critical field, $B_{c3}(\Theta)$, rather than $B_c(\Theta)$, and, second, at $\Theta = 0$ the first derivative of $B_c(\Theta)$ curve has a discontinuity. The approximate $B_{c3}(\Theta)$ dependence was derived in[23], using the interpolation formula for Θ dependence on the spatial coordinate describing the position of the nucleation of the superconducting phase.

For superconductors with two flat surfaces the precise results were obtained in two limiting cases (2D and 3D modes). Tinkham derived the formula for the $B_c(\Theta)$ dependence in the thin film 2D limit $(d \ll \xi)$[19], which gives a discontinuous derivative $\partial B_c(\Theta)/\partial\Theta$ at $\Theta = 0$,

$$\left|\frac{B_c^{2D}(\Theta)\sin\Theta}{B_c(\pi/2)}\right| + \left[\frac{B_c^{2D}(\Theta)\cos\Theta}{B_c(0)}\right]^2 = 1 \qquad (2)$$

At the same time Saint-James established that the $B_c(\Theta)$ shape and the nucleation position of the superconducting phase are mutually related to each other. He has formulated the perturbation theory for the GL equations solution[20] and obtained the slope $(1/B_c) \times \partial B_c / \partial \Theta \big|_{\Theta=0}$ of the $B_c(\Theta)$ characteristic as a function of dimensionless parameter $h = L(2\pi H/\Phi_0)^{1/2}$, where L is the sample thickness and Φ_0 is the flux quantum (see also[24]). This result includes as a limiting cases the regimes of very thin film[19] and of semi infinite sample[18]. In the thin film limit $h \ll h_1$ (h_1 being the parameter at which the second derivative of B_c with respect to Θ has a discontinuity) the nucleation always starts at the center of the slab. For large sample thickness $h \gg h_1$ the nucleation starts apart from the center and reaches the familiar surface superconductivity situation[18].

2. $B_c(\Theta)$ dependence in layered structures

Layered structure of superconducting and non superconducting films of finite thickness creates additional difficulties for understanding the $B_c(\Theta)$ dependences. Only qualitative change of the angular dependence in 2D and 3D modes has been observed in Josephson coupled artificial multilayers[22,24-26], in high-T_c (HTS) compounds with different anisotropy[27,28] and in proximity coupled S/N multilayers[2-4]. For example, in S/N multilayers in 2D mode (according to the temperature dependence of $B_{c\parallel}(T) \equiv B_c(T, \Theta = 0)$) the $B_c(\Theta)$ points fall faster with angle growth with respect to the Tinkham expression (2) for thin film[2,3]. Moreover, in the temperature region of linear $B_{c\parallel}(T)$ dependence (where the 3D case is realized), the angular dependences sometimes reveal the pronounced cusp[2,22], which is a feature of 2D regime. To resolve these discrepancies between theoretical descriptions of homogeneous superconductors and experiments on multilayers, Glazman proposed to take account of both the discrete structure of multilayer and of the finite force of the coupling between individual S layers in the sample[29]. As a result, for certain values of superconducting layer thickness and the thickness of interlayers separating them, the dimensional crossover could be observed on the $B_c(\Theta)$ dependence when the 3D mode at $\Theta = \pi/2$ changes smoothly into 2D whereas the angle Θ decreases to zero. Experimentally this kind of crossover was observed on S/I multilayers[26], HTS[28], S/M multilayers[12]. The main result of 2D-3D angular crossover could be easily applied to different types of multilayers due to the general character of the physical discussion performed in[29].

For Josephson coupled superlattices (infinite structures) the problem of reconstruction of the full angular dependence of B_c was considered in[30,31]. The attempt to describe the full angular dependence of B_c of layered superconductors, stimulated by the success in studying the properties of HTS, was also performed by Schnieder and Schmidt[32]. Their approach presents the modified Lawrence-Doniach (LD) model[21] and is based on the calculation of B_c of a periodic stack of superconducting slabs of finite thickness coupled through Josephson junctions[33]. The amplitude of the superconducting wave function was supposed to be constant over the film thickness, while the phase was varied between the neighboring S layers. Obviously, it is difficult to apply this model to proximity coupled structures. Attempt to resolve the discrepancy for infinite proximity coupled superlattices between the measured 3D behavior according to $B_{c\parallel}(T)$ dependence and 2D according to $B_c(\Theta)$ data has been done in[34]. It was shown that the main features of $B_c(\Theta)$ dependences of S/N multilayers remain the same as for isolated slabs. In particular, the $B_c(\Theta)$ curve has a cusp at $\Theta=0$ if the order parameter has a preferential position in the layer structure, which is essential for the real multilayers experimentalists usually deal with.

Calculations of the full $B_c(\Theta)$ dependences in proximity coupled S/N multilayers were performed in[35] on the base of the continuous anisotropic GL model with smooth periodic coefficient functions[36]. The key parameter of this model is the quantity $\Lambda/\xi_\perp$ (Λ being the period of the structure and $\xi_\perp$ being the perpendicular coherence length), which determines the character of B_c versus Θ dependence. The main results are that for layered S/N superconductors, as for thin films, in 2D regime the angular dependence of the critical field reveals a cusp at $\Theta = 0$, and in 3D regime $B_c(\Theta)$ takes the form of a bell-shaped curve similar to that in the LD model[21]. Moreover, the angular dependence of B_c in the region away from $\Theta = 0$ is not related to the dimensionality of $B_{c\parallel}(T)$. At the same time the model gives only qualitative explanation of the $B_c(\Theta)$ behavior in S/N multilayers. It has too many fitting parameters. In order to obtain the agreement with the experimental data it is necessary to resolve each time numerically the boundary task for differential equations.

The main reason for the discrepancy in $B_c(\Theta)$ behavior between the film and multilayers could be related to two facts: the real layered materials are often in the region of dimensional 2D-3D crossover, which usually is widely spread in temperature, and both models[19,21] are not applicable in this case and/or the integral effect of a large number of S/N interfaces influences significantly the nucleation position and spatial shape of the superconducting wave function, which, as a result, produces a change of an analytical expression for $B_c(\Theta)$ dependence. The solution of the first problem for S/I structures was proposed in[37], where the general analytical expression for the $B_c(\Theta)$ curve containing both expressions[19,21] as the limiting cases was derived. For proximity coupled metallic S/N multilayers both tasks were not analyzed up to now.

In this work, we present the results of measurements of the $B_c(\Theta)$ dependences for Nb/Pd multilayers, and we analyze the difference with the existing theoretical descriptions in order to deepen the understanding of the nature of the proximity coupling. We performed angular measurements in the temperature range where, according to $B_{c\|}(T)$ result, samples are in 2D mode. The measured $B_c(\Theta)$ data were not described by the 2D model (Eq.2). We determined the angle Θ^* at which the dimensional crossover occurs, and we obtained the temperature dependence of Θ^*. Finally, we show that for our S/N multilayers the actual temperature region of the dimensional crossover is widely spread due to the strong proximity effect between Nb and Pd.

3. Results

Here we discuss the results related to Nb/Pd sample (NP) containing 10 bilayers with $d_{Pd} = d_{Nb}/2 = 10$ nm. All the details about the sample preparation and its characterization could be found elsewhere[11,38].

In Ref. 38 we published the results of the detailed study of the B-T phase diagrams for this sample. We established that it reveals the usual $B_{c\|}(T)$ dependence for S/N multilayers having $d_S \sim d_N \sim \xi$, i.e. the presence of temperature dimensional crossover at temperature $T^* = 3.60$ K (Fig. 2a of Ref. 38). At $T < T^*$, $B_{c\|}(T) \sim (T_c^{2D} - T)^{1/2}$ (T_c^{2D} being the critical temperature in the 2D mode), which is a signature of the bi-dimensionality, while at $T > T^*$ multilayer behaves like a 3D system and $B_{c\|}(T) \sim (T_c - T)$.

We performed angular measurements in the temperature range 1.8 K $< T < T^*$. The B_c values were extracted from the $R(B)$ curves taken at different angles. Example of such curves for $T = 2.30$ K is shown in Fig. 1. In Fig. 2 $B_c(\Theta)$ measured at another temperature, 3.45 K, is presented. The general features of all the curves in this temperature range is the presence of a cusp at $\Theta = 0$. At a first glance, this result is a consequence of the bi-dimensional behavior of $B_{c\|}(T)$ at $T < T^*$, but the Tinkham`s expression (2) does not fit the experimental data (dashed curve in Fig. 2). As it was shown by Glazman for S/I superlattices[29], the reason is the incorrect application of Eq. 2 for describing the *whole* $B_c(\Theta)$ dependence. In fact, the $B_c(\Theta)$ curve is divided in two parts which are related to two different dimensionalities of the sample even at $T < T^*$. In other words, it exists an angle Θ^*, such that when $\Theta < \Theta^*$ the experimental dependence is well described by the Tinkham's expression, in which, however, the $B_{c\perp}$ value is a free parameter and it is smaller than the measured value of the perpendicular critical magnetic field. In our case this fit is shown as a solid line in Fig. 2. For $\Theta > \Theta^*$, in the case of S/I superlattices, the 3D mode takes place and the angular dependence of B_c is described by the Eq. 1[26]. The 3D-2D angular crossover occurs in this case smoothly. In our sample, as it is seen from Fig. 2, the situation is different. At $T = 3.45$ K a sharp transition takes place at the angle Θ^*. Actually the same is valid also for another temperatures[39]. We suppose that this angular crossover has the same physical meaning of the temperature crossover. The peculiarity is related to the unusual 3D mode at $\Theta > \Theta^*$. We tried to do fit both using Eq. 1 (short-dotted line in Fig. 2) and Eq. 2 (dash-dotted line in Fig. 2) with free parameter $B_{c\|}$ and we obtain the practically ideal correspondence to the experimental data of the Tinkham's formula. Physical reason for this result will be discussed later.

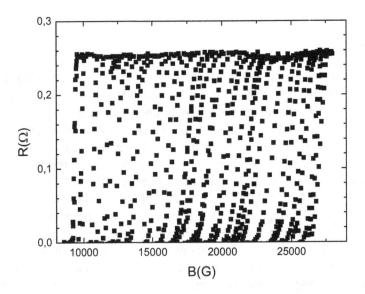

Figure 1. R(B) curves for Θ values from $0°$ to $90°$ at $T = 2.30$ K.

The obtained angle Θ^* is a function of temperature, which is shown in Fig. 3. It is interesting to note, that the crossover angle Θ^* could be obtained in a primitive way. In Fig. 2 the result for Tinkham formula is shown by dashed lines. The deviation of the experimental data from the Tinkham's behavior could be illuminated better by plotting the ratio $\delta = [B_c(\Theta)/B_c^{2D}(\Theta)]^2$ as a function of Θ as it is shown in Fig. 4. When magnetic field is directed along the basal plane of the substrate or at the perpendicular direction, $\delta = 1$, since the experimental values of $B_{c\|}$ and $B_{c\perp}$ were used as parameters for the Tinkham's formula (2).

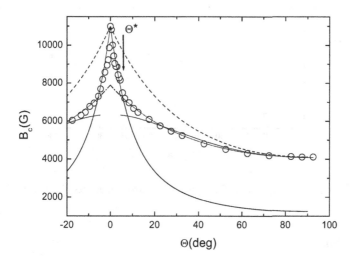

Figure 2. Angular dependence of the upper critical magnetic field at $T = 3.45$ K. Arrow indicates the value of Θ^*.

The shape of the deviations as a function of angle is similar for all temperatures, but the angle Θ_{max}, at which the deviation $\Delta\delta(\Theta) = 1 - \delta(\Theta)$ reaches its maximum value $\Delta\delta_{max}$, is temperature dependent, as it is shown in Fig. 3. As it is seen, the values of Θ_{max} coincide well with Θ^*. Coincide also their temperature dependences.

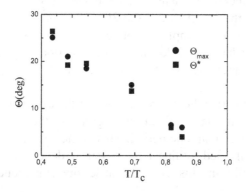

Figure 3. Temperature dependences of Θ^* (squares) and Θ_{max} (circles).

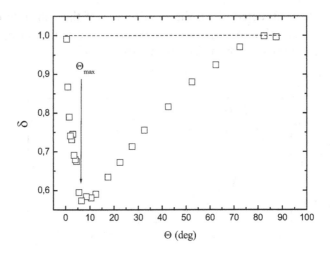

Figure 4. Relative deviation δ as a function of Θ. Arrow indicates the value of Θ_{max}. $T = 3.45$ K.

4. Discussion

The obtained 2D mode at $\Theta < \Theta^*$ for the sample NP is related to the coherence length values in the temperature range 2.0K...3.6K. Indeed, from $B_{c\perp}(T)$ dependence we may obtain the GL coherence length at zero temperature $\xi(0)$, which is equal to 12.6 nm. From this we get that $\xi(T) = \xi(0)(1-T/T_c)^{-1/2}$ changes from ≈18 nm at 2K up to ≈33 nm at 3.60 K, i.e. it is order of d_{Nb} and less or order of the period of the multilayer structure ($\Lambda = d_{Nb} + d_{Pd}$). This means that for $T < T^*$ in parallel magnetic field the superconducting nucleus is really localized in one period of the S/N structure, and in this temperature range the wave function of the superconducting condensate changes relatively weakly on the thickness of the S layer[40]. Both these facts mean the 2D nature of superconductivity in parallel field. When the sample is rotated from $\Theta = 0$ towards Θ^* the superconducting nucleus is still localized in one period of the S/N structure and the S/N interfaces favorite the localization of the superconducting nucleus.

For $\Theta > \Theta^*$ the perpendicular component of the external magnetic field has an influence on the localization degree of the order parameter, "spreading" it on more than one period. This is the characteristic feature of 3D object as is confirmed by linear $B_{c\perp}$ versus T dependence. But the LD formula is not valid in this case, because the superconducting order parameter, remaining relatively homogeneous in one S layer, changes significantly in the whole sample. This probably explains the applicability of the Tinkham formula (2) also at $\Theta > \Theta^*$. So, we get, from one side, at $\Theta > \Theta^*$ structure with bound S layers (i.e. 3D structure), and from the other, at $\Theta < \Theta^*$, we get an analog of the homogeneous thin film and at Θ^* the change of the topology of the superconducting nucleus takes place. Therefore the B_c angular dependence measurements look like a new useful tool for the investigation of dimensionality effects in anisotropic superconductors.

Within the above picture it is possible to explain the smaller $B_{c\perp}$ values obtained from the fitting procedure using Eq. 2 at $\Theta < \Theta^*$. In this angular range our sample is reduced to a simple 1 bilayer S/N structure with smaller T_c value with respect to the 3D multilayer structure (which has larger total S layer thickness). The reduction of Θ^* with temperature (Fig. 3) can also be explained within the proposed description. In fact, while increasing the temperature the $\xi(T)$ increases, and this reduces the localization degree of the superconducting nucleus in parallel field and causes its delocalization at lower Θ^* values.

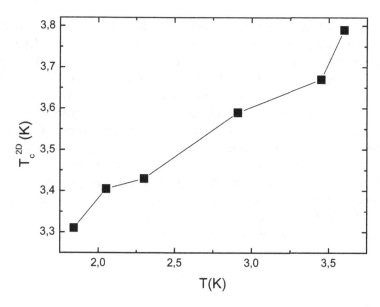

Figure 5. Temperature dependence of T_c^{2D}.

The obtained results help us to better understand the mechanism of the proximity coupling in S/N multilayers. Indeed, the reduction of the multilayer to 1 period at $\Theta < \Theta^*$ means that the $B_c(\Theta)$ dependence in this angle range is described by Eq. 2, but with the fitting parameter $B_{c\perp}^{2D} < B_{c\perp}$, where $B_{c\perp}^{2D}$ is the critical field of one Nb layer located between two Pd layers. The value of $B_{c\perp}^{2D}$ can be related to $B_{c\perp}$ as[26]

$$B_{c\perp}^{2D} = B_{c\perp}\left(T_c^{2D} - T\right)/\left(T_c - T\right) \tag{3}$$

where T_c^{2D} is the critical temperature of one S layer imbedded between two N layers. For a pure 2D system this value should be constant. In Fig. 5 we plotted the T_c^{2D} values obtained from the fitting procedure for different temperatures. It is clearly seen, that T_c^{2D} grows with temperature. It means, that for S/N proximity coupled multilayers even at $T < T^*$ and $\Theta < \Theta^*$ thin superconducting films are not in the pure 2D mode. Proximity coupling still influences the dimensional properties of multilayers and due to the final interface transparency the real superconducting thickness depends on the temperature[41, 42].

5. Conclusion

In conclusion, we investigated the dimensionality of proximity coupled multilayers by means of angle measurements. The dimensionality depends on the orientation of the magnetic field. We established the angle $\Theta^*(T)$ of dimensional crossover. At $\Theta < \Theta^*$ multilayer behaves like 2D and at $\Theta > \Theta^*$ it is in 3D mode. Strong proximity coupling between Nb and Pd causes the peculiarities of the bi-dimensionality in multilayers which is related to the change of the effective superconducting thickness of S layer located between two N layers with temperature. That is the reason that the separation of the S films is essential, but not sufficient feature for the observation of the "pure" 2D state in S/N multilayers. The proposed novel tool of investigation of the dimensional effects and crossover phenomena also could be applied for new layered superconductors, like magnesium diboride, in which the dimensional effects play an important role[43].

References

1.   G. Blatter, M.V. Feigelman, V.B. Geshkenbein, A.I. Larkin, and V.M. Vinokur, Vortices in high temperature superconductors, *Rev. Mod. Phys.* 66(4), 1125-1388 (1994).

2. I. Banerjee, Q.S. Yang, C.M. Falco, and I.K. Schuller, Anisotropic critical fields in superconducting superlattices, *Phys. Rev. B* 28(9), 5037-5040 (1983).

3. C.S.L. Chun, G.G. Zheng, J.L. Vicent, and I.K. Schuller, Dimensional crossover in superlattice superconductors, *Phys. Rev. B* 29(9), 4915-4920 (1984).

4. I. Banerjee and I.K. Schuller, Transition temperatures and critical fields of Nb/Cu superlattices, *J. Low Temp. Phys.* 54(5/6), 501-517 (1984).

5. J. Guimpel, M.E. de la Cruz, F. de la Cruz, H.J. Fink, O. Laborde, and J.C. Villigier, Size dependence of the superconducting critical temperature and fields of Nb/Al multilayers, *J. Low Temp. Phys.* 63(1/2), 151-165 (1986).

6. K. Kanoda, H. Mazaki, T. Yamada, N. Hosoito, and T. Shinjo, Dimensional crossover and commensurability effect in V/Ag superconducting multilayers, *Phys. Rev. B* 33(3), 2052-2055 (1986).

7. J.P. Locquet, W. Sevenhans, Y. Bruynseraede, H. Homma, and I.K. Schuller, Nature of coupling and dimensional crossover in superconducting multilayers, *IEEE Trans. Magn.* 23(2), 1393-1396 (1987).

8. P.R. Broussard and T.H. Geballe, Critical fields of Nb-Ta multilayers, *Phys. Rev. B* 35(4), 1664-1668 (1987).

9. V.I. Dediu, V.V. Kabanov, and A.A. Sidorenko, Dimensional effects in V/Cu superconducting superlattices, *Phys. Rev. B* 49(6), 4027-4032 (1994).

10. C. Coccorese, C. Attanasio, L.V. Mercaldo, M. Salvato, L. Maritato, J.M. Slaughter, C.M. Falco, S.L. Prischepa, and B.I. Ivlev, Vortex properties in superconducting Nb/Pd multilayers, *Phys. Rev. B* 57(13), 7922-7929 (1998).

11. C. Cirillo, C. Attanasio, L. Maritato, L.V. Mercaldo, S.L. Prischepa, and M. Salvato, Upper critical fields of Nb/Pd multilayers, *J. Low Temp. Phys.* 130(5/6), 509-528 (2003).

12. C. Attanasio, C. Coccorese, L.V. Mercaldo, M. Salvato, L. Maritato, A.N. Lykov, S.L. Prischepa, and C.M. Falco, Angular dependence of the upper critical field in Nb/CuMn multilayers, *Phys. Rev. B* 57(10), 6056-6060 (1998).

13. S. Takahashi and M. Tachiki, Theory of the upper critical field of superconducting superlattices, *Phys. Rev. B* 33(7), 4620-4631 (1986).

14. R.T.W. Koperdraad and A. Lodder, Magnetic-coherence-length scaling in metallic multilayers, *Phys. Rev. B* 54(1), 515-522 (1996).

15. C. Ciuhu and A. Lodder, Influence of the boundary resistivity on the proximity effect, *Phys. Rev. B* 64(22), 224526 (2001).

16. R.A. Klemm, A. Luther, and M.R. Bealsey, Theory of the upper critical field in layered superconductors, *Phys. Rev. B* 12(3), 877-891 (1975).

17. B.Y. Jin and J.B. Ketterson, Artificial metallic superlattices, *Adv. Phys.* 38(3), 189-366 (1989).

18. D. Saint-James and P.G. De Gennes, Onset of superconductivity in decreasing fields, *Phys. Lett.* 7(5), 306-308 (1963).

19. M. Tinkham, Effect of fluxoid quantization on transition of superconducting films, *Phys. Rev.* 129(6), 2413-2422 (1963).

20. D. Saint-James, Angular dependence of the upper critical field of type II superconductors ; theory, *Phys. Lett.* 16(3), 218-220 (1965).

21. W.E. Lawrence and S. Doniach, Theory of layer structure superconductors, Proceedings of the 12th International Conference on Low Temperature Physics, Kyoto, 1970, edited by E. Kanada (Keigaki, Tokyo, 1971) p. 361-362.

22. S.T. Ruggiero, T.W. Barbee Jr., and M.R. Beasley, Superconducting properties of Nb/Ge metal semiconductor multilayers, *Phys. Rev. B* 26(9), 4894-4908 (1982).

23. K. Yamafuji, E. Kusayanagi, and F. Irie, On the angular dependence of the surface superconducting critical field, *Phys. Lett.* 21(1), 11-13 (1966).

24. R.S. Thompson, Tilted vortices in type II superconductors, *Zh. Eksp. Teor. Fiz.* 69(6), 2249 (1975) [*Sov. Phys. JETP* 42(6), 1144 (1975)].

25. B.Y. Jin, J.B. Ketterson, E.J. McNiff, S. Foner, and I.K. Schuller, Anisotropic upper critical fields of disordered $Nb_{0.53}Ti_{0.47}$-Ge multilayers, *J. Low Temp. Phys.* 69(1/2), 39-51 (1987).

26. 26 .L. Tovazhnyanskii, V.G. Cherkasova, and N.Ya. Fogel', Angular dependence of the critical field of superconducting superlattices: experiment, *Zh. Eksp. Teor. Fiz.* 93(4), 1384-1393 (1987) [*Sov. Phys. JETP* 66(4), 787-797 (1987)].

27. M.J. Naughton, R.C. Yu, P.K. Davies, J.E. Fischer, R.V. Chamberlin, Z.Z. Wang, T.W. Jing, N.P. Ong, and P.M. Chaikin, Orientational anisotropy of the upper critical field in single crystal $YBa_2Cu_3O_7$ and $Bi_{2.2}CaSr_{1.9}Cu_2O_{8+x}$, *Phys. Rev. B* 38(13), 9280-9283 (1988).

28. R. Fastampa, M. Giura, R. Marcon, and E. Silva, 2D to 3D crossover in Bi-Sr-Ca-Cu-O: comparison with synthetic multilayered superconductors, *Phys. Rev. Lett.* 67(13), 1795-1798 (1991).

29. L.I. Glazman, Angular dependence of the critical field of superconducting superlattices: theory, *Zh. Eksp. Teor. Fiz.* 93(4), 1373-1383 (1987) [*Sov. Phys. JETP* 66(4), 780-786 (1987)].

30. E.V. Minenko and I.O. Kulik, Tilted vortices in superconductors, *Fiz. Nizk. Temp.* 5(11), 1237 (1979) [*Sov. J. Low Temp. Phys.* 5, 583 (1979)].

31. V.M. Gvozdikov, Critical magnetic fields of superconducting superlattices, *Fiz. Nizk. Temp.* 16(1), 5-10 (1990) [*Sov. J. Low Temp. Phys.* 16(1), 1-6 (1990)].

32. T. Schneider and A. Schmidt, Dimensional crossover in the upper critical field of layered superconductors, *Phys. Rev B* 47(10), 5915-5921 (1993).

33. G. Deutscher and O. Entin-Wohlman, Critical fields of weakly coupled superconductors, *Phys. Rev. B* 17(3), 1249-1252 (1978).

34. K. Takanaka, Angular dependence of superconducting critical fields, *J. Phys. Soc. Jpn.* 58(2), 668-672 (1989).

35. N. Takezawa, T. Koyama, and M. Tachiki, Angular dependence of the upper critical field in layered superconductors, *Physica C* 207(3/4), 231-238 (1993).

36. T. Koyama, N. Takezawa, Y. Naruse, and M. Tachiki, New continuous model for intrinsic layered superconductors, *Physica C* 194(1/2), 20-30 (1992).

37. V.P. Mineev, General expression for the angular dependence of the critical field in layered superconductors, *Phys. Rev. B* 65(1), 012508 (2001).

38. V.N. Kushnir, S.L. Prischepa, C. Cirillo, M.L. Della Rocca, A. Angrisani Armenio, L. Maritato, M. Salvato, and C. Attanasio, Nucleation of superconductivity in finite metallic multilayers: Effect of the symmetry, *Europ. Phys. J. B* 41(4), 439-444 (2004).

39. S.L. Prischepa, C. Cirillo, V.N. Kushnir, E.A. Ilyina, M. Salvato, and C. Attanasio, Effect of geometrical symmetry on the angular dependence of the critical magnetic field in superconductor/normal metal multilayers, *Phys. Rev. B* 72(2), 024535 (2005).

40. V.N. Kushnir, S.L. Prischepa, M.L. Della Rocca, M. Salvato, and C. Attanasio, Effect of symmetry on the resistive characteristics of proximity coupled metallic multilayers, *Phys. Rev. B* 68(21), 212505 (2003).

41. A. Tesauro, A.Aurigemma, C. Cirillo, S.L. Prischepa, M. Salvato, and C. Attanasio, Interface transparency and proximity effect in Nb/Cu triple layers realized by sputtering and molecular beam epitaxy, *Supercond. Sci. Technol.* 18(1), 1-8 (2005).

42. C. Cirillo, S.L. Prischepa, M. Salvato, and C. Attanasio, Interface transparency of Nb/Pd layered systems, *Europ. Phys. J. B* 38(1), 59-64 (2004).

43. A.S. Sidorenko, L.R. Tagirov, A.N. Rossolenko, N.S. Sidorov, V.I. Zdravkov, V.V. Ryazanov, M. Klemm, S. Horn, and R. Tidecks, Fluctuation conductivity in superconducting MgB_2, *Pis`ma Zh. Eksper. Teor. Fiz.* 76(1), 20-24 (2002) [*JETP Letters* 76(1), 17-20 (2002)].

ANDREEV STATES AND SPONTANEOUS SPIN CURRENTS IN SUPERCONDUCTOR-FERROMAGNET PROXIMITY SYSTEMS

James F. Annett and Balazs L. Gyorffy
University of Bristol, H.H. Wills Physics Laboratory, Tyndall Avenue, Bristol, BS8 1TL, UK

M. Krawiec
Institute of Physics and Nanotechnology Center, M. Curie-Sklodowska University, pl. M. Curie-Sklodowskeij 1, PL 20-031 Lublin, Poland

Abstract. The proximity effect between a ferromagnet and a superconductor is examined within the framework of the self-consistent Bogoliubov de Gennes equations. It is shown that for a thin ferromagnetic layer on a semi-infinite superconductor, a spontaneous supercurrent may flow parallel to the interface. The current turns on or off repeatedly as a function of the exchange splitting, E_{ex} for a fixed thickness film, d. It is shown that the current state is stabilized by a lowering of the total quasiparticle kinetic energy, and this is interpreted in terms of the Andreev bound states in the ferromagnetic layer.

Key words: Ferromagnet-superconductor Interface, Proximity Effect, Andreev States, FFLO State, Spontaneous Supercurrent

1. Introduction

The proximity effects between conventional s-wave superconductors (S) and itinerant ferromagnets (F) have recently led to a number of surprising new experimental results. These include: the appearance of π-junction behaviour in Josephson junctions with a ferromagnetic tunnel barrier (Kontos, 2001; Ryazanov, 2001; Frolov, 2004), an unusual giant proximity effect (Petrashov, 1999), and negative refraction in F/S superlattices (Pimenov, 2005). These and other experiments have stimulated a considerable body of theoretical work, which was recently summarised in two major review articles (Lyuksyutov and Pokrovsky, 2005; Buzdin, 2005). Even more recently there have even been dramatic new demonstrations of long ranged proximity effects in S/F/S junctions in which it is believed that the supercurrents can only carried by spin triplet Cooper pairs, which could be generated by spin-orbit interactions at the S/F interfaces (Keizer, 2006; Krivoruchko, 2006; Sosnin et al., 2006).

K. Scharnberg and S. Kruchinin (eds.),
Electron Correlation in New Materials and Nanosystems, 317–324.
© 2007 *Springer.*

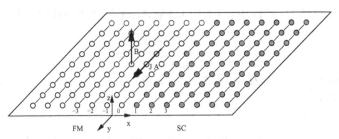

Figure 1. The model system we consider. The ferromagnet (FM) consists of d atomic layers ($x = -d, \ldots 0$), and is in contact with a semi-infinite superconductor ($x = 1, 2, \ldots$). A current, J, may flow along y, parallel to the surface. This current leads to a magnetic vector potential A in the same direction, and to a B-field in the z direction perpendicular to the current, as shown.

The fundamental interaction which is relevant to the proximity effect in superconductor (S) ferromagnet (F) interfaces is the exchange splitting of up and down spin states at the Fermi level. In bulk superconductors this exchange splitting leads to the famous Fulde-Ferrel (Fulde Ferrell, 1964) Larkin-Ovhinnikov (Larkin Ovchinnikov, 1964) state (FFLO) in which the superconducting order parameter oscillates in real space. In the S-F proximity effect the pairing amplitude $\chi(\mathbf{r}) = \langle c_{\mathbf{r}\downarrow} c_{\mathbf{r}\uparrow} \rangle$ becomes non-zero within the ferromagnet, and also acquires an oscillatory character,

$$\chi(\mathbf{r}) \equiv \langle c_{\mathbf{r}\downarrow} c_{\mathbf{r}\uparrow} \rangle \propto \frac{\sin(x/\xi_F)}{(x/\xi_F)} \tag{1}$$

where x is the distance into the ferromagnet from the superconducting interface, $\xi_F = 2t/E_{ex}$ is the ferromagnetic coherence length, t is the hopping, and E_{ex} is the exchange splitting in the ferromagnet. It is these oscillations which lead to sign changes in the S-F-S Josephson effect, and hence to the π-junction behaviour observed by Kontos *et al.* (Kontos, 2001) and others.

In a number of previous papers we have examined the self-consistent solutions to the Bogoliubov de Gennes equations for a simple S-F proximity system(Krawiec, 2002; Krawiec, 2003a; Krawiec, 2003b). The model system is shown in Fig. 1, consisting of a semi-infinite superconductor in contact with a small ferromagnetic layer of finite thickness. We found that for certain combinations of the thickness of the ferromagnetic layer, d, and exchange splitting, E_{ex}, the self-consistent solution is not simply the oscillating solution in Eq. 1, but there is an additional effect of a spontaneous current parallel to the interface, as indicated by J, in Fig. 1. The current flows as a quasiparticle current flowing in one direction in the ferromagnetic region and with a return current in the opposite direction in the superconductor. This current circulation generates an effective magnetic flux, indicated in the direction **B** in Fig. 1.

Interestingly the current is also partially spin-polarized in the ferromagnet, leading to a net spin-current. Although we originally found this result within the solution of the Bogoliubov de Gennes equations, in later work we were able to show that the same effect is also present when one uses a quasiclassical transport approach (Krawiec, 2004). This latter approach also allowed us to examine the stability of the spontaneous current to impurity scattering and to reduced interface transparency of the F-S junction.

In this paper we show that the finite current solution is indeed the global ground state energy of the system, and we describe the relative importance of the various contributions to the total energy. A short report of these results was published earlier (Annett, 2006). Here we provide additional details and discuss the physical interpretation of the results.

2.   Model System

The model Hamiltonian for the system shown in Fig. 1 is given by

$$H = \sum_{ij\sigma}[t_{ij} + (\varepsilon_{i\sigma} - \mu)\delta_{ij}]c_{i\sigma}^{+}c_{j\sigma} + \sum_{i\sigma}\frac{U_i}{2}\hat{n}_{i\sigma}\hat{n}_{i-\sigma} \qquad (2)$$

where t_{ij} are nearest neighbour hopping integrals, given by

$$t_{ij} = -te^{-ie\int_{\vec{r}_i}^{\vec{r}_j}\vec{A}(\vec{r})\cdot d\vec{r}}, \qquad (3)$$

where $\vec{A}(\vec{r})$ is the magnetic vector potential. In Eq. 2 $\varepsilon_{i\sigma}$ are spin dependent on-site energies, and U_i is an effective on-site attractive interaction. We model the ferromagnet-superconductor interface in Fig. 1 by simply taking $U_i = 0$ for layers $-d, \ldots, 0$, and $U_i = U$ for layers $1, 2, \ldots$. The ferromagnetic exchange energy is simply modelled by a spin-splitting $\varepsilon_{i\sigma} = \pm\frac{1}{2}E_{ex}$ for the ferromagnetic layers $-d, \ldots, 0$ and zero otherwise.

We solve this model numerically by computing the self-consistent layer dependent order parameter

$$\Delta_n = U_n \sum_{k_y}\langle c_{n\downarrow}(k_y)c_{n\uparrow}(k_y)\rangle = -U_n \sum_{ky}\int d\omega\frac{1}{\pi}\text{Im}G_{nn}^{12}(\omega, k_y)f(\omega) \qquad (4)$$

where k_y is the crystal momentum parallel to the interface, $G_{nn}^{\alpha\beta}(\omega, k_y)$ is the (4×4) Nambu-Gorkov Green function and $f(\omega)$ is the Fermi function. We choose a gauge where Δ_n is real, and where the vector potential is $A_y(n)$ oriented in the y direction in layer n. The vector potential is computed self-consistently from the discretized Maxwell equation

$$A_y(n + 1) - 2A_y(n) + A_y(n - 1) = -4\pi J_y(n) \qquad (5)$$

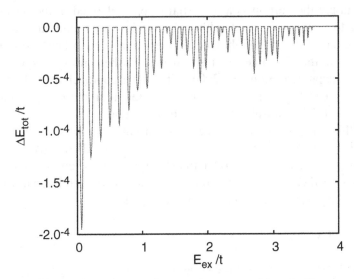

Figure 2. The difference in total energy between the system with and without a non-zero current $J_y(n)$, for a system with $d = 20$ ferromagnetic layers. Here we used $U = -2.345t$ in the superconductor and temperature $T = 0.01t$.

where the current in layer n, as indicated in Fig. 1, is obtained from

$$J_{y\uparrow(\downarrow)}(n) = -2et \sum_{k_y} sin(k_y - eA_y(n)) \int d\omega \frac{1}{\pi} Im G_{nn}^{11(33)}(\omega, k_y)f(\omega). \quad (6)$$

3. Finite-current solution

Solving the above set of equations self-consistently we find two distinct types of solutions, with either zero current J_y, or finite current. Both are saddle points of the non-linear set of equations. Typically only one or the other solution is stable numerically, for example allowing a finite J_y in the starting conditions will sometimes result in a stable solution with finite J_y and for other parameter values will result in a self-consistent solution with no-current, $J_y = 0$ (Krawiec, 2002; Krawiec, 2003a; Krawiec, 2003b).

In order to confirm which solution is the true ground state it is necessary to compute the total energy for both, and then the physical solution is obviously the one with the lowest total energy. Fig. 2 shows the difference in computed total energy between the non-zero current and zero-current solutions for a typical system with $d = 20$ ferromagnetic layers. We see that as a function of the ferromagnetic exchange splitting E_{ex} there are a series of dramatic

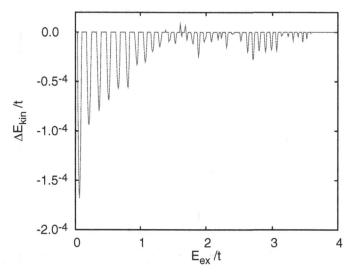

Figure 3. The difference in quasiparticle kinetic energy between the system with and without a non-zero current $J_y(n)$, for a system with $d = 20$ ferromagnetic layers. Here we used $U = -2.345t$ in the superconductor and temperature $T = 0.01t$. The values of exchange splitting, E_{ex}, vary from 0 (i.e. a paramagnet) to $4t$, corresponding to a fully spin polarized ferromagnetic insulator.

minima, corresponding to parameter values where the finite current solution is stable. Between minima there are finite ranges of the parameter E_{ex} where the zero-current solution is stable, corresponding to zero total energy difference in Fig. 2. For these values there is no stable saddle point solution with finite current.

As we have shown elsewhere (Krawiec, 2002; Krawiec, 2003a; Krawiec, 2003b) the finite current state appears whenever the parameters E_{ex} and d are such that a pair of Andreev bound states cross zero energy simultaneously. This degeneracy at the Fermi level can be removed by introducing the supercurrent. Therefore we might expect that the quasiparticle kinetic energy plays a large role in the formation of the stable current-carrying solution. This expectation is confirmed in Fig. 3, where we see that the quasiparticle kinetic energy change is almost always negative, and of the same magnitude as the change in total energy whenever the finite current solution is stable.

In Fig. 4 we examine two contributions to the kinetic energy, corresponding to hopping terms t_{ij} either perpendicular to, or parallel to, the F-S interface in Fig. 1. Interestingly these are not only an order of magnitude bigger that the total change in kinetic energy in Fig. 3, but they are also of opposite sign. At each critical value of E_{ex} where the finite current solution first

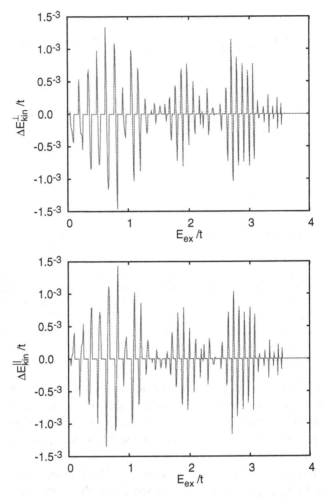

Figure 4. The difference in parallel and perpendicular quasiparticle kinetic energy (relative to the F-s interface) between the system with and without a non-zero current $J_y(n)$ as shown in Fig. 2. Note that the overall scale is an order of magnitude larger than the total kinetic energy in Fig. 3, and that the two contributions here are always of opposite sign.

becomes stable we see first a dramatic positive increase in $\Delta E_{kin}^{\perp}$ matched by a corresponding negative change in $\Delta E_{kin}^{\parallel}$. Upon further increasing E_{ex}, to the point where the current carrying solution is turning off, we then see the opposite effect of a large positive change in $\Delta E_{kin}^{\parallel}$ and a corresponding negative $\Delta E_{kin}^{\perp}$. It seems that in order to lower the total system energy the superconductor sometimes pay a cost of increased kinetic energy in part of the system, for example in the bulk supercurrent flow parallel to the surface.

However any such increases in energy are compensated for with larger gains in energy, by avoiding the Fermi level degeneracy between Andreev states.

4. Conclusions

The Andreev bound states in a thin ferromagnetic multi-layer on the surface of a superconductor are shown to lead to a state of spontaneous supercurrent parallel to the F-S interface. This current turns on or off repeatedly as a function of the exchange splitting E_{ex} for a finite ferromagnetic thickness d. Alternatively, for fixed E_{ex} we expect the same switching on and off as a function of thickness d. The current carrying state is stable whenever a pair of Andreev bound states cross the Fermi energy. We have shown here that the formation of the current carrying solution leads to a decrease in overall quasiparticle kinetic energy, although the contributions to the kinetic energy from parallel and perpendicular motion (relative to the F-S interface) can be either positive or negative.

Acknowledgements

We thank the ESF network AQDJJ for partial support for this work.

References

T. Kontos, M. Aprili, J. Lesueur and X. Grison, Phys. Rev. Lett. **86**, 304 (2001).

V.V. Ryazanov, V. A. Oboznov, A. Yu. Rusanov, A. V. Veretennikov, A. A. Golubov, and J. Aarts, Phys. Rev. Lett. **86**, 2427 (2001).

S.M. Frolov, D. J. Van Harlingen, V. A. Oboznov, V. V. Bolginov, and V. V. Ryazanov, Phys. Rev. B **70**, 144505-1 (2004).

V.T. Petrasov, I. A. Sosnin, I. Cox, A. Parsons, and C. Troadec, Phys. Rev. Lett. **83**, 3281 (1999).

A. Pimenov, A. Loidl, P. Przyslupski and B. Dabrowski, Phys. Rev. Lett. **95**, 247009 (2005).

I.F. Lyuksyutov IF and V.L. Pokrovsky, Adv. Phys. **54**, 67 (2005).

A. I. Buzdin, Rev. Mod. Phys. **77**, 935-976 (2005).

R.S. Keizer, S.T.B. Goennenwein, T.M. Klapwijk, G. Miao, G. Xiao, A. Gupta, Nature **439**, 825 (2006).

V.N. Krivoruchko, V.Y. Tarenkov, A.I. D'yachenko and V.N. Varyukhin VN, Europhys. Lett. **75**, 294 (2006).

I. Sosnin, H. Cho, V.T. Petrashov and A.F. Volkov, Phys. Rev. Lett. **96**, 157002 (2006).

P. Fulde and R.A. Ferrell, Phys. Rev. **135**, A550 (1964).

A.I. Larkin and Yu.N. Ovchinnikov, Zh. Exp. Teor. Fiz. **47**, 1136 (1964) [Sov. Phys. JETP **20**, 762 (1965)].

M. Krawiec, B.L. Gyorffy and J.F. Annett, Phys. Rev. B **66**, 172505 (2002).

M. Krawiec, B.L. Gyorffy and J.F. Annett, Physica C **387**, 7 (2003).

M. Krawiec, B.L. Gyorffy and J.F. Annett, Eur. Phys. J. B **32**, 163 (2003).
M. Krawiec, B. L. Gyorffy, J. F. Annett, Phys. Rev. B **70** 134519 (2004).
J.F. Annett, M. Krawiec and B. L. Gyorffy, Physica C **437-438**, 7 (2006).

PART III

Spintronics

KONDO EFFECT IN MESOSCOPIC SYSTEM

A. N. Rubtsov[a], M. I. Katsnelson[b], E. N. Gorelov[c, d],
and A. I. Lichtenstein[c] (alichten@physnet.uni-hamburg.de)

[a] Department of Physics, Moscow State University, 119992 Moscow, Russia

[b] Institute for Molecule and Materials, University of Nijmegen,
6525 ED Nijmegen, The Netherlands

[c] I. Institut für Theoretische Physik, Universität Hamburg,
20355 Hamburg, Germany

[d] Theoretical Physics and Applied Mathematics Department,
Urals State Technical University, Mira Street 19, 620002 Ekaterinburg,
Russia

Abstract. We discuss numerically exact continuous-time Quantum Monte Carlo algorithms for electronic structure calculations of nanosystems with a general non-local in space-time interaction. The problem of three interacting Kondo impurities on metallic substrate has been investigated. The suppression of the Kondo resonance by interatomic exchange interactions for different cluster geometries is investigated. It is shown that a drastic difference between the Heisenberg and Ising cases appears for an antiferromagnetically coupled trimer. Effects of magnetic frustrations in the trimer are also studied.

Key words: Kondo effect, Mesoscopic systems, Multi-band model, Quantum Monte Carlo algorithm, Kondo trimer, Non-local interaction in space and time

1. Introduction

Most newly discovered physical phenomena stimulate our imagination from the very beginning and give us a strong feeling of novel and exciting opportunities. Superconductivity is probably the most famous example in solid state physics. However, sometimes the situation is very different. An effect does not look very impressive, the perspectives of practical applications are far from obvious, and the only motivation for scientists to work in this field is curiosity. Then, after half a century of efforts, the effect becomes a hotly

K. Scharnberg and S. Kruchinin (eds.),
Electron Correlation in New Materials and Nanosystems, 327–341.
© 2007 Springer.

discussed and widely explored phenomenon which excites the minds of a broad range of scientists. This is, in brief, what happened to the surprising phenomenon which is now called "Kondo effect". It was discovered in the 1930th as a result of routine measurements of the temperature dependent resistivity for different metals by the Dutch experimentalist G.C. van den Berg and co-authors in the famous Kamerlingh Onnes laboratory in Leiden.

Recently, the interest in Kondo physics has been renewed due to developments in scanning tunneling microscopy (STM), which allows to image conducting surfaces with atomic resolution, move individual atoms, and measure the local electronic structure. In particular, STM gives the unique opportunity to investigate an essentially many-body phenomenon like the Kondo effect (Madhavan et al., 1998; Li and Schneider, 1998; Manoharan et al., 2000; Kolesnichenko et al., 2002) directly. In comparison to previous studies, where only indirect methods such as the analysis of temperature and magnetic field dependencies of thermodynamic and transport properties were available (Grüner and Zawadowski, 1974; Hewson, 1993), the STM technique permits physicists to "see" the surfaces and thereby study in detail the formation of the Kondo effect on individual magnetic atoms.

The first results on the observation of the Kondo effect around a single magnetic impurity positioned on a non-magnetic surface have been reported almost simultaneously in 1998 by two groups: Michael Crommie's group at Boston University (Madhavan et al., 1998) and Wolf-Dieter Schneider's group at the University of Lausanne. (Li and Schneider, 1998) Since then, more exciting investigations by this very powerful technique have followed. Probably the most convincing demonstration of the essentially quantum nature of the Kondo effect has been given by Don Eigler and colleagues at IBM Almaden (Manoharan et al., 2000) with the so called quantum mirage. In these experiments, an ellipse of atoms around a cobalt impurity, which was placed at one of the two focal points of the ellipse, was built. Then, a STM was used to measure the energy spectrum on the cobalt impurity, revealing the existence of a feature that corresponds to the Kondo resonance. The symmetry of an ellipse is such that electron waves passing through one focus inevitably converge at the second one, thus creating a mirror image of the Kondo resonance. The energy spectrum measured at the second focus also has a Kondo-like feature, in spite of the fact that there is no magnetic impurity at that point.

The electronic structure of adatoms and clusters on surfaces constitutes one of the most fascinating subjects in condensed matter physics and modern nanotechnology (For a review see the special issue of Surf. Sci., especially (Plummer et al., 2002), (Schen and Kirschner, 2002)). The scanning tunneling microscopy or spectroscopy technique allows the study of atomic structure, the electronic energy spectrum, and magnetic properties of different surfaces at an atomic scale.

Recently STM studies of small transition metal nanoclusters on different surfaces have been performed, including Co dimers (Chen et al., 1999) and Cr trimers (Jamneala et al., 2001) on a Au surface, and Co clusters on carbon nanotubes (Odom et al., 2000). The electron spectrum of these nanosystems, in particular the existence of the Kondo resonance, turns out to be very sensitive to the geometry of the clusters as well as to the type of magnetic adatoms. The latter can be important for nanotechnological fine tuning of surface electronic structure.

The "quantum-corral" type of STM-experiments provides an unique opportunity to investigate in detail an interplay between the single-impurity Kondo effect and interatomic magnetic interactions in nanoclusters. The interaction between itinerant electrons and localized ones leads to the screening of the impurity moment; on the other hand, the RKKY exchange interaction between localized spins suppress the Kondo resonance at the Fermi level. Quantum critical points at the boundary of different phases is a subject of special interest (Si et al., 2001; Zhu et al., 2003). At the same time, due to the extreme complexity of the problem, theoretical investigations of electronic structure for several Kondo centers usually involve some uncontrollable approximations, such as a replacement of the Heisenberg interatomic exchange interactions by the Ising ones (Zhu et al., 2003) or a variational approach based on a simple trial function (Kudasov and Uzdin, 2002).

In this paper we use the continuous time quantum Monte Carlo (CT-QMC) scheme to investigate correlation effects in complex nanosystems and give results of a numerically exact solution of the Kondo trimer problem. For the antiferromagnetic (AFM) interatomic exchange interaction, in contrast to the ferromagnetic (FM) one, the results for the Heisenberg and Ising systems differ substantially.

2. Exact scheme for correlated nanosystems

The quantum Monte Carlo (QMC) scheme is the most universal tool for the numerical study of quantum many-body systems with strong correlations. So-called determinantal Quantum Monte Carlo scheme for fermionic systems appeared more than 20 years ago (Scalapino and Sugar, 1981; Blankenbecler et al., 1981; Hirsch, 1983; Hirsch, 1985). This scheme is now widely used for the numerical investigation of physical models with strong interactions, as well as for quantum chemistry and nanoelectronics. Although the first numerical attempts were made for a model Hamiltonians with local interactions, real systems are described by a many-particle action of a more general form.

For example, many non-local matrix elements of the Coulomb interaction do not vanish in the context of quantum chemistry (White, 2002) and solid state physics (Zhang and Krakauer, 2003). For a realistic description of Kondo impurities like cobalt atoms on a metallic surface it is of crucial importance to use the spin and orbital rotationally invariant Coulomb vertex in the non-perturbative investigation of electronic structure. The recently developed Dynamical Mean-Field theory (DMFT) (Georges et al., 1996) for correlated materials brings a non-trivial frequency-dependent bath Green function on the scene, and its extension (Sun and Kotliar, 2002) deals with an action that is non-local in time. Moreover, the same frequency dependent single-electron Green-function and retarded electron-electron interaction naturally appear in any electronic subsystems where the rest of system is integrated out.

The determinantal grand-canonical auxiliary-field scheme (Scalapino and Sugar, 1981; Blankenbecler et al., 1981; Hirsch, 1983; Hirsch, 1985) is extensively used for interacting fermions, since other known QMC schemes (like stochastic series expansion in powers of Hamiltonian (Sandvik and Kurkijärvi, 1991) or worm algorithms (Prokof'ev et al., 1996)) suffer from an unacceptably bad sign problem. The following two points are essential for the determinantal QMC approach: first, the imaginary time is artificially discretized, and the Hubbard-Stratonovich transformation (Hubbard, 1959; Stratonovich, 1957) is performed to decouple the fermionic degrees of freedom. After the decoupling, fermions can be integrated out, and Monte Carlo sampling should be performed in the space of auxiliary Hubbard-Stratonovich fields. Hirsch (Hirsch, 1983) proposed a so-called discrete Hubbard-Stratonovich transformation to improve the efficiency of the original scheme. It is worth noting that for a system of N atoms the number of auxiliary field scales $\propto N$ for the local (short-range) interaction and as N^2 for the long-range one. This makes the calculation rather ineffective for the non-local case. In fact, the scheme is developed for the local interaction only.

The problem of systematic errors arising from the time discretization was addressed in several studies. For bosonic quantum systems, continuous time loop algorithm (Beard and Wiese, 1996), worm diagrammatic world line Monte Carlo scheme (Prokof'ev et al., 1996) and continuous time path-integral QMC (Kornilovitch, 1998) overcame these problems. Recently, a continuous-time modification of the fermionic QMC algorithm was proposed (Rombouts et al., 1999). It is based on a series expansion for the partition function in powers of the interaction. The scheme is free of time-discretization errors, but the Hubbard-Stratonovich transformation still needs to be invoked. Therefore, the number of auxiliary fields scales similarly to the discrete scheme, so that the scheme remains local.

New developments in the field of interacting fermion systems (Savrasov et al., 2001) clearly urge the construction of a new type of QMC scheme suitable for non-local, time-dependent interaction. One can consider the partition function $Z = \mathrm{Tr} T e^{-S}$ for the nanosystem with pair interaction in the most general case and split S into two parts: an unperturbed action S_0 of Gaussian form and an interaction W.

$$S = S_0 + W, \tag{1}$$

$$S_0 = \int \int t_r^{r'} c_{r'}^\dagger c^r \, dr \, dr'$$

$$W = \int \int \int \int w_{r_1 r_2}^{r_1' r_2'} c_{r_1'}^\dagger c^{r_1} c_{r_2'}^\dagger c^{r_2} \, dr_1 \, dr_1' \, dr_2 \, dr_2'$$

Here, T is a time-ordering operator, $r = \{\tau, s, i\}$ is a combination of the continuous imaginary-time variable τ, spin projection s and the discrete index i numbering the single-particle orbital states in nanosystems. Integration over dr implies the integral over τ, and the sum over all lattice states and spin projections: $\int dr \equiv \sum_i \sum_s \int_0^\beta d\tau$.

Now we switch to the interaction representation and make the perturbation series expansion for the partition function Z assuming S_0 as an unperturbed action:

$$Z = \sum_{k=0}^{\infty} Z_k = \sum_{k=0}^{\infty} \int dr_1 \int dr_1' ... \int dr_{2k} \int dr_{2k}' \Omega_k(r_1, r_1', ..., r_{2k}, r_{2k}'), \tag{2}$$

$$\Omega_k = Z_0 \frac{(-1)^k}{k!} w_{r_1 r_2}^{r_1' r_2'} ... w_{r_{2k-1} r_{2k}}^{r_{2k-1}' r_{2k}'} D_{r_1' r_2' ... r_{2k}'}^{r_1 r_2 ... r_{2k}}.$$

Here $Z_0 = \mathrm{Tr}(T e^{-S_0})$ is a partition function for the unperturbed system and

$$D_{r_1' ... r_{2k}'}^{r_1 ... r_{2k}} = < T c_{r_1'}^\dagger c^{r_1} ... c_{r_{2k}'}^\dagger c^{r_{2k}} > . \tag{3}$$

Hereafter the triangular brackets denote the average over the unperturbed system: $< A > = Z_0^{-1} \mathrm{Tr}(T A e^{-S_0})$. Since S_0 is Gaussian, one can apply Wick theorem to decouple Eq. (3). Thus $D_{r_1' ... r_{2k}'}^{r_1 ... r_{2k}}$ is the determinant of a $2k \times 2k$ matrix which consists of the bare two-point Green functions $g_{r'}^r = < T c_{r'}^\dagger c^r >$:

$$D_{(2k)} \equiv D_{r_1' r_2' ... r_{2k}'}^{r_1 r_2 ... r_{2k}} = \det |g_{r_j'}^{r_i}| \tag{4}$$

Now we can express the interacting two-point Green function for the system (1) using the perturbation series expansion (2). It reads:

$$
G_{r'}^r \equiv Z^{-1} < T c_{r'}^\dagger c^r e^{-W} >
$$
$$
= Z^{-1} \sum_k \int dr_1 \int dr_1' ... \int dr_{2k}' G_{r'}^r(r_1, r_1', ..., r_{2k}') \Omega_k(r_1, r_1', ..., r_{2k}') \quad (5)
$$

where $G_{r'}^r(r_1, r_1', ..., r_{2k}')$ is defined as

$$
G_{r'}^r(r_1, r_1', ..., r_{2k}') = \frac{< T c_{r'}^\dagger c^r c_{r_1'}^\dagger c^{r_1} ... c_{r_{2k}'}^\dagger c^{r_{2k}} >}{< T c_{r_1'}^\dagger c^{r_1} ... c_{r_{2k}'}^\dagger c^{r_{2k}} >}
$$

This is nothing else than the ratio of two determinants: $D_{(2k+1)}/D_{(2k)}$. Similarly, one can write formulae for other averages, for example the two-particle Green function which is related to four-point correlation functions and contains important information about magnetic excitations in nanosystems:

$$
\chi_{r'r'''}^{rr''} \equiv Z^{-1} < T c_{r'}^\dagger c^r c_{r'''}^\dagger c^{r''} e^{-W} >
$$
$$
= Z^{-1} \sum_k \int dr_1 \int dr_1' ... \int dr_{2k}' \chi_{r'r'''}^{rr''}(r_1, r_1', ..., r_{2k}') \Omega_k(r_1, r_1', ..., r_{2k}'). \quad (6)
$$

Here

$$
\chi_{r'r'''}^{rr''}(r_1, r_1', ..., r_{2k}') = \frac{< T c_{r'}^\dagger c^r c_{r'''}^\dagger c^{r''} c_{r_1'}^\dagger c^{r_1} ... c_{r_{2k}'}^\dagger c^{r_{2k}} >}{< T c_{r_1'}^\dagger c^{r_1} ... c_{r_{2k}'}^\dagger c^{r_{2k}} >}
$$

is the ratio of two determinants: $D_{(2k+2)}/D_{(2k)}$.

An important property of the above formulae is that the integrands remain unchanged under the permutations $r_i, r_{i'}, r_{i+1}, r_{i'+1} \leftrightarrow r_j, r_{j'}, r_{j+1}, r_{j'+1}$ with any i, j. Therefore it is possible to introduce a quantity K, which we call "state of the system" and that is a combination of the perturbation order k and an *unnumbered set* of k tetrades of coordinates. Now, denote $\Omega_K = k!\Omega_k$, where the factor $k!$ reflects all possible permutations of the arguments. For the Green functions, $k!$ in the nominator and denominator cancel each other, so that $g_K = g_k$. In this notation,

$$
Z = \int \Omega_K D[K], \quad (7)
$$
$$
G_{r'}^r = Z^{-1} \int G_K \Omega_K D[K],
$$

where $\int D[K]$ means the summation over k and integration over all possible realizations of the above-mentioned unnumbered set at each k. One can check, that the factorial factors are indeed taken into account correctly with this definition.

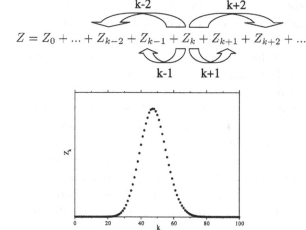

Figure 1. Schematic picture of random walks in the space of $k; r_1, r'_1, ..., r_{2k}, r'_{2k}$ according to perturbation series expansion (2) and an example of the histogram for the perturbation order k.

The important thing to notice is that the series expansion for an exponential *always* converges for finite fermionic systems. A mathematically rigorous proof can be constructed for Hamiltonian systems. Indeed, many-body fermionic Hamiltonians H_0 and W have a finite number of eigenstates that is $2^{N_{orb}}$, where N_{orb} is the total number of electronic orbitals in the system. Now one can observe that $\Omega_k < const \cdot W_{max}^k$, where W_{max} is the eigenvalue of W with maximal modulus. This proves the convergence of (2) because $k!$ in the denominator grows faster than the numerator. In our calculations for non-Hamiltonian systems we did not observe any indications of the divergence, either.

Although formula (7) looks rather formal, it exactly corresponds to the idea of the proposed QMC scheme. We simulate a Markov random walk in a space of all possible K with a probability density $P_K \propto |\Omega_K|$ to visit each state. Two kinds of the trial steps are necessary: one should try either to increase or to decrease k by 1, and, respectively, to add or to remove the corresponding tetrad of "coordinates". Then the standard Metropolis acceptance criterion can be constructed using the ratio

$$\frac{\|w\|}{k+1} \cdot \left| \frac{D_{r'_1 \dots r'_{2k+2}}^{r_1 \dots r_{2k+2}}}{D_{r'_1 \dots r'_{2k}}^{r_1 \dots r_{2k}}} \right|. \tag{8}$$

for the incremental steps and its inverse for the decremental ones.

In general, one may also want to add/remove several tetrades simultaneously. Such a random walk is shown in Fig. 1. The same figure presents a typical distribution diagram for a perturbation order k in the QMC calculation.

The most time consuming operation of the algorithm is a calculation of the ratio of determinants and Green-function matrices. It is necessary for the calculation of MC weights, as well as for the Green functions. There exist so called fast-update formulae for the calculation of the ratios of determinants and Green-function matrices. The usual procedure takes N^3 operations, while the fast-update technique requires only N^2 or less operations, where N is a matrix size. Usually, the two types of steps ($k + 1$ and $k \to k - 1$) considered are sufficient. However, the steps $k \to k \pm 2$ can be also employed in certain cases and are useful for two-particle Green functions.

The only matrix which one needs to keep during MC steps is the inverse matrix of the inverse bare Green functions: $M_{(2k)} = g^{-1}_{(2k)}$. In the following formulae, the matrix $M_{(2k)}$ is enlarged to a $(2k + 1) \times (2k + 1)$ matrix with $M_{2k+1,2k+1} = 1$ and $M_{2k+1,i} = 0$, $M_{i,2k+1} = 0$. This does not change the ratio of determinants.

It is easy to obtain fast-update formulae for the step $k \to k + 1$. The expression for the matrix $M^{(2k+1)}$ has the following form (Rubtsov et al., 2005):

$$M_{(2k+1)} = \begin{pmatrix} \cdots & \cdots & \cdots & -L_{1,2k+1}\lambda^{-1} \\ \cdots & M'_{i,j} & \cdots & \cdots \\ \cdots & \cdots & \cdots & -L_{2k,2k+1}\lambda^{-1} \\ -\lambda^{-1}R_{2k+1,1} & \cdots & -\lambda^{-1}R_{2k+1,k} & \lambda^{-1} \end{pmatrix}$$

where $M'_{i,j} = M^{(2k)}_{i,j} + L_{i,2k+1}\lambda^{-1}R_{2k+1,j}$, $R_{i,j} = \sum_n g_{i,n}M^{(2k)}_{n,j}$ and $L_{i,j} = \sum_n M^{(2k)}_{i,n} g_{n,j}$ and λ is equal to the ratio of two determinants:

$$\frac{\det D_{(2k+1)}}{\det D_{(2k)}} = g_{2k+1,2k+1} - \sum_{i,j=1}^{2k} g_{2k+1,i} M^{(2k)}_{i,j} g_{j,2k+1} = \lambda$$

For the step $k \to k - 1$ (removal of the column and row n) the fast update formulae for matrix $M^{(2k-1)}$ and the ratio of determinants are as follows:

$$M^{(2k-1)}_{i,j} = M^{(2k)}_{i,j} - \frac{M^{(2k)}_{i,n} M^{(2k)}_{n,j}}{M^{(2k)}_{n,n}}, \tag{9}$$

$$\frac{\det D_{(2k-1)}}{\det D_{(2k)}} = M^{(2k)}_{n,n}.$$

In the same manner one can obtain fast-update formulae for steps ± 2. Introducing the 2×2 matrix λ:

$$\lambda_{q,q'} = g_{q,q'} - \sum_{i,j=1}^{2k} g_{q,i} M^{(2k)}_{i,j} g_{j,q'}, \tag{10}$$

where $\{q, q'\} = 2k + 1, 2k + 2$, the ratio of two determinants can be written as

$$\frac{\det D_{(2k+2)}}{\det D_{(2k)}} = \det |\lambda|$$

Using the fast update formula for M, the Green function can be obtained both in imaginary time and at Matsubara frequencies:

$$G^{\tau}_{\tau'} = g^{\tau}_{\tau'} - \sum_{i,j} g^{\tau}_{\tau_i} M_{i,j} g^{\tau_j}_{\tau'} \tag{11}$$

$$G(\omega) = g(\omega) - g(\omega)[\frac{1}{\beta} \sum_{i,j} M_{i,j} e^{i\omega(\tau_i - \tau_j)}] g(\omega).$$

Here $g^{\tau}_{\tau'}$ and $g(\omega)$ are the bare Green function in imaginary time and Matsubara spaces, respectively.

In order to reduce and in some cases avoid the sign problem in CT-QMC we introduce additional quantities $\alpha^r_{r'}$, which, in principle, can be functions of time, spin and the number of lattice states. Thus, up to an additive constant we have the new separation of our action:

$$S_0 = \int \int \left(t^{r'}_r + \int \int \alpha^{r_2}_{r'_2} (w^{r'r'_2}_{rr_2} + w^{r'_2 r'}_{r_2 r}) dr_2 dr'_2 \right) c^{\dagger}_{r'} c^r \, dr dr',$$

$$W = \int \int \int \int w^{r'_1 r'_2}_{r_1 r_2} (c^{\dagger}_{r'_1} c^{r_1} - \alpha^{r_1}_{r'_1})(c^{\dagger}_{r'_2} c^{r_2} - \alpha^{r_2}_{r'_2}) dr_1 dr'_1 dr_2 dr'_2.$$

In this case, the determinants $D^{r_1...r_{2k}}_{r'_1...r'_{2k}}$ of $2k \times 2k$ matrices have the following form:

$$D^{r_1 r_2 ... r_{2k}}_{r'_1 r'_2 ... r'_{2k}} = \det |g^{r_i}_{r'_j} - \alpha^{r_i}_{r'_j} \delta_{ij}| \tag{12}$$

where $i, j = 1, ..., 2k$.

A proper choice of α can completely remove the sign problem. In the case of Hubbard model, the choice $\alpha_{\uparrow} = 1 - \alpha_{\downarrow}$ eliminates the sign problem for repulsive systems with particle-hole symmetry. Note however, that the proper choice of the α's depends on the particular system under consideration.

3. Multi-band models for nanosystems

An important advantage of the CT-QMC algorithm is that it allows one to consider multi-band problems with interactions of the most general form:

$$\hat{U} = \frac{1}{2} \sum_{ijkl;\sigma\sigma'} U_{ijkl} c^{\dagger}_{i\sigma} c^{\dagger}_{j\sigma'} c^{k\sigma'} c^{l\sigma}. \tag{13}$$

We apply the continuous time QMC to the rotationally invariant three-band impurity problem with semicircular density of states (DOS) with the half-bandwidth $D = 1$. We assume an average occupancy of 4 on the impurity site. The interaction matrix U_{ijkl} is parametrised in the standard way (Lichtenstein and Katsnelson, 1998) with $U = 1$ and $J = 0.3$ at $\beta = 10$ including spin-flip and pair-hopping terms. In Fig. 2 we show the imaginary time Green function for the rotationally invariant form of the interaction and a diagonal density-density approximation, which is normally used in the standard Hirsch MC-formalism. Because the system is away from half-filling, the Green function is non-symmetric with respect to $\beta/2$. The maximum-entropy analytical continuation for the density of states is presented in the inset of Fig. 2. We also include the case of $D = 0$ in order to compare with the exact solution for the atomic limit. One can see that the effect of spin-flip interactions decreases when the fermionic bath bandwidth is increased.

In order to investigate the effects of spin-flip interactions we calculate the corresponding two-band models at half-filling with $U = 4$ and $J = 1$ for $\beta = 4$. Results for the local and non-local in time spin-flip are shown in Fig. 3. It is interesting to note that Green function is almost insensitive to the details of spin-flip retardation. Both Green functions are very similar and correspond to qualitatively the same density of states. To demonstrate the effects due to retardation we calculated the four-point quantity $\chi(\tau)$. It turns out that a switch to the non-local in time exchange modifies $\chi(\tau)$ dramatically. The local in time exchange results in a pronounced peak of $\chi(\tau)$ at $\tau \approx 0$, whereas the non-local spin-flip results in almost time-independent spin-spin correlations (Fig. 4).

4. The Kondo trimer

We start with the system of three correlated impurity sites with Hubbard repulsion U in a metallic bath and with an effective exchange interaction J_{ij} between them, a minimal model which however includes all relevant interactions necessary to describe magnetic nanoclusters on a metallic surface. The effective action for such cluster in a metallic bath has the following form:

$$S_0 = -\int_0^\beta \int_0^\beta d\tau d\tau' \sum_{i,j;\sigma} c_{i\sigma}^\dagger(\tau) \mathcal{G}_{ij}^{-1}(\tau - \tau') c_{j\sigma}(\tau'),$$

$$W = \int_0^\beta d\tau \left(U \sum_i n_{i\uparrow}(\tau) n_{i\downarrow}(\tau) + \sum_{i,j} J_{ij} \mathbf{S}_i(\tau) \mathbf{S}_j(\tau) \right).$$

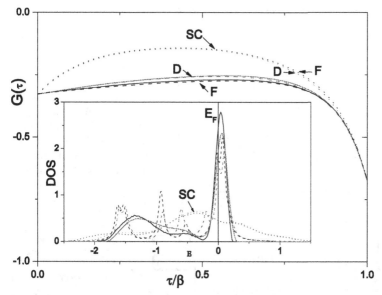

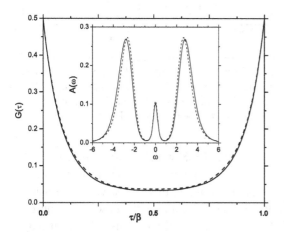

Figure 2. Imaginary-time Green Function for the rotationally-invariant three-band model: the dot (or full) lines are QMC results for full (F) and diagonal (D) interaction vertex for semicircular (SC) bath (or atomic limit), the dashed lines are the exact diagonalisation results for the atomic limit. The inset shows the corresponding density of states.

Figure 3. Imaginary-time Green Function for the rotationally-invariant two-band model. Solid and dot lines correspond to the static and to the nonlocal in time spin-flip, respectively. The inset shows DOS estimated from the Green function.

The last term in the right-hand-side of Eq. (1) allows us to consider the most important "Kondo lattice" feature, that is, the mutual suppression of the Kondo screening and intersite exchange interactions (Doniach, 1977; Irkhin and Katsnelson, 1997; Irkhin and Katsnelson, 1999; Irkhin and Katsnelson, 2000). Another factor, the coherence of the resonant Kondo scattering, is taken into account by introducing inter-impurity hopping integrals t_{ij} to the bath Green function, which is supposed to be $\mathcal{G}_{ij}^{-1} = \mathcal{G}_i^{-1}\delta_{ij} - t_{ij}$. Here

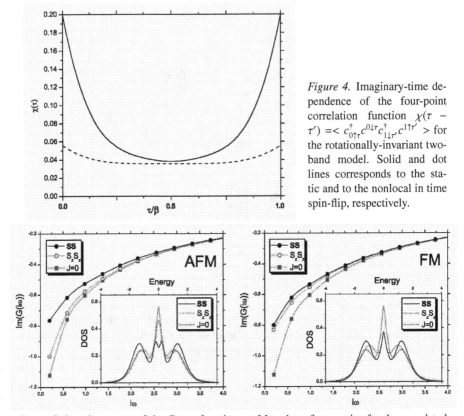

Figure 4. Imaginary-time dependence of the four-point correlation function $\chi(\tau - \tau') =< c^\dagger_{0\uparrow\tau}c^{0\downarrow\tau}c^\dagger_{1\downarrow\tau'}c^{1\uparrow\tau'} >$ for the rotationally-invariant two-band model. Solid and dot lines corresponds to the static and to the nonlocal in time spin-flip, respectively.

Figure 5. Imaginary part of the Green functions at Matsubara frequencies for the correlated adatom equilateral triangle in the metallic bath for AFM (left panel) and FM (right panel) types of effective exchange interaction. Parameters: $U = 2, J = \pm 0.2, t = 0, \beta = 16, \mu = U/2$. There are three dependencies on each picture for **SS**, $S_z S_z$ and $J = 0$ (which corresponds to single atom in the metallic bath) types of interaction. The insets show DOS.

$\mathcal{G}_i^{-1}(i\omega_n)$ corresponds to the semicircular DOS with band-width 2. For real adatom clusters the exchange interactions dependend on the electronic structure of both adatoms and host metal. To simulate this effect we will consider J_{ij} as independent parameters. We will concentrate on the case of an equilateral triangle for which $J_{ij} = J$. To check an approximation used by (Zhu et al., 2003) we will also investigate the case when all spin-flip exchange terms are ignored.

Let us discuss the electronic structure of correlated adatom trimer in the metallic bath. First we show that **SS** type of interactions suppress the resonance in AFM case. We study the equilateral triangle at half-filling. The Green functions at the Matsubara frequencies and corresponding DOS are presented in Fig. (5). The case of $J = 0$ corresponds to the single Kondo

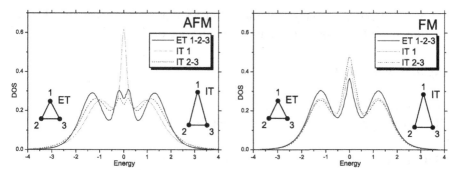

Figure 6. DOS for equilateral triangle (ET) and isosceles triangle (IT) geometries with AFM (left panel) and FM (right panel) types of effective exchange interaction. Parameters are the same as in Fig. 5. Values of the effective exchange integrals for IT are as follows: $J_{23} = J, J_{12} = J/3, J_{13} = J/3$. There are two dependencies in case of IT: one for adatom 1 and another for equivalent adatoms 2 and 3. All adatoms are equivalent in the case of ET (one dependence).

impurity in the bath. One can see that there is no essential difference between $S_z S_z$ and SS types of interaction in the FM case for a wide range of the model parameters while for the AFM case this difference is very important. The AFM SS interaction leads to a pronounced suppression of the Kondo resonance at the Fermi level for physically relevant values of J (Fig. 5).

In order to describe the experimental situation we changed the geometry of the adatom trimer. An observation of the Kondo resonance reconstruction was reported for an isosceles arrangement of three Cr atoms on a gold surface (Jamneala et al., 2001). We therefore study isosceles triangles with AFM and FM types of effective exchange interactions. To model the experimental system we have chosen $J_{23} = J, J_{12} = J/3, J_{13} = J/3$. The computational results are presented in Fig. (6) where one can see the reconstruction of resonance in AFM and FM cases in accordance with experimental data. Note that the Kondo resonance appears only for the more weakly bonded adatoms in the AFM case. In order to describe specific experimental results quantitatively the method developed here should be combined with first-principle calculations (Lichtenstein and Katsnelson, 1998).

5.   Concluding remarks

In conclusion, we have discussed the general computational scheme for correlated nanosystems within a continuous time quantum Monte Carlo method for interactions which are non-local in space and time. We have shown that the electronic structure of the correlated Kondo trimer on a metallic surface drastically depends on the type of the exchange interactions. In the AFM case, the effective exchange interaction of **SS** type leads to a more efficient suppression of the Kondo resonance than the one of $S_z S_z$ type.

Acknowledgements

This research was supported by the DFG-research grant (SFB 668-A3) and Russian Scientific Schools Grant 96-1596476.

References

Beard, B. B. and Wiese, U.-J. (1996) Simulations of Discrete Quantum Systems in Continuous Euclidean Time, *Phys. Rev. Lett.* **77**, 5130.

Blankenbecler, R., Scalapino, D. J., and Sugar, R. L. (1981) Monte Carlo calculations of coupled boson-fermion systems. I, *Phys. Rev. D* **24**, 2278.

Chen, W., Jamneala, T., Madhavan, V., and Crommie, M. F. (1999) Disappearance of the Kondo resonance for atomically fabricated cobalt dimers, *Phys. Rev. B* **60**, R8529.

Doniach, S. (1977) The Kondo lattice and weak antiferromagnetism, *Physica B* **91**, 231.

Georges, A., Kotliar, G., Krauth, W., and Rozenberg, M. (1996) Dynamical mean-field theory of strongly correlated fermion systems and the limit of infinite dimensions, *Rev. Mod. Phys.* **68**, 13.

Grüner, G. and Zawadowski, A. (1974) Magnetic Impurities in Non-magnetic Metals, *Rep. Prog. Phys.* **37**, 1497.

Hewson, A. C. (1993) *The Kondo Problem to Heavy Fermions*, Cambridge University Press, Cambridge.

Hirsch, J. E. (1983) Discrete Hubbard-Stratonovich transformation for fermion lattice models, *Phys. Rev. B* **28**, 4059.

Hirsch, J. E. (1985) Two-dimensional Hubbard model: Numerical simulation study, *Phys. Rev. B* **31**, 4403.

Hubbard, J. (1959) Calculation of Partition Functions, *Phys. Rev. Lett.* **3**, 77.

Irkhin, V. Y. and Katsnelson, M. I. (1997) Scaling picture of magnetism formation in the anomalous f-electron systems: Interplay of the Kondo effect and spin dynamics, *Phys. Rev. B* **56**, 8109.

Irkhin, V. Y. and Katsnelson, M. I. (1999) Scaling theory of magnetic ordering in the Kondo lattices with anisotropic exchange interactions, *Phys. Rev. B* **59**, 9348.

Irkhin, V. Y. and Katsnelson, M. I. (2000) Non-Fermi-liquid behavior in Kondo lattices induced by peculiarities of magnetic ordering and spin dynamics, *Phys. Rev. B* **61**, 14640.

Jamneala, T., Madhavan, V., and Crommie, M. F. (2001) Kondo Response of a Single Antiferromagnetic Chromium Trimer, *Phys. Rev. Lett.* **87**, 256804.

Kolesnichenko, O. Y., de Kort, R., Katsnelson, M. I., Lichtenstein, A. I., and van Kempen, H. (2002) Real-space imaging of the orbital Kondo resonance on the Cr(001) surface, *Nature* **415**, 507.

Kornilovitch, P. E. (1998) Continuous-Time Quantum Monte Carlo Algorithm for the Lattice Polaron, *Phys. Rev. Lett.* **81**, 5382.

Kudasov, Y. B. and Uzdin, V. M. (2002) Kondo State for a Compact Cr Trimer on a Metallic Surface, *Phys. Rev. Lett.* **89**, 276802.

Li, J. and Schneider, W.-D. (1998) Kondo Scattering Observed at a Single Magnetic Impurity, *Phys. Rev. Lett.* **80**, 2893.

Lichtenstein, A. I. and Katsnelson, M. I. (1998) Ab initio calculations of quasiparticle band structure in correlated systems: LDA++ approach, *Phys. Rev. B* **57**, 6884.

Madhavan, V., Chen, W., Jamneala, T., Crommie, M. F., and Wingreen, N. S. (1998) Tunneling into a Single Magnetic Atom: Spectroscopic Evidence of the Kondo Resonance, *Science* **280**, 567.

Manoharan, H. C., Lutz, C. P., and Eigler, D. M. (2000) Quantum mirages formed by coherent projection of electronic structure, *Nature* **403**, 512.

Odom, T. W., Huang, J.-L., Cheung, C. L., and Lieber, C. M. (2000) Magnetic Clusters on Single-Walled Carbon Nanotubes: The Kondo Effect in a One-Dimensional Host, *Science* **290**, 1549.

Plummer, E. W., Ismail, Matzdorf, R., Melechko, A. V., Pierce, J., and Zhang, J. (2002) Surfaces: A Playground for Physics with Broken Symmetry in Reduced Dimensionality, *Surf. Sci.* **500**, 1.

Prokof'ev, N. V., Svistunov, B. V., and Tupitsyn, I. S. (1996) Exact quantum Monte Carlo process for the statistics of discrete systems, *Pis'ma Zh. Eksp. Teor. Fiz.* **64**, 853, JETP Lett. **64**, 911.

Rombouts, S. M. A., Heyde, K., and Jachowicz, N. (1999) Quantum Monte Carlo Method for Fermions, Free of Discretization Errors, *Phys. Rev. Lett.* **82**, 4155.

Rubtsov, A. N., Savkin, V. V., and Lichtenstein, A. I. (2005) Continuous-time quantum Monte Carlo method for fermions, *Phys. Rev. B* **72**, 035122.

Sandvik, A. W. and Kurkijärvi, J. (1991) Quantum Monte Carlo simulation method for spin systems, *Phys. Rev. B* **43**, 5950.

Savrasov, S. Y., Kotliar, G., and Abrahams, E. (2001) Correlated electrons in δ-plutonium within a dynamical mean-field picture, *Nature* **410**, 793.

Scalapino, D. J. and Sugar, R. L. (1981) Method for Performing Monte Carlo Calculations for Systems with Fermions, *Phys. Rev. Lett.* **46**, 519.

Schen, J. and Kirschner, J. (2002) Tailoring magnetism in artificially structured materials: the new frontier, *Surf. Sci.* **500**, 300.

Si, Q., Rabello, S., Ingersent, K., and Smith, J. L. (2001) Locally critical quantum phase transitions in strongly correlated metals, *Nature* **413**, 804.

Stratonovich, R. L. (1957), *Dokl. Akad. Nauk SSSR* **115**, 1097, translation: Soviet Phys. Doklady 2, 416 (1958).

Sun, P. and Kotliar, G. (2002) Extended dynamical mean-field theory and GW method, *Phys. Rev. B* **66**, 085120.

White, S. R. (2002) Numerical canonical transformation approach to quantum many-body problems, *J. of Chem. Phys.* **117**, 7472.

Zhang, S. W. and Krakauer, H. (2003) Quantum Monte Carlo Method using Phase-Free Random Walks with Slater Determinants, *Phys. Rev. Lett.* **90**, 136401.

Zhu, J.-X., Grempel, D. R., and Si, Q. (2003) Continuous Quantum Phase Transition in a Kondo Lattice Model, *Phys. Rev. Lett.* **91**, 156404.

1/f NOISE AND TWO-LEVEL SYSTEMS IN JOSEPHSON QUBITS

Alexander Shnirman[1] (shnirman@tfp.uni-karlsruhe.de), Gerd Schön[1], Ivar Martin[2], and Yuriy Makhlin[3]

[1] *Institut für Theoretische Festkörperphysik, Universität Karlsruhe, D-76128 Karlsruhe, Germany*

[2] *Theoretical Division, Los Alamos National Laboratory, Los Alamos, NM 87545, USA*

[3] *Landau Institute for Theoretical Physics, Kosygin st. 2, 119334 Moscow, Russia*

Abstract. Quantum state engineering in solid-state systems is one of the most rapidly developing areas of research. Solid-state building blocks of quantum computers have the advantages that they can be switched quickly, and they can be integrated into electronic control and measuring circuits. Substantial progress has been achieved with superconducting circuits (qubits) based on Josephson junctions. Strong coupling to the external circuits and other parts of the environment brings, together with the advantages, the problem of noise and, thus, decoherence. Therefore, the study of sources of decoherence is necessary. Josephson qubits themselves are very useful in this study: they have found their first application as sensitive spectrometers of the surrounding noise.

Key words: Josephson qubits, $1/f$ noise

1. Introduction

Josephson junction based systems are among the promising candidates for quantum state engineering with solid state systems. In recent years great progress was achieved in this area. After initial breakthroughs of the groups in Saclay and NEC (Tsukuba) in the late 90's, there are now many experimental groups worldwide working in this area, many of them with considerable previous experience in nano-electronics. By now the full scope of single-qubit (NMR-like) control is possible. One can drive Rabi oscillations, observe Ramsey fringes, apply composite pulses and echo technique (Collin et al., 2004). The goal of 'single shot' measurements has almost been achieved (Astafiev et al., 2004b; Siddiqi et al., 2004). There are first reports about 2-bit operations (Yamamoto et al., 2003). The decoherence times have reached microseconds, which would allow for hundreds of gates. Finally, a setup equivalent to cavity QED was realized in superconducting circuits (Wallraff

K. Scharnberg and S. Kruchinin (eds.),
Electron Correlation in New Materials and Nanosystems, 343–356.
© 2007 *Springer.*

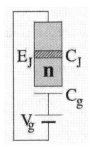

Figure 1. Charge Qubit.

et al., 2004). We refer the reader to the recent reviews (Esteve and Vion, 2005; Wendin and Shumeiko, 2005).

Despite the great progress decoherence remains the limiting factor in solid state circuits. Since one wants to manipulate and measure the qubits, some decoherence is unavoidable. There are, however, noise sources which are purely intrinsic, i.e., they are not related to any controlling or measuring circuitry. Eliminating those sources as much as possible is therefore of greatest importance. The main intrinsic source of decoherence in most superconducting qubits is $1/f$ noise of either the charge, the flux or the critical Josephson current.

On the other hand, the full control of 1-qubit circuits opens the possibility to use qubits as efficient noise detectors (Aguado and Kouwenhoven, 2000; Schoelkopf et al., 2003). The idea is to measure the decoherence times of the qubit while changing its parameters and extract from the data the noise in the qubit's environment. An experiment of this type was performed by (Astafiev et al., 2004a). Further information about the noise was obtained in recent studies (Ithier et al., 2005; Martinis et al., 2005). In this paper we give a short overview of new and improved understanding of the nature of $1/f$ noise.

2. Charge qubit and charge noise

To introduce the basic concepts we consider the simplest charge qubit. The system is shown in Fig. 1. Its Hamiltonian reads

$$H = \sum_n \left[E_{\text{ch}}(n, V_g)|n\rangle\langle n| + \frac{E_J}{2}|n\rangle\langle n \pm 1| \right], \tag{1}$$

where the charging energy in given by

$$E_{\text{ch}}(n, V_g) = \frac{(2ne - Q_g)^2}{2(C_g + C_J)}, \tag{2}$$

and the induced gate charge is $Q_g = C_g V_g$. Near $Q_g = e$ one can consider the two lowest energy charge states. In the spin-1/2 representation one obtains the following Hamiltonian

$$H = -\frac{1}{2}\Delta E_{ch}(V_g)\, \hat{\sigma}_z - \frac{1}{2}E_J\, \hat{\sigma}_x \,, \tag{3}$$

Introducing an angle $\eta(V_g)$ such that $\tan \eta = E_J/E_{ch}(V_g)$ we rewrite the Hamiltonian as

$$H = -\frac{1}{2}\Delta E\, (\cos \eta\, \hat{\sigma}_z + \sin \eta\, \hat{\sigma}_x) \,. \tag{4}$$

We now assume that the gate charge has a noisy component, i.e., $Q_g = C_g V_g + \delta Q$. Then the charging energy fluctuates and we obtain

$$H = -\frac{1}{2}\Delta E\, (\cos \eta\, \hat{\sigma}_z + \sin \eta\, \hat{\sigma}_x) - \frac{1}{2}X\hat{\sigma}_z \,, \tag{5}$$

where $X = e\delta Q/(C_g + C_J)$. In the eigenbasis of the qubit this gives

$$H = -\frac{1}{2}\Delta E\, \hat{\sigma}_z - \frac{1}{2}X(\cos \eta\, \hat{\sigma}_z - \sin \eta\, \hat{\sigma}_x) \,. \tag{6}$$

For sufficiently weak noise with regular spectrum $S_X(\omega)$, the Bloch-Redfield theory (Bloch, 1957; Redfield, 1957) gives the dissipative rates. The relaxation (spin flip) rate is given by

$$\Gamma_1 \equiv \frac{1}{T_1} = \frac{1}{2}\sin^2 \eta\, S_X(\omega = \Delta E) \,, \tag{7}$$

while the dephasing rate

$$\Gamma_2 \equiv \frac{1}{T_2} = \frac{1}{2}\Gamma_1 + \Gamma_\varphi \,, \tag{8}$$

with

$$\Gamma_\varphi = \frac{1}{2}\cos^2 \eta S_X(\omega = 0) \,. \tag{9}$$

is a combination of spin-flip effects (Γ_1) and of the so called 'pure' dephasing, characterized by the rate $\Gamma_\varphi = 1/T_2^*$. The pure dephasing is usually associated with the inhomogeneous level broadening in ensembles of spins, but occurs also for a single spin due to the 'longitudinal' (coupling to σ_z) low-frequency noise.

We now consider the situation where the noise X is characterized by the spectral density

$$S_X(\omega) = \frac{\alpha}{|\omega|} \tag{10}$$

in the interval of frequencies $\omega_{ir} < \omega < \omega_c$. In this case Eq. (9) is clearly inapplicable. Several models of $1/f$ noise and pure dephasing were developed in the literature (Cottet, 2002; Shnirman et al., 2002; Paladino et al., 2002; Galperin et al., 2004a). In all of them the T_2-decay of the coherences (i.e. of the off-diagonal elements of the density matrix) is given by decay law $e^{-\Gamma_1 t/2} f(t)$. The pure decoherence described by the function $f(t)$ depends on the statistics of the noise. For our purposes here a very rough estimate is enough. When deriving the Bloch-Redfield results, e.g., Eq. (9), one realizes that $S(\omega = 0)$ should actually be understood as the noise power averaged over the frequency band of width $\sim \Gamma_\varphi$ around $\omega = 0$. We thus obtain a time scale of the pure dephasing from the self-consistency condition $\Gamma_\varphi = S_X(\Gamma_\varphi)$. This gives

$$\Gamma_\varphi \approx \sqrt{\alpha}\,\cos\eta\,. \tag{11}$$

For the cases of "strongly non-Gaussian" statistics (Galperin et al., 2004a), α and Γ_φ should be understood as typical rather than ensemble averaged quantities. From the study of many examples we came to the conclusion that the relation (11) is universal irrespective of the noise statistics as long as $\Gamma_\varphi > \omega_{ir}$.

3. Analysis of the NEC experiments

(Astafiev et al., 2004a) measured the T_1 and T_2^* time scales in a charge qubit. As the energy splitting ΔE and the angle η were independently controlled, they could extract the noise power $S(\omega)$ in the GHz range using Eq. (7). In addition they were able to determine the strength of the $1/f$ noise, α, using Eq. (11). The results suggested a connection between the strengths of the Ohmic high-frequency noise, responsible for the relaxation of the qubit (T_1-decay), and the low-frequency $1/f$ noise, which dominates the dephasing (T_2-decay). The noise power spectra, extrapolated from the low- and high-frequency sides, turn out to cross at ω of order T. Expressing the high-frequency noise at $\omega > T$ as $S_X(\omega) = a\omega$, they found that the strength of the low-frequency noise scales as $\alpha = aT^2$ (see Fig. 2). The T^2 dependence of the low-frequency noise power was observed earlier for the $1/f$ noise in Josephson devices (Wellstood, 1988; Kenyon et al., 2000). Further evidence for the T^2 behavior was obtained recently (Astafiev, 2004; Wellstood et al., 2004). But the fact that the two parts of the spectrum are characterized by the same constant a was surprising.

4. Resonances in phase qubits

Additional information was obtained from experiments with phase qubits (current biased large area Josephson junction) by (Simmonds et al., 2004).

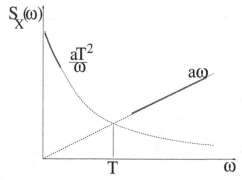

Figure 2. Asymptotic behavior of noise at low and high frequencies.

These experiments revealed the presence of spurious quantum two-level systems with strong effects on the high-frequency (~10 GHz) qubit dynamics. In a phase qubit one controls the energy splitting between the qubit states by changing the bias current. It turned out that at certain values of the bias current the system ceased to behave as a two-level system but showed rather a 4-level dynamics. This phenomenon can be attributed to the existence of a collection of coherent two-level fluctuators in the oxide of the tunnel barrier. When the energy splitting of the qubit coincides with that of one of the fluctuators a pair of states $|g_{\text{qubit}}\rangle|e_{\text{fluctuator}}\rangle$ and $|e_{\text{qubit}}\rangle|g_{\text{fluctuator}}\rangle$ are degenerate. Here $|g_{\text{qubit}}\rangle$ and $|g_{\text{fluctuator}}\rangle$ denote the ground states of the qubit and the fluctuator respectively, while $|e_{\text{qubit}}\rangle$ and $|e_{\text{fluctuator}}\rangle$ are the excited states. This degeneracy is lifted by the qubit-fluctuator interaction, which leads to a gap (avoided crossing) in the spectroscopy of the qubit. Most surprising was the observation that the two-level fluctuators are more coherent than the qubit. Hence, the decoherence of the 4-level system in a resonant situation is dominated by the decoherence of the qubit.

5.   High- and low-frequency noise from coherent TLF's

Motivated by the experiments mentioned above we have pointed out (Shnirman et al., 2005) that a set of *coherent* two-level systems may produce both high- and low-frequency noise with strengths that are naturally related. As a model we consider a set of coherent two-level systems described by the Pauli matrices $\sigma_{p,j}$, where $p = x, y, z$, and j labels the particular TLF. We write the Hamiltonian of the set in the basis such that their contributions to the relevant fluctuating quantity, e.g., the gate charge, are $X \equiv \sum_j v_j \sigma_{z,j}$. Then

$$H_{\text{TLS}} = \sum_j \left[-\frac{1}{2} \left(\varepsilon_j \sigma_{z,j} + \Delta_j \sigma_{x,j} \right) + H_{\text{diss},j} \right]. \tag{12}$$

Here, in the language of tunneling TLSs (TTLS), ε_j are the bias energies and Δ_j the tunnel amplitudes between two states. Each individual TLS, j, is subject to dissipation due to its own bath with Hamiltonian $H_{\text{diss},j}$. We do not specify $H_{\text{diss},j}$, but only assume that it produces the usual relaxation (T_1) and dephasing (T_2) processes. We assume that all the TLSs are under-damped, with $\Gamma_{1,j} \equiv T_{1,j}^{-1} \ll E_j$ and $\Gamma_{2,j} \equiv T_{2,j}^{-1} \ll E_j$. Here $E_j \equiv \sqrt{\epsilon_j^2 + \Delta_j^2}$ is the energy splitting.

Our goal in the following is to investigate the noise properties of the fluctuating field X. For that reason we evaluate the (unsymmetrized) correlator

$$C_X(\omega) \equiv \int dt \left\{ \langle X(t)X(0) \rangle - \langle X \rangle^2 \right\} e^{i\omega t}. \tag{13}$$

For independent TLSs the noise is a sum of individual contributions, $C_X = \sum_j v_j^2 C_j$, where

$$C_j(\omega) \equiv \int dt \left\{ \langle \sigma_{z,j}(t)\sigma_{z,j}(0) \rangle - \langle \sigma_{z,j} \rangle^2 \right\} e^{i\omega t}. \tag{14}$$

To obtain C_j we first transform to the eigenbasis of the TLS. This gives

$$H_{\text{TLS}} = \sum_j \left\{ -\frac{1}{2} E_j \rho_{z,j} + H_{\text{diss},j} \right\}, \tag{15}$$

and

$$X = \sum_j v_j \left(\cos\theta_j \rho_{z,j} - \sin\theta_j \rho_{x,j} \right), \tag{16}$$

where $\tan\theta_j \equiv \Delta_j/\epsilon_j$. Proceeding in the spirit of the Bloch-Redfield theory (Bloch, 1957; Redfield, 1957) we readily find

$$C_j(\omega) \approx \cos^2\theta_j \left[1 - \langle \rho_{z,j} \rangle^2 \right] \frac{2\Gamma_{1,j}}{\Gamma_{1,j}^2 + \omega^2}$$

$$+ \sin^2\theta_j \left[\frac{1 + \langle \rho_{z,j} \rangle}{2} \right] \frac{2\Gamma_{2,j}}{\Gamma_{2,j}^2 + (\omega - E_j)^2}$$

$$+ \sin^2\theta_j \left[\frac{1 - \langle \rho_{z,j} \rangle}{2} \right] \frac{2\Gamma_{2,j}}{\Gamma_{2,j}^2 + (\omega + E_j)^2}. \tag{17}$$

In thermal equilibrium we have $\langle \rho_{z,j} \rangle = \tanh(E_j/2T)$. The first term, due to the longitudinal part of the coupling, describes random telegraph noise of a thermally excited TLS. We have assumed $\Gamma_{1,j} \ll T$, so that this term is symmetric (classical). The second term is due to the transverse coupling and describes absorption by the TLS, while the third term describes the transitions of the TLS with emission. We observe that TLSs with $E_j \gg T$ contribute to C_X only at the (positive) frequency $\omega = E_j$. Indeed their contribution at $\omega = 0$ is suppressed by the thermal factor $1 - \langle \rho_{z,j} \rangle^2 = 1 - \tanh^2(E_j/2T)$. The negative frequency (emission) contribution at $\omega = -E_j$ is also suppressed. These high-energy TLSs remain always in their ground state. Only the TLSs with $E_j < T$ are thermally excited, performing real random transitions between their two eigenstates, and contribute at $\omega = \pm E_j$ and at $\omega = 0$. Note that the separation of the terms in Eq. (17) into low- and high-frequency noise is meaningful only if the typical width $\Gamma_{1,j}$ of the low-ω Lorentzians is lower than the high frequencies of interest, which are defined, e.g., by the qubit's level splitting or the temperature.

For a dense distribution of the parameters ϵ, Δ, and v we can evaluate the low- and high-frequency noise. For positive high frequencies, $\omega \gg T$, we obtain

$$C_X(\omega) \approx \sum_j v_j^2 \sin^2 \theta_j \frac{2\Gamma_{2,j}}{\Gamma_{2,j}^2 + (\omega - E_j)^2}$$

$$\approx N \int d\epsilon d\Delta dv \, P(\epsilon, \Delta, v) \, v^2 \sin^2 \theta \cdot 2\pi \delta(\omega - E) \,, \qquad (18)$$

where N is the number of fluctuators, $P(\epsilon, \Delta, v)$ is the distribution function normalized to 1, $E \equiv \sqrt{\epsilon^2 + \Delta^2}$, and $\tan \theta = \Delta/\epsilon$. Without loss of generality we take $\epsilon \geq 0$ and $\Delta \geq 0$.

At negative high frequencies ($\omega < 0$ and $|\omega| > T$) the correlator $C_X(\omega)$ is exponentially suppressed. On the other hand, the total weight of the low-frequency noise (up to $\omega \approx \Gamma_{1,\max}$, where $\Gamma_{1,\max}$ is the maximum relaxation rate of the TLSs) follows from the first term of (17). (Since we have assumed $\Gamma_{1,j} \ll E_j$ we can disregard the contribution of the last two terms of (17).) Each Lorentzian contributes 1. Thus we obtain

$$\int_{\text{low freq.}} \frac{d\omega}{2\pi} C_X(\omega)$$

$$\approx \int_{\text{low freq.}} \frac{d\omega}{2\pi} \sum_j v_j^2 \cos^2 \theta_j \left[1 - \langle \rho_{z,j} \rangle^2 \right] \frac{2\Gamma_{1,j}}{\Gamma_{1,j}^2 + \omega^2}$$

$$\approx N \int d\epsilon d\Delta dv \, P(\epsilon, \Delta, v) \, v^2 \cos^2 \theta \frac{1}{\cosh^2 \frac{E}{2T}} \,. \qquad (19)$$

Equations (18) and (19) provide the general framework for further discussion.

Next we investigate possible distributions for the parameters ϵ, Δ, and v. We consider a log-uniform distribution of tunnel splittings Δ, with density $P_\Delta(\Delta) \propto 1/\Delta$ in a range $[\Delta_{min}, \Delta_{max}]$. This distribution is well known to provide for the $1/f$ behavior of the low-frequency noise (Dutta and Horn, 1981a). It is natural for TTLSs as Δ is an exponential function of, e.g., the tunnel barrier height (Phillips, 1972), which is an almost uniformly distributed parameter. The relaxation rates are, then, also distributed log-uniformly, $P_{\Gamma_1}(\Gamma_1) \propto 1/\Gamma_1$, and the sum of many Lorentzians of width Γ_1 centered at $\omega = 0$ adds up to the $1/f$ noise.

The distribution of v is rather arbitrary. We only assume that it is uncorrelated with ε and Δ. Finally we have to specify the distribution of ϵs. First, we assume that the temperature is lower than Δ_{max}. For the high-frequency part, $T < \omega < \Delta_{max}$, we find, after taking the integral over Δ in Eq. (18), that

$$C_X(\omega) \propto \frac{1}{\omega} \int_0^\omega P_\varepsilon(\varepsilon) d\varepsilon. \tag{20}$$

This is consistent with the observed Ohmic behavior $C_X \propto \omega$ only for a linear distribution $P_\varepsilon(\varepsilon) \propto \varepsilon$.

Remarkably, this distribution, $P(\varepsilon, \Delta) \propto \varepsilon/\Delta$, produces at the same time the $T^2 \ln(T/\Delta_{min})$ behavior of the low-frequency weight (19), observed in several experiments (Wellstood, 1988; Kenyon et al., 2000; Astafiev, 2004; Wellstood et al., 2004). If the low-frequency noise has a $1/f$ dependence, the two parts of the spectrum would cross around $\omega \sim T$ (Astafiev et al., 2004a).

In the opposite limit, $T \gg \Delta_{max}$, the high-frequency noise depends on the detailed shape of the cutoff of $P_\Delta(\Delta)$ at Δ_{max}. As an example, for a hard cutoff the Ohmic spectral density implies that $P_\varepsilon \propto \varepsilon^3$, and the low-frequency weight scales with T^4. For a $1/f$ low-frequency behavior, the spectra would cross at $\omega \sim T^2/\Delta_{max} \gg T$, which is not in agreement with the result of Ref. (Astafiev et al., 2004a).

A remark is in order concerning the crossing at $\omega \approx T$ discussed above. It is not guaranteed that the spectrum has a $1/f$ dependence up to $\omega \sim T$. Rather the high-frequency cutoff of the low-frequency $1/f$ noise is given by the maximum relaxation rate of the TLSs, $\Gamma_{1,max} \ll T$, as we assumed. Then the *extrapolations* of the low-frequency $1/f$ and high-frequency Ohmic spectra cross at this $\omega \sim T$.

We would like to emphasize that the relation between low- and high-frequency noise is more general, i.e., it is not unique to an ensemble of two-level systems. Consider an ensemble of many-level systems with levels $|n\rangle$ and energies E_n such that the coupling is via an observable which has both transverse and longitudinal components. By 'transverse component' we

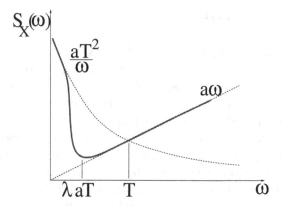

Figure 3. Noise spectrum in a selfconsistent model.

mean the part constructed with operators $|n\rangle\langle m|$, where $n \neq m$, while the 'longitudinal component' is built from the projectors $|n\rangle\langle n|$. If the system is under-damped, that is, if the absorption and emission lines are well defined, the correlator of such an observable will have Lorentzian-like contributions at $\omega = E_n - E_m$ as well as at $\omega = 0$. An example is provided by an ensemble of an-harmonic oscillators with $X = \sum_j v_j x_j$, where x_j are the oscillator's coordinates. Due to the anharmonicity x_j acquires a longitudinal component, in addition to the usual transverse one. Thus a relation between the low- and high-frequency noise would emerge naturally with details depending on the ensemble statistics.

6.   Self-consistent model

In this section we consider a possibility that the Γ_1 decay of each individual TLS is caused by the other TLSs. This model explains further details of the behavior of $S_X(\omega)$. We assume that each individual fluctuator "feels" the same charge noise as the qubit, however reduced by a factor $\lambda < 1$ due to the small size of the fluctuators. That is we assume that the relaxation rate of the fluctuators is given by

$$\Gamma_{1,j} = \frac{\lambda}{2} \sin \theta_j^2 \, S_X(\omega = E_j) \,. \tag{21}$$

As only the fluctuators with $E_j \leq T$ contribute to the $1/f$ noise, we estimate the maximum possible relaxation rate of the fluctuators to be $\Gamma_{1,\mathrm{max}} \sim \lambda a T$. This leads to a crossover from $1/f$ to $1/f^2$ dependence around $\omega \sim \Gamma_{1,\mathrm{max}} \sim \lambda a T$ as indicated in Fig. 3. We note that such a crossover (soft cut-off) is compatible with the recent experimental data (Ithier et al., 2005).

7. Relation to other work

It is useful to relate our phenomenological results to the recent work of (Faoro et al., 2005), (de Sousa et al., 2005), (Grishin et al., 2005), and (Faoro and Ioffe, 2005), where physical models of the fluctuators, coupling to and relaxing the qubit, were considered. In Ref. (Faoro et al., 2005) three models were studied: (I) a single electron trap in tunnel contact with a metallic gate, (II) a single electron occupying a double trap, and (III) a double trap that can absorb/emit a Cooper pair from the qubit or a superconducting gate ('Andreev fluctuator'). In all models a uniform distribution of the trap energy levels was assumed. Then one can show that the distribution for the two-level systems corresponding to the models II and III are linear in the energy level splitting, $P(\epsilon) \propto \epsilon$. Since the switching in these models is tunneling dominated, we find that $P(\Delta) \propto 1/\Delta$. Therefore, both models II and III are characterized by distribution $P(\epsilon, \Delta) \propto \epsilon/\Delta$, introduced above, and hence can naturally account for the experimentally observed low- and high-frequency noises. In contrast, as shown in (de Sousa et al., 2005; Grishin et al., 2005), single electron traps do not behave as coherent two-level systems. Depending on the ratio between the hybridization with the metal and the temperature a single trap at the Fermi energy can either show the random telegraph noise or, when the hybridization dominates, it can make the qubit feel the Ohmic particle-hole spectrum of the metal.

It was argued recently (Faoro and Ioffe, 2005) that one needs an unphysically high density of fluctuators in order to explain the experimental findings. This argument is based on the assumption that the traps' energies are distributed homogeneously over the energy band of order of the Fermi energy (of order 1 eV). (Faoro and Ioffe, 2005) proposed an alternative scenario where the low energy scale needed for qubit relaxation is provided by Kondo physics.

8. Mechanisms of coupling between Josephson qubits and two-level systems

In this section we discuss possible mechanisms of coupling between the TLS and the Josephson qubits in light of the recent experimental findings (Simmonds et al., 2004; Astafiev et al., 2004a; Martinis et al., 2005). For qubits with strong domination of the charging energy, like those of Ref. (Astafiev et al., 2004a), the charge noise considered in Section 2 is the main source of dissipation. In these qubits the fluctuators are most likely located away from the tunnel junction. This follows from the fact that a single charge of e jumping back and forth a distance of a single atomic bond (~ 0.1nm)

across a 2nm wide tunnel junction (Martinis et al., 2005) would create charge
noise with amplitude (defined as square root of integrated power) of order
$10^{-2}e$. A typical amplitude of the charge noise in single electron devices
is smaller, $\sim 10^{-3}e$. If a fluctuator of this type would happen to be in the
tunnel junction, such an "unfortunate" sample would, probably, be discarded
by experimentalists. On the other hand, as we have already mentioned in
Section 4, the coherent two-level systems have been observed in large area
Josephson junctions. There are at least two types of interactions which could
be responsible for lifting the degeneracy between the states $|g_{\text{qubit}}\rangle |e_{\text{fluctuator}}\rangle$
and $|e_{\text{qubit}}\rangle |g_{\text{fluctuator}}\rangle$ (Martin et al., 2005). One corresponds to a situation in
which a two-level fluctuator blocks a conducting channel in one of its states
and, thus, influences the Josephson energy of the junction (Simmonds et al.,
2004). This can be expressed by substituting the Josephson energy as

$$E_J \rightarrow E_J \left[1 + \frac{1}{2} \sum_j v_j \sigma_{z,j} \right], \tag{22}$$

where $v_j = \delta G_j / G$ is the ratio of the conductance variation δG_j due to the
motion of the fluctuator j to the total tunnel conductance G of the junction.
The other coupling mechanism (Martin et al., 2005) arises due to a dipole
moment of the fluctuator interacting with the electric field in the junction,
i.e., it is directly related to the charge noise. The interaction energy of this
mechanism can be written as $E_{\text{int}} = (1/2) eV \sum_j \tilde{v}_j \sigma_{z,j}$, where V is the (operator of) voltage across the junction. The coupling constants $\tilde{v}_j$ are given
this time by $\tilde{v}_j = d_j/l$, where l is the width of the junction and d_j is the
distance the fluctuator moves across the junction. Recent studies (Martinis
et al., 2005) point towards this coupling mechanism. Thus it is plausible
that the fluctuators producing the charge noise in the charge qubits and the
coherent fluctuators in the phase qubits are of the same origin. The former
ones are located in the oxide away from the tunnel junction, while the latter
ones reside in the oxide of the tunnel junction.

9. Conclusions

Josephson qubits have found their first application as sensitive meters of their
environment. Measurements of qubit relaxation produced new surprising information about the properties of $1/f$ noise. Motivated by these experiments,
we have shown that an ensemble of coherent two-level systems with the
distribution function, $P(\epsilon, \Delta) \propto \epsilon/\Delta$, produces Ohmic high-frequency noise
and, at the same time, $1/f$ low-frequency noise with strength which scales
with temperature as T^2. The two branches of the noise power spectrum cross

at $\omega \sim T$ in accordance with the experimental observation (Astafiev et al., 2004a). Thus, recent experimental findings (Astafiev et al., 2004a; Martinis et al., 2005) shed new light on the nature of the low frequency fluctuations in mesoscopic systems.

Acknowledgements

This work was supported by the EU IST Project EuroSQIP and by the CFN (DFG), and the grant MD-2177.2005.2 (YM).

References

Aguado, R. and Kouwenhoven, L. P. (2000) Double Quantum Dots as Detectors of High-Frequency Quantum Noise in Mesoscopic Conductors, *Phys. Rev. Lett.* **84**, 1986.

Anderson, P. W., Halperin, B. I., and Varma, C. M. (1972) Anomalous Low-Temperature Thermal Properties of Glasses and Spin Glasses, *Phylos. Mag.* **25**, 1.

Astafiev, O. (2004), *private communication*.

Astafiev, O., Pashkin Yu. A., Nakamura, Y., Yamamoto, T., and Tsai, J. S. (2004a) Quantum Noise in the Josephson Charge Qubit, *Phys. Rev. Lett.* **93**, 267007.

Astafiev, O., Pashkin, Y. A., Yamamoto, T., Nakamura, Y., and Tsai, J. S. (2004b) Single-Shot Measurement of the Josephson Charge Qubit., *Phys. Rev. B* **69**, 180507.

Bernamont, J. (1937) Fluctuations de potentiel aux bornes d'un conducteur metallique de faible volume parcouru par un courant., *Ann. Phys. (Leipzig)* **7**, 71.

Black, J. L. (1981) Low-Energy Excitations in Metallic Glasses, In H.-J. Güntherodt and H. Beck (eds.), *Glassy metals*, Berlin, Springer-Verlag.

Black, J. L. and Halperin, B. I. (1977) Spectral Diffusion, Phonon Echoes, and Saturation Recovery in Glasses at Low-Temperatures, *Phys. Rev. B* **16**, 2879.

Bloch, F. (1957) Generalized Theory of Relaxation, *Phys. Rev.* **105**, 1206.

Collin, E., Ithier, G., Aassime, A., Joyez, P., Vion, D., and Esteve, D. (2004) NMR-like Control of a Quantum Bit Superconducting Circuit., *Phys. Rev. Lett.* **93**, 157005.

Cottet, A. (2002), *PhD thesis, Université Paris VI*.

de Sousa, R., Whaley, K. B., Wilhelm, F. K., and von Delft, J. (2005) Ohmic Noise from a Single Defect Center Hybridized with a Fermi Sea., *Phys. Rev. Lett.* **95**, 247006.

Dutta, P. and Horn, P. M. (1981a) Low-Frequency Fluctuations in Solids, *Rev. Mod. Phys.* **53**, 497.

Dutta, P. and Horn, P. M. (1981b) Low-Frequency Fluctuations in Solids: $1/f$ Noise, *Rev. Mod. Phys.* **53**, 497.

Esteve, D. and Vion, D. (2005) Solid State Quantum Bits, *cond-mat/0505676*.

Falci, G., D'Arrigo, A., Mastellone, A., and Paladino, E. (2005) Initial Decoherence in Solid State Qubits, *Phys. Rev. Lett.* **94**, 167002.

Faoro, L., Bergli, J., Altshuler, B. A., and Galperin, Y. M. (2005) Models of Environment and T_1 Relaxation in Josephson Charge Qubits, *Phys. Rev. Lett.* **95**, 046805.

Faoro, L. and Ioffe, L. B. (2005) Quantum Two Level Systems and Kondo-like Traps as Possible Sources of Decoherence in Superconducting Qubits., *cond-mat/0510554*.

Feng, S., Lee, P. A., and Stone, A. D. (1986) Sensitivity of the Conductance of a Disordered Metal to the Motion of a Single Atom: Implications for 1/f Noise, *Phys. Rev. Lett.* **56**, 1960.

Galperin, Y. M., Altshuler, B. L., and Shantsev, D. V. (2004a) Low-Frequency Noise as a Source of Dephasing of a Qubit, In I. V. Lerner, B. L. Altshuler, and Y. Gefen (eds.), *Fundamental Problems of Mesoscopic Physics*, Dordrecht, Boston, London, Kluwer Academic Publishers, cond-mat/0312490.

Galperin, Y. M., Kozub, V. I., and Vinokur, V. M. (2004b) Low-Frequency Noise in Tunneling through a Single Spin, *Phys. Rev. B* **70**, 033405.

Grishin, A., Yurkevich, I. V., and Lerner, I. V. (2005) Low-Temperature Decoherence of Qubit Coupled to Background Charges, *Phys. Rev. B* **72**, 060509.

Imry, Y., Fukuyama, H., and Schwab, P. (1999) Low-Temperature Dephasing in Disordered Conductors: The Effect of "$1/f$" Fluctuations, *Europhys. Lett.* **47**, 608.

Ithier, G., Collin, E., Joyez, P., Meeson, P. J., Vion, D., Esteve, D., Chiarello, F., Shnirman, A., Makhlin, Y., Schriefl, J., and Schön, G. (2005) Decoherence in a Superconducting Quantum Bit Circuit., *Phys. Rev. B* **72**, 134519.

Kenyon, M., Lobb, C. J., and Wellstood, F. C. (2000) Temperature Dependence of Low-Frequency Noise in Al-Al2O3-Al Single-Electron Transistors, *J. Appl. Phys.* **88**, 6536.

Kogan, S. M. and Nagaev, K. E. (1984) On the Low-Frequency Current Noise in Metals, *Solid State Comm.* **49**, 387.

Korotkov, A. N. and Averin, D. V. (2001) Continuous Weak Measurement of Quantum Coherent Oscillations, *Phys. Rev. B* **64**, 165310.

Ludviksson, A., Kree, R., and Schmid, A. (1984) Low-Frequency 1/f Fluctuations of Resistivity in Disordered Metals, *Phys. Rev. Lett.* **52**, 950.

Makhlin, Y. and Shnirman, A. (2004) Dephasing of Solid-State Qubits at Optimal Points, *Phys. Rev. Lett.* **92**, 107001.

Martin, I., Bulaevskii, L., and Shnirman, A. (2005) Tunneling Spectroscopy of Two-level Systems Inside a Josephson Junction., *Phys. Rev. Lett.* **95**, 127002.

Martinis, J. M., Cooper, K. B., McDermott, R., Steffen, M., Ansmann, M., Osborn, K., Cicak, K., Oh, S., Pappas, D. P., Simmonds, R., and Yu, C. C. (2005) Decoherence in Josephson Qubits from Dielectric Loss., *Phys. Rev. Lett.* **95**, 210503.

Nakamura, Y., Pashkin Yu. A., Yamamoto, T., and Tsai, J. S. (2002) Charge Echo in a Cooper-Pair Box, *Phys. Rev. Lett.* **88**, 047901.

Paladino, E., Faoro, L., Falci, G., and Fazio, R. (2002) Decoherence and 1/f noise in Josephson Qubits, *Phys. Rev. Lett.* **88**, 228304.

Phillips, W. A. (1972) Tunneling States in Amorphous Solids, *J. Low. Temp. Phys.* **7**, 351.

Rabenstein, K., Sverdlov, V. A., and Averin, D. V. (2004) Qubit Decoherence by Gaussian Low-Frequency Noise, *JETP Lett.* **79**, 783.

Redfield, A. G. (1957) On the theory of relaxation processes, *IBM J. Res. Dev.* **1**, 19.

Schoelkopf, R. J., Clerk, A. A., Girvin, S. M., Lehnert, K. W., and Devoret, M. H. (2003) Qubits as Spectrometers of Quantum Noise, In Y. V. Nazarov (ed.), *Quantum Noise in Mesoscopic Physics*, Dordrecht, Boston, pp. 175–203, Kluwer Academic Publishers, cond-mat/0210247.

Schriefl, J. (2005), *PhD Thesis, University of Karlsruhe*.

Shnirman, A., Makhlin, Yu., and Schön, G. (2002) Noise and Decoherence in Quantum Two-Level Systems, *Physica Scripta* **T102**, 147.

Shnirman, A., Mozyrsky, D., and Martin, I. (2004) Output Spectrum of a Measuring Device at Arbitrary Voltage and temperature, *Europhys. Lett.* **67**, 840.

Shnirman, A., Schön, G., Martin, I., and Makhlin, Y. (2005) Low- and High-Frequency Noise from Coherent Two-Level Systems, *Phys. Rev. Lett.* **94**, 127002.

Siddiqi, I., Vijay, R., Pierre, F., Wilson, C. M., Metcalfe, M., Rigetti, C., Frunzio, L., and Devoret, M. H. (2004) RF-Driven Josephson Bifurcation Amplifier for Quantum Measurement, *Phys. Rev. Lett.* **93**, 207002.

Simmonds, R. W., Lang, K. M., Hite, D. A., Nam, S., Pappas, D. P., and Martinis, J. M. (2004) Decoherence in Josephson Phase Qubits from Junction Resonators, *Phys. Rev. Lett.* **93**, 077003.

VanHarlingen, D. J., Robertson, T. L., Plourde, B. L. T., Reichardt, P. A., Crane, T. A., and Clarke, J. (2004) Decoherence in Josephson-junction qubits due to critical current fluctuations, *Phys. Rev. B* **70**, 064517.

Vion, D., Aassime, A., Cottet, A., Joyez, P., Pothier, H., Urbina, C., Esteve, D., and Devoret, M. H. (2002) Manipulating the quantum state of an electrical circuit, *Science* **296**, 886.

Wallraff, A., Schuster, D. I., Blais, A., Frunziol, L., Huang, R.-S., Majer, J., Kumar, S., Girvin, S. M., and Schoelkopf, R. J. (2004) Strong Coupling of a Single Photon to a Superconducting Qubit using Circuit Quantum Electrodynamics, *Nature* **431**, 162.

Wellstood, F. C. (1988), *PhD thesis, University of California, Berkeley.*

Wellstood, F. C., Urbina, C., and Clarke, J. (2004) Flicker (1/f) Noise in the Critical Current of Josephson Junctions at 0.09-4.2 K, *Appl. Phys. Lett.* **85**, 5296.

Wendin, G. and Shumeiko, V. S. (2005) Superconducting Quantum Circuits, Qubits and Computing, *cond-mat/0508729.*

Yamamoto, T., Pashkin, Y. A., Astafiev, O., Nakamura, Y., and Tsai, J. S. (2003) Demonstration of Conditional Gate Operation using Superconducting Charge Qubits., *Nature* **425**, 941.

Zimmerli, G., Eiles, T. M., Kautz, R. L., and Martinis, J. M. (1992) Noise in the Coulomb Blockade Electrometer, *Appl. Phys. Lett.* **61**, 237.

Zorin, A. B., Ahlers, F.-J., Niemeyer, J., Weimann, T., Wolf, H., Krupenin, V. A., and Lotkhov, S. V. (1996) Background Charge Noise in Metallic Single-Electron Tunneling Devices, *Phys. Rev. B* **53**, 13682.

SINGLE-ELECTRON PUMP: DEVICE CHARACTERIZATION
AND LINEAR-RESPONSE MEASUREMENTS

R. Schäfer (`roland.schaefer@ifp.fzk.de`), B. Limbach, P. vom Stein
and C. Wallisser
*Forschungszentrum Karlsruhe, Institut für Festkörperphysik,
Postfach 3640, 76021 Karlsruhe, Germany*

Abstract. We report experimental investigations on single-electron pumps, which by definition consist of two metallic islands formed by three tunnel junctions in series. Here, we focus
on device characterization and show that all parameters needed to model our samples can be
determined. A similar study for the single-electron transistor—a much simpler device than the
two-island system studied here—has been very successful and revealed the amazingly close
match between the standard Hamiltonian of single-electron devices and the actual samples it
describes. The most valuable quantity for a careful comparison between theory and experiment
turns out to be the linear-response conductance, which is presented as the main result of our
measurements. A comparison of these data to theoretical models does not rely on any fitting
parameter. We find strong quantum-fluctuation phenomena manifested by a $\log\left(k_B T/(2E_{co})\right)$
reduction of the maximal conductance, where E_{co} measures the coupling strength between the
islands.

Key words: Single-electron devices, Coulomb blockade, Tunneling

1. Introduction

Single-electron tunneling (SET) devices (see e. g. Grabert and Devoret (1992))
are made up from a number of tunneling contacts connecting several small
metallic islands among themselves as well as to external reservoirs. These
tunneling contacts make a change of the charge on the otherwise isolated
islands possible. In addition, gate electrodes can be used to tune the electrostatic potential of the islands by a purely capacitive coupling. Single-electron
effects occur at a given temperature T if the total capacity C of a island obeys
$C \lesssim e^2/k_B T$. This relation sets a size restriction on the nanometer scale for
the tunneling contacts even at dilution refrigerator temperatures of $\sim 20\,\text{mK}$.

Here we report measurements on SET devices of a common type produced by the so called "*shadow evaporation technique*" (Niemeyer, 1974):
A micro-structured PMMA mask held at some distance from the substrate is

K. Scharnberg and S. Kruchinin (eds.),
Electron Correlation in New Materials and Nanosystems, 357–369.
© 2007 *Springer.*

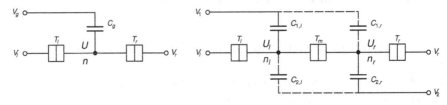

a. single-electron transistor   b. single-electron pump

Figure 1. Schematic representation of two SET devices with one and two islands. Tunneling contacts are labeled T_l, T_m, and T_r (each contact is characterized by its capacitance C_i and conductance $(e^2/h)g_i$ ($i \in \{l,m,r\}$). The islands are formed between the contacts. n, n_l, n_r count the number of electrons by which the island charge differs from neutrality. Both devices are biased by $V_r - V_l$. The electrostatic potentials on the islands (U, U_l, and U_r) can be tuned by V_g (transistor) or V_1 and V_2 (pump). V_g couples to the single island of the device by C_g. V_1 and V_2 couples directly to the two pump islands by the capacitance C_{1l} and C_{2r}. C_{2l} and C_{1r} represent the experimentally unavoidable stray capacitances of the two island device.

used to pattern an aluminum film of about 50 nm thickness during evaporation. Two identical aluminum patterns transposed with respect to each other are produced by choosing different evaporation angles in two consecutive evaporation steps. By a proper alignment of the angles an overlap between the two patterns can be achieved and restricted to very small areas. Between the two evaporation steps a natural aluminum oxide layer is grown on the first aluminum film by flooding the high vacuum evaporation chamber with a small amount of oxygen. The tunneling barriers arise from the oxide layer in the overlap regions.

The single-electron islands of these devices are made out of aluminum. Their size is still so large that the electronic states form a continuum at the Fermi level (the mean level spacing is small compared to the thermal energy scale at typical temperatures of order of $T \sim e^2/(100 k_B C)$). For each tunneling contact many electronic states on the left and right bank of the barrier exhibit a small tunneling matrix element leading to an overall finite but small transparency.

Electrical transport properties of such SET devices can be studied in the linear-response region by slightly biasing the external reservoirs. In this case, a device can be completely characterized by a rather small set of $N_T + (N^2 + 3N)/2$ parameters. Here N_T counts the tunneling contacts and N the single-electron islands. The first N_T parameters are given by the Ohmic conductance $G_i \equiv (e^2/h)g_i, i \in [1, N_T]$, each contact would exhibit in a voltage biased configuration. The additional parameter describe the electrostatic response of the SET device. $(N^2 + N)/2$ parameters stem from the symmetric charging-energy matrix $\mathbf{E} = (e^2/2)\mathbf{C}^{-1}$ of rank N, where $\mathbf{C}$ is the capacity matrix of the islands. The remaining N parameters describe the influence of the gates (including the influence of the biasing reservoirs on the electrostatic potential of the islands connected to them). This influence can be cast into a vector of charges $q_i, i \in [1, N]$ (or even more convenient into dimensionless quantities $n_i = q_i/e$) describing an offset charge q_i on each island.

1.1. A SINGLE ISLAND DEVICE: THE SET TRANSISTOR

To get more specific we mention the normal conducting SET transistor, which is probably the best studied SET device (see Fig. 1a). According to the above rule we need four parameters to characterize this device. The linear-response conductance is correctly described (see below) by a Hamiltonian $H_{tr} = H_{qp} + H_{ch} + H_T$. H_{qp} describes the quasiparticle states of the two reservoirs and of the island in the limit of vanishing coupling between these three parts. H_{ch} describes the electrostatic charging energy of the island charged up by n electrons, which can be written as $E_{ch} = E_c(n - n_1)^2$. The charging-energy matrix is a scalar, $E_c = e^2/(2C_\Sigma)$ and represents our first parameter in the present case. C_Σ is the sum of all capacities in Fig. 1a. In the relation for E_{ch} the rôle of the offset charge en_1 (our second parameter) is apparent: For $n = n_1$ the charging energy is minimal. Note however that n is integer (since charge is quantized in units of e), while $n_1 \approx C_g V_g$ is a continuous quantity depending on the gate voltage V_g and capacity C_g (In general, it depends on the voltage of the biasing reservoirs as well. In the limit of linear response these terms will vanish and the last relation becomes exact). The final part of the Hamiltonian, H_T, describes the tunneling through the two contacts. It is understood that the conductance in the present situation depends on an averaged tunneling matrix element only, which is proportional to the Ohmic conductance $G_i \equiv (e^2/h)g_i, i \in \{l, r\}$. These last two quantities complete our set of parameters. Naturally, the conductance is measured as a function of temperature T and of n_1, parameters which easily can be controlled by external means.

For more complex SET devices with more than one island we still split the Hamiltonian into the parts H_{qp} (describing the quasiparticles on the various parts of the device), H_{ch} (describing the electrostatics by a charging-energy matrix and one offset charge for each island), and H_T (describing each tunneling junction with its conductance parameter).

1.2. SELF HEATING AND THE ADVANTAGE OF LINEAR-RESPONSE

In principle, the complete Hamiltonian should include terms describing the exchange of energy with the environment during inelastic tunneling events (e. g. Ingold and Nazarov (1992)). In most practical cases, however, it is sufficient to include elastic tunneling events only. While in general e.g. the

Figure 2. Scanning electron micrograph of a SEP in a special layout. The middle tunneling contact is seen in the center. It is produced in a different oxidation process than the outer contacts, which are rotated by 90° with respect to the inner contact. The islands are seen adjacent to the middle junction on top and bottom. Each island is connected by two outer contacts to the corresponding reservoir in the final experiment. The gates can be seen in the midway of the top and bottom edge of the picture.

finite resistance of the leads between the reservoirs and the tunneling junction allow an exchange of energy during the tunneling events, this has little influence as long as external impedances are small compared to the internal junction resistances. Since the junctions are low conducting objects anyhow, this assumption is justified in most cases.

One should keep in mind, however, that ordinarily elastic tunneling events result in a non-equilibrium situation: empty states exceeding the Fermi energy by as much as the applied bias voltage get occupied by the tunneling electrons and a subsequent dissipative relaxation process is needed to restore equilibrium. In general, the energy dissipation after the tunneling events is a major problem in operating SET devices since it leads to severe heating of the electronic system (Kautz et al., 1993; Korotkov et al., 1994; Krupenin et al., 1999). In many cases it is possible to keep the phononic system at base temperature by taking proper action. Then, the bottleneck of the relaxation process is the energy transfer between the electronic and the phononic system due to fundamental reasons. In general, the performance of SET devices "degrades" quickly with increasing electronic temperature. We restrict ourselves to the investigation of the linear-response regime, where the tunneling rates, as well as the excess energy of tunneling electrons, is small. It is only in this regime that the electrons on single-electron islands can be cooled down to dilution refrigerator temperatures.

The theory predicts many interesting features in the non-linear transport regime at higher bias voltages. However, these features are difficult to investigate experimentally in a quantitative manner as a result of the energy dissipation just discussed. The most annoying fact about self heating is that the resulting electronic temperature is practically unknown. It depends on

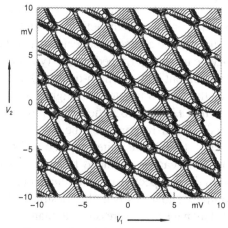

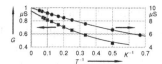

b. High temperature conductance. From the fits (solid lines) g_0 and E_c are determined (see text).

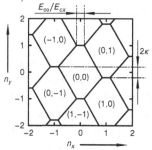

a. Contour plot of the current through a SEP as a function of the two gate voltages V_1 and V_2. The thin lines represent currents from 20 pA to 220 pA in steps of 20 pA. Also included are the stability lines (thicker solid hexagons). They correspond to the hexagons of Fig. 3c, but are linearly transformed due to the different coordinate systems of figures 3a and 3c

c. Stability diagram of the SEP. The hexagonal cells mark regions where the indicated charge state (n_l, n_r) possesses the lowest energy $E_{ch}(n_l, n_r)$.

Figure 3.

the coupling between the phononic and the electronic systems which enters into the description as a fitting parameter. This makes an unambiguous comparison between experiment and theoretical calculations almost impossible. Furthermore, at elevated electronic temperatures most features tend to wash out and are thus difficult to investigate.

1.3. SET-TRANSISTOR PHYSICS IN THE LINEAR-RESPONSE REGIME

For the SET transistor it had been shown that the Hamiltonian H_{tr} describes the linear-response conductance with amazingly high accuracy (Joyez et al., 1997). The most sensitive test (Wallisser et al., 2002) depended strongly on a careful device characterization. In the most natural transistor design, with exactly one contact between the island and each of the two reservoirs, the two conductance parameters, $g_i, i \in \{l, r\}$, cannot be determined independently. Only the serial conductance $g_0 = 1/(g_l^{-1} + g_r^{-1})$ is accessible in this case. For an unambiguous description the parallel conductance $g = g_l + g_r$ is needed as well. Wallisser et al. (2002) solved the problem of measuring g with a special transistor layout with more than two tunneling contacts connected to the island. In the final experiment all contacts had been connected to only two different reservoirs so as to operate several contacts in parallel. This arrangement is equivalent to a transistor with tunneling conductances which equal the sum of the conductances of the contacts connected in parallel. The individual junction conductances can be determined in this type of device. Wallisser et al. found perfect agreement between high-quality Quantum Monte Carlo (QMC) simulations of the SET transistor and high-resolution experimental data giving confidence that H_{tr} describes the whole physics of the single-electron device at least in the linear-response regime.

The QMC studies in Wallisser et al. (2002) are quite involved and there is little hope that simulation of that kind can be extended to systems with more than two islands. However, from the perspective of functionalizing SET devices, a detailed modeling of complex situations is highly desirable. Perturbative methods are known to yield manageable expressions in many cases. The lowest-order perturbative theory in H_T for the present case is known as the sequential tunneling model (see e. g. Ingold and Nazarov (1992)). This very useful method gives reliable information on the qualitative device performance even in rather complicated situations. However, in most practical cases large corrections to the sequential behavior are observed in experiments. Considerable effort has been made to extend the perturbative expansion beyond the lowest order. However, the data in Wallisser et al. (2002) suggest that even for the transistor perturbation theory is limited to rather low values of g (which is the natural expansion parameter for the SET transistor) and fails altogether for special values of n_1. To get further insight into the relation between Monte Carlo data and perturbative methods as well as experimental data on real SET devices, we have provided data on two island systems (Limbach et al., 2005). Guided by our experiments on the transistor, we place special emphasis on device characterization. In the major part of this paper we give details about the characterization process giving the full set of parameters needed two write down the two-island analog of H_{tr}.

2. Device characterization for two island systems

For a two-island system the electrostatic charging energy in the Hamiltonian is:

$$E_{ch} = E_l(n_l - n_{0l})^2 + E_r(n_r - n_{0r})^2 + 2E_{co}(n_l - n_{0l})(n_r - n_{0r}). \quad (1)$$

Here n_i denotes the number of elementary charges and en_{0i} the offset charge on the left ($i = l$) and right ($i = r$) island. $E_i, i \in \{l, r\}$ labels the charging energy of the two islands. In addition we have a coupling term with energy scale E_{co}. In accordance with our formula of Sec. 1 E_{ch} contains five parameters. Among the different two-island devices we choose the single-electron pump (SEP), which is especially simple to characterize in terms of linear-response measurements, for our investigation. For this device the two islands are coupled by a tunneling contact and each island is connected by a tunneling contact to its own reservoir (see Fig. 1b).

2.1.   DETERMINATION OF THE CONDUCTANCE PARAMETERS

In the most natural layout of a SEP, constituted by exactly three junctions and two gates, the individual conductances of the tunneling contacts are inaccessible. We have seen a similar situation for the SET transistor already. Also, for a simple SEP only the serial conductance—now given by $g_0 = 1/(g_l^{-1} + g_r^{-1} + g_m^{-1})$—can be measured directly. The layout shown in Fig. 2 remedies this deficiency. Here current-voltage (IV) characteristics can be measured along different routes across one or both islands. From these characteristics the conductances of different serial combinations of contacts can be determined. With a little algebra the individual tunnel conductances can be determined.

However, further comments are required. The conductance parameter is defined as the conductance of the junction in a voltage biased configuration (see Sec. 1). In this case, a linear dependence between voltage and current would be observed, and the conductance could be measured in the linear-response limit. It is not possible to voltage bias a junction even in the layout shown in Fig. 2. All current routes across several junctions include at least one of the islands. As a consequence, non-Ohmic IV characteristics with a strong Coulomb blockade are observed at low temperatures. But even at relatively high temperatures the Coulomb blockade renormalizes the slope of the IV characteristic at low bias voltage V. At large positive and negative voltages V the characteristic is expected to approach a constant slope, and it is this slope which can be identified with the serial conductance of the barriers crossed by the specific current route. Note however that this picture is rather naive. A more sophisticated treatment shows that at high voltages environmental effects (of the type briefly mentioned in Sec. 1.2) get more and more important (Wahlgren et al., 1995). This alters the slope of the IV curve at high biases. It is an experimental fact that even from a thorough inspection of the IV characteristic a region of constant slope cannot be extracted. As a consequence, the determination of the sequential conductance is limited to about 5%. This error propagates through the algebra to the final estimates of the individual junction conductances. While this accuracy seems rather low at a first sight, it turns out to be sufficient for the purpose of a detailed comparison with theoretical predictions (Wallisser et al., 2002).

Another way to determine the serial conductance g_0 is presented in Sec. 2.3. Both methods give consistent values. In Sec. 3 we present data

TABLE I. SEP parameters

samp.	$\dfrac{E_l}{k_B}$ (K)	$\dfrac{E_r}{k_B}$ (K)	$\dfrac{E_{co}}{k_B}$ (K)	$\dfrac{E_{cx}}{k_B}$ (K)	$\dfrac{E_m}{k_B}$ (K)	$\dfrac{E_c}{k_B}$ (K)	g_l	g_r	g_m
1	2.6	2.2	0.9	1.6	3.0	2.5	0.52	0.83	1.32
2	2.6	2.6	0.3	1.5	4.7	4.5	0.73	0.57	0.026

samp.	C_l (aF)	C_r (aF)	C_m (aF)	C_{1l} (aF)	C_{2r} (aF)	C_{1r} (aF)	C_{2l} (aF)	G_0 (µS)	κ
1	181	236	173	50.5	58.6	18.0	21.5	10.0	0.1054
2	244	242	43	50.9	53.6	18.6	18.5	0.95	0.0013

for two samples. The corresponding conductance parameters determined as explained here are given in Tab. I.

2.2. DETERMINATION OF THE OFFSET CHARGE

As apparent from Eq. 1 the behavior of the SEP is periodic in the offset charges $n_{0i}, i \in \{l, r\}$. If we shift, e.g. n_{0l} by an integer number ($n_{0l} \rightarrow n'_{0l} = n_{0l} + m$), this change can be absorbed in a rescaling of n_l ($n_l \rightarrow n'_l = n_l - m$) leaving the difference terms in Eq. 1 unchanged ($n'_l - n'_{0l} = n_l - n_{0l}$). Since the number of electrons can be freely adjusted by tunneling events, the behavior of the SEP at n_{0l} is identical to the behavior at n'_{0l}. In practice, two gates (with voltages V_1 and V_2, see Fig. 1b) are used which both couple to the islands with different capacities. Tuning V_1 and V_2 makes it possible to scan the (n_{0l}, n_{0r}) plane (see Fig. 3a): $en_{0i} = C_{1i}V_1 + C_{2i}V_2, i \in \{l, r\}$. Thus, it is easy to extract $C_{ij}, i \in \{1, 2\}, j \in \{l, r\}$ from data of the type shown in Fig. 3a by utilizing the apparent periodicity (see Tab. I).

A general problem of SET devices is the randomness of the offset charges. They are not solely determined by controllable gate voltages, but equally influenced by trapped charges in the substrate and the tunneling barriers. These so-called background charges produce an image charge on the island shifting the electrostatic potential on the islands just as a deliberately tuned gate would. Even worse, the background charge configuration is not completely stable but jumps from time to time even at very low temperatures (e. g. Zorin et al., Furlan et al. (1996, 2000)). For this reason the (V_1, V_2) contour plot in Fig. 3a is not perfectly periodic but display jumps. These jumps are relatively seldom for our samples (on the time scale of one to several hours). It should be clear from Fig. 3a that it is possible to scan a representative cell of the (n_{0l}, n_{0r}) plane with sufficient accuracy which is not disturbed by a background charge redistribution.

It is convenient to shift from the coordinate system (n_{0l}, n_{0r}) to a different one defined by $n_x = n_{0l} + n_{0r}$ and $n_y = n_{0r} - n_{0l} - \kappa n_x$, where $\kappa = (E_l - E_r)/E_m$ is an asymmetry parameter with $E_m = E_l + E_r - 2E_{co}$. The charging energy can be expressed as $E_{ch} = E_{cx}(n_x - n_s)^2 + E_m(n_y + \Delta n + \kappa n_s)^2$, with $E_{cx} = (E_l E_r - E_{co}^2)/E_m$. The advantage of this transformation becomes clear by looking at the tiling of the (n_x, n_y) plane shown in Fig. 3b and by noting that the coordinate n_x is associated with a change of the total charge number $n_s = n_l + n_r$, while n_y redistributes the charge between both islands ($\Delta n = n_l - n_r$). Each hexagonal shaped tile in Fig. 3c corresponds to a region where a specific charge state (n_l, n_r), characterized by the numbers of elementary charges on both islands, has the lowest charging energy $E_{ch}(n_l, n_r)$ of all states. In what follows we will specifically focus on measurements along the line $0 < n_x < 2, n_y = -\kappa$.

2.3. DETERMINATION OF THE CHARGING-ENERGY MATRIX

At very low temperatures the linear-response conductance vanishes except for the triple points in the (n_x, n_y) plane, where three adjacent charge states are degenerate. The first step in determining the charging-energy matrix is to measure these points. Then it is straight forward to extract the ratio E_{co}/E_{cx} and the asymmetry parameter κ (see Fig. 3c). This procedure fixes the capacity matrix up to an overall scaling factor, which has to be determined independently.

In principle, it would be possible to extract this last parameter from the IV characteristic measured across both islands, which has been used already to determine the conductance parameters. The high-voltage asymptotes at positive and negative voltages (in the naive picture they have a slope corresponding to the serial conductance of the three SEP junctions) are separated (in the same rather naive picture) by twice the threshold voltage $V_{th} = (E_l + E_r + E_m)/e$. However, while we get the serial conductance from the asymptotic behavior with about 5% accuracy, their separation cannot be determined with an accuracy better than about 15%. This is unsatisfactory. A similar observation has been made for the SET transistor. As with the transistor, a more accurate determination of the charging-energy matrix utilizes the temperature dependence of the device at high temperatures. At elevated temperatures the conductance does not depend on the gate voltages V_1 and V_2. Still, it is renormalized due to the Coulomb blockade in the linear-response regime. The conductance can be described by the formula $g = g_0(1 + E_c/(3k_bT) + c/T^2)$ (see Fig. 3b) in analogy to the SET transistor. It can be shown (Hirvi et al.,

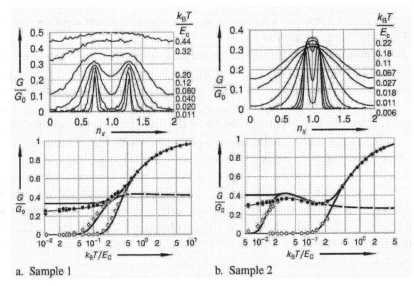

a. Sample 1   b. Sample 2

Figure 4. Top row: Conductance as a function of n_x at $n_y = -\kappa$. Temperatures of the measurements are indicated at the right. Bottom row: The minimal and maximal Conductance as a function of temperature. Also show is the conductance G_m at $(n_x, n_y) = (1, -\kappa)$, a point of especially high symmetry. The experimental data are marked by symbols. Thick solid lines indicated the result of the sequential tunneling model. The dashed line and the thin line are discussed in the text.

1995; Limbach, 2002) that the relation

$$E_{cx} = \frac{E_c}{g_0} \left(\frac{e_+}{g_l} + \frac{e_-}{g_r} + \frac{e_+ + e_- - 2E_{co}/Ecx}{g_m} \right)^{-1}$$

holds, where $e_\pm = ((E_{co}/E_{cx})(\kappa \pm 1) \mp 2)/(\kappa \mp 1)$. Thus, the determination of E_c from the fits shown in Fig. 3b fixes the charging-energy matrix and completes our device characterization (Tab. I).

3. Results of the measurements

In Fig. 4 we show results of our measurements. The linear-response conductance is measured as a function of V_1 and V_2 at fixed temperature. At low T we find very narrow peaks of finite conductance at the position of the degeneracy points of three charge states (see Fig. 3c). They are grouped in pairs. Normally we restrict the area of our measurement to a single pair. At the top of Fig. 4 we present data corresponding to a cut through the (V_1, V_2) plane crossing the maxima of the peaks. The cut line runs along the n_x axis at $n_y = -\kappa$ (see Sec. 2.2). In Fig. 4 we scaled the abscissa to n_x coordinates.

As we increase the temperature the peaks broaden and start to merge. In an intermediate temperature range the conductance G_m midway between the peaks is finite, while the minimal conductance stays zero. If the temperature is further raised the peaks merge completely, and the conductance displays its maximum at the center position ($n_x = 1, n_y = -\kappa$). Now the conductance is finite in the whole (V_1, V_2) plane. At even higher temperatures (see Fig. 3b) the conductance does not depend on V_1 or V_2 anymore.

The bottom part of Fig. 4 shows the minimal G_{min} and maximal G_{max} conductance as well as G_m as extracted from the top part of the figure. As explained in Sec. 2 the two samples are completely characterized. We have included in the lower part of Fig. 4 the outcome of a calculation in the framework of the sequential tunneling model using the parameters of Tab. I (thick solid lines). The physics of the SEP is described by the sequential model in a qualitative manner; e.g. the position and strength of the local minimum observed in $G_{max}(T)$ of sample 2 is reproduced with sufficient accuracy. For $T < E_c/k_B$ only four states contribute to the sequential model (see dashed lines in Fig. 4). Strong corrections to the sequential model are clearly observed, most pronounced for G_{max} at low temperatures. The corrections are fitted by the thin lines in the figure. They are achieved by adding a term of the form $\alpha \log(k_B T/(2E_{co}))$ to the sequential model.

A similar correction term is found for the single-electron transistor. It is related to quantum fluctuations of the charge states on the transistor island. The logarithmic dependence hints to a connection with Kondo physics, and, indeed, it has been shown (Matveev, 1991) that H_{tr} can be mapped onto a multichannel Kondo model in the low temperature limit near a degeneracy point. Here only two charge states are involved in transport, which can be identified with a pseudo spin coupled to the lead states by the large number of tunneling channels. Pohjola et al. (Pohjola et al., 1999) analyzed the linear response of the SEP by renormalization group methods and have found a logarithmic behavior of the low temperature conductance in qualitative agreement with our experimental result. However, quantitative results were obtained only for certain limiting cases, characterized by highly symmetric system parameters. For a detailed comparison with our experiment a calculation using our sample parameters is highly desirable. In consideration of the success of QMC studies (Göppert et al., 2000; Wallisser et al., 2002) for the transistor, similar efforts for the current situation appear quite promising (Theis and Srivilai, 2005). We expect that the interplay between our data and the various theoretical techniques will lead to further insight into the applicability range of the perturbation expansion.

For the single-electron transistor the most astonishing deviations were found for G_{min} at low temperature. In this regime perturbation theory does not have to deal with degenerate levels, is not burdened by divergencies of the Kondo type and has been explored quite early (Averin and Odintsov, 1989; Averin and Nazarov, 1992). The deviations between G_{min} and the sequential model are less pronounced for the SEP. This is expected, since in the double island structure, cotunneling contributions are strongly suppressed as compared to the single-electron transistor (Averin and Nazarov, 1992). G_m, however, is significantly influenced by cotunneling contributions and is, in this respect, closely related to the transistor's G_{min}. A careful look at Fig. 4 reveals significant deviations between G_m and the sequential predictions. In our opinion, this is the most sensitive region to look for deviations between QMC simulations and theories exploring perturbative series.

Acknowledgements

We acknowledge useful discussions with C. Theis, G. Göppert, P. Srivilay, G. Johannson, P. Joyez, J. König, H. Pothier, C. Meingast and H. v. Löhneysen. This work has been carried out as part of the Center for Functional Nanostructures (CFN) supported by DFG.

References

Averin, D. and Nazarov, Y. V. (1992) Macroscopic Quantum Tunneling of Charge and Co-Tunneling, In (Grabert and Devoret, 1992), chapter 6.

Averin, D. V. and Odintsov, A. A. (1989) Macroscopic quantum tunneling of the electric charge in small tunnel junctions, *Phys. Lett. A* **140**, 251–257.

Furlan, M., Heinzel, T., Jeanneret, B., Lotkhov, S. V., and K., E. (2000) Non-Gaussian distribution of nearest-neighbour Coulomb peak spacings in metallic single-electron transistors, *Euro. Phys. Lett.* **49**, 369–375.

Göppert, G., Hüpper, B., and Grabert, H. (2000) Conductance of the single-electron transistor for arbitrary tunneling strength, *Phys. Rev. B* **62**, 9955–9958.

Grabert, H. and Devoret, M. H. (eds.) (1992) *Single Charge Tunneling —Coulomb Blockade Phenomena in Nanostructures*, No. Vol. 294 in NATO ASI series. Series B, Physics, New York, Plenum Press.

Hirvi, K. P., Kauppinen, J. P., Korotkov, A. N., Paalanen, M. A., and Pekola, J. P. (1995) Arrays of normal metal tunnel junctions in weak Coulomb blockade regime, *Appl. Phys. Lett.* **67**, 2096–2098.

Ingold, G.-L. and Nazarov, Y. V. (1992) Charge Tunneling Rates in Ultrasmall Junctions, In (Grabert and Devoret, 1992), chapter 2.

Joyez, P., Bouchiat, V., Esteve, D., Urbina, C., and Devoret, M. H. (1997) Strong tunneling in the single electron transistor, *Phys. Rev. Lett.* **79**, 1349–1352.

Kautz, R. L., Zimmerli, G., and Martinis, J. M. (1993) Self-heating in the Coulomb-blockade electrometer, *J. Appl. Phys.* **73**, 2386–2396.

Korotkov, A. N., Samuelsen, M. R., and Vasenko, S. A. (1994) Effects of overheating in a single-electron transistor, *J. Appl. Phys.* **76**, 3623–3631.

Krupenin, V. A., Lotkhov, S. V., Scherer, H., Weimann, T., Zorin, A. B., Ahlers, F.-J., Niemeyer, J., and Wolf, H. (1999) Charging and heating effects in a system of coupled single-electron tunneling devices, *Phys. Rev. B* **59**, 10778–10784.

Limbach, B. (2002) Metallische Doppelinselstrukturen mit hohen Tunnelleitwerten, Technical Report FZKA 6791, Forschungszentrum Karlsruhe.

Limbach, B., vom Stein, P., Wallisser, C., and Roland, S. (2005) Coulomb blockade in two-island systems with highly conductive junctions, *Phys. Rev. B* **72**, 045319.

Matveev, K. A. (1991) Quantum fluctuations of the charge of a metal particle under the Coulomb blockade conditions, *Sov. Phys. JETP* **72**, 892–899.

Niemeyer, J. (1974) Eine einfache Methode zur Herstellung kleiner Josephson-Elemente, *PTB-Mitteilungen* **84**, 251.

Pohjola, T., König, J., Schoeller, H., and Schön, G. (1999) Strong tunneling in double-island structures, *Phys. Rev. B* **59**, 7579–7589.

Theis, C. and Srivilai, P. (2005), private communication.

Wahlgren, P., Delsing, P., and Haviland, D. B. (1995) Crossover from global to local rule for the Coulomb blockade in small tunnel junctions, *Phys. Rev. B* **52**, R2293–R2296.

Wallisser, C., Limbach, B., vom Stein, P., Schäfer, R., Theis, C., Göppert, G., and Grabert, H. (2002) Conductance of the single-electron transistor: A comparison of experimental data with Monte Carlo calculations, *Phys. Rev. B* **66**, 125314.

Zorin, A. B., Ahlers, F.-J., Niemeyer, J., Weimann, T., Wolf, H., Krupenin, V. A., and Lotkhov, S. V. (1996) Background charge noise in metallic single-electron tunneling devices, *Phys. Rev. B* **53**, 13682–13687.

ZERO-BIAS CONDUCTANCE THROUGH SIDE-COUPLED

DOUBLE QUANTUM DOTS

J. Bonča (janez.bonca@ijs.si)
J. Stefan Institute, SI-1000 Ljubljana, and Department of Physics, FMF, University of Ljubljana, SI-1000 Ljubljana, Slovenia

R. Žitko (rok.zitko@ijs.si)
J. Stefan Institute, SI-1000 Ljubljana, Slovenia

Abstract. Low temperature zero-bias conductance through two side-coupled quantum dots is investigated using Wilson's numerical renormalization group technique. A low-temperature phase diagram is computed. Near the particle-hole symmetric point localized electrons form a spin-singlet associated with weak conductance. For weak inter-dot coupling we find enhanced conductance due to the two-stage Kondo effect when two electrons occupy quantum dots . When quantum dots are populated with a single electron, the system enters the Kondo regime with enhanced conductance. Analytical expressions for the width of the Kondo regime and the Kondo temperature in this regime are given.

Key words: Quantum dots, Kondo effect, Two-stage Kondo effect, Fano resonance

1. Introduction

The interplay between the Kondo effect (which involves coupling of the local moment to the conduction electrons) and the magnetic ordering of the moments has been studied in the context of bulk materials such as heavy-fermion metals (Jones and Varma, 1987). Recent advances in nanotechnology have enabled studies of transport through single as well as coupled quantum dots where Kondo physics and magnetic interactions play an important role at low temperatures. A double-dot system represents the simplest possible generalization of a single-dot system which has been extensively studied in the past. Recent experiments demonstrate that an extraordinary control over the physical properties, such as the intra-dot coupling, can be achieved in multiple dot systems (Jeong et al., 2001; Craig et al., 2004; Holleitner et al., 2002; Chen et al., 2004). This enables direct experimental investigations of the competition between the Kondo effect and the exchange interaction between localized moments on the dots. One manifestation of this competition

371

K. Scharnberg and S. Kruchinin (eds.),
Electron Correlation in New Materials and Nanosystems, 371–380.
© 2007 *Springer.*

is a two stage Kondo effect (Hofstetter and Schoeller, 2003). Experimentally, it manifests itself as a sharp drop in the conductance vs. gate voltage (van der Wiel et al., 2002).

The Fano resonance is a characteristic of noninteracting electrons for which the shape of the Fano resonance can be analytically determined. However, the influence of interactions on the appearance of Fano resonances remains an open question. It has been recently observed in experiments on rings with embedded quantum dots (Kobayashi et al., 2002) and quantum wires with side-coupled dots (Kobayashi et al., 2004). The interplay between Fano and Kondo resonance was investigated using equation of motion (Bulka and Stefanski, 2001; Stefanski et al., 2004) and Slave boson techniques (Lara et al., 2004).

We study a double quantum dot (DQD) in a side-coupled configuration (Fig. 1), connected to a single conduction-electron channel. Similar systems were studied previously with different techniques, such as: non-crossing approximation (Kim and Hershfield, 2001), embedding technique (Apel et al., 2004), and slave-boson mean field theory (Kang et al., 2001; Lara et al., 2004). Numerical renormalization group (NRG) calculations were also performed recently (Cornaglia and Grempel, 2005), where only narrow regimes of enhanced conductance were found at low temperatures and no Fano resonances were reported. We will present a low-temperature phase diagram of a DQD as a function of intra-dot coupling strengths and gate-voltage potential indicating regions with enhanced conductance due to Kondo effect and regions of nearly zero conductance. In particular, when the intra-dot overlap is large and DQD occupancy is one, wide regimes of enhanced conductance as a function of gate-voltage exist at low temperatures due to Kondo effect, separated by the regimes where localized spins on DQD are antiferromagnetically (AFM) coupled. Kondo temperatures T_K follow a prediction based on the poor-man's scaling and Schrieffer-Wolff transformation. In the limit when the dot a is only weakly coupled, the system enters the "two stage" Kondo regime (Vojta et al., 2002a; Cornaglia and Grempel, 2005), where we again find a wide regime of enhanced conductivity under the condition that the high- and the low- Kondo temperatures (T_K and T_K^0, respectively) are well separated and the temperature of the system T is in the interval $T_K^0 \ll T \ll T_K$.

2. Model and Method

The Hamiltonian that we study reads

$$H = \delta(n_d - 1) + \delta(n_a - 1) - t_d \sum_\sigma \left(d_\sigma^\dagger a_\sigma + a_\sigma^\dagger d_\sigma \right)$$

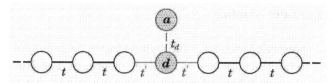

Figure 1. Side-coupled configuration of quantum dots.

$$+ \frac{U}{2}(n_d - 1)^2 + \frac{U}{2}(n_a - 1)^2 \qquad (1)$$

$$+ \sum_{k\sigma} \epsilon_k c_{k\sigma}^\dagger c_{k\sigma} + \sum_{k\sigma} V_d(k)\left(c_{k\sigma}^\dagger d_\sigma + d_\sigma^\dagger c_{k\sigma}\right)$$

$$- J_{ad}\mathbf{S}_a \cdot \mathbf{S}_d,$$

where $n_d = \sum_\sigma d_\sigma^\dagger d_\sigma$ and $n_a = \sum_\sigma a_\sigma^\dagger a_\sigma$. Operators $d_\sigma^\dagger$ and $a_\sigma^\dagger$ are creation operators for an electron with spin σ on site d or a. On-site energies of the dots are defined by $\epsilon = \delta - U/2$. For simplicity, we choose the on-site energies and Coulomb interactions to be equal on both dots. Coupling between the dots is described by the inter-dot tunnel coupling t_d. Dot d couples to both leads with equal hopping t'. Operator $c_{k\sigma}^\dagger$ creates a conduction band electron with momentum k, spin σ and energy $\epsilon_k = -D \cos k$, where $D = 2t$ is the half-bandwidth. Spin operator $\mathbf{S} = \sum_{s,s'} c_s^\dagger \vec{\sigma}_{s,s'} c_{s'}$ is defined using Pauli matrices and J_{ad} represents additional Heisenberg exchange interaction. The momentum-dependent hybridization function is $V_d(k) = -(2/\sqrt{N+1})\, t' \sin k$, where N in the normalization factor is the number of conduction band states.

We use Meir-Wingreen's formula for conductance in the case of proportionate coupling (Meir and Wingreen, 1992), which is known to apply under very general conditions (for example, the system need not be in a Fermi-liquid ground state), with spectral functions obtained using the NRG technique (Wilson, 1975; Krishna-Murthy et al., 1980; Costi et al., 1994; Costi, 2001; Hofstetter, 2000). At zero temperature, the conductance is

$$G = G_0 \pi \Gamma \rho_d(0), \qquad (2)$$

where $G_0 = 2e^2/h$, $\rho_d(\omega)$ is the local density of states of electrons on site d and $\Gamma/D = (t'/t)^2$.

The NRG technique consists of logarithmic discretization of the conduction band, mapping onto a one-dimensional chain, and iterative diagonalization of the resulting Hamiltonian (Wilson, 1975). Only low-energy part of the spectrum is kept after each iteration step; in our calculations we kept 1200 states, not counting spin degeneracies, using discretization parameter $\Lambda = 1.5$.

3. Strong intra-dot coupling

In Fig. 2a we present conductance through a double quantum dot at different values of intra-dot couplings vs. δ. Due to formation of Kondo correlations, conductance is enhanced, reaching the unitary limit in a wide range of δ. Regimes of enhanced conductance appear in the intervals approximately given by $\delta_1 < |\delta| < \delta_2$, where $\delta_1 = t_d(2\sqrt{1 + (U/4t_d)^2} - 1)$ and $\delta_2 = (U/2 + t_d)$. These estimates were obtained from the lowest energies of states with one and two electrons on the isolated double quantum dot, i.e. $E_1 = U/2 - \delta - t_d$ and $E_2 = -2t_d\sqrt{1 + (U/4t_d)^2} + U/2$, respectively.

Using the above estimates we present a phase diagram in Fig. 3. In the gray region, called Kondo regime, with border lines given by $\delta_1(t_d), \delta_2(t_d)$, the Kondo effect is responsible for an enhanced conductance. Kondo plateaus in Fig. 2 fall in this regime. The Conductance is zero everywhere outside this region, except in the limit when $t_d \to 0$, where a two-stage Kondo effect two stage is responsible for enhanced conductance, as discussed further in the text.

To gain further physical insight, we focus on various correlation functions, defined within the DQD system. In Fig. 2b we show S, calculated from expectation value $\langle \mathbf{S}_{tot}^2 \rangle = S(S + 1)$, where $\mathbf{S}_{tot} = \mathbf{S}_a + \mathbf{S}_d$ is the total spin operator. S reaches value $1/2$ in the Kondo regime where $G/G_0 = 1$. Enhanced conductance is thus followed by the local moment formation. This is further supported by the average double-dot occupancy $\langle n \rangle$, where $n = n_a + n_d$, which in the regime of enhanced conductivity approaches odd-integer values, i.e. $\langle n \rangle = 1$ and 3 (see Fig. 2c and 3). Transitions between regimes of nearly

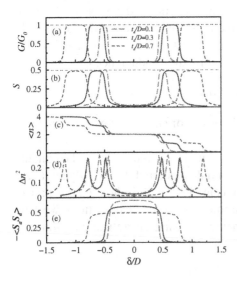

Figure 2. Conductance and correlation functions of DQD vs. δ. Besides different values of t_d, indicated in the figure, other parameters of the model are $\Gamma/D = 0.03, U/D = 1$ and $J_{ad} = 0$. Temperature T is chosen to be far below T_K, i.e. $T \ll T_K$.

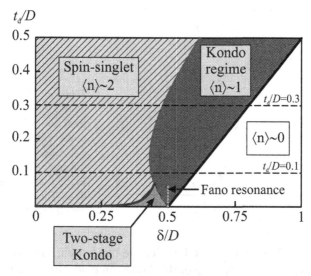

Figure 3. Phase diagram of a DQD for $U/D = 1$ and $J_{ad} = 0$, obtained using analytical estimates as given in the text. Grey areas represent Kondo regimes where $S \sim 1/2$, $\langle n \rangle \sim 1$, and $G/G_0 \sim 1$. In shaded area, called spin - singlet regime, where $S \sim 0$ and $\langle n \rangle \sim 2$, there is enhanced spin-spin correlation function, *i.e.* $\langle \mathbf{S}_a \cdot \mathbf{S}_d \rangle \lesssim -0.5$ and $G/G_0 \sim 0$. The two-stage Kondo regime is explained further in the text.

integer occupancies are rather sharp; they are visible as regions of enhanced charge fluctuations measured by $\Delta n^2 = \langle n^2 \rangle - \langle n \rangle^2$, as shown in Fig. 2d. Finally, we show in Fig. 2e spin-spin correlation function $\langle \mathbf{S}_a \cdot \mathbf{S}_d \rangle$. Its value is negative between two separated Kondo regimes where conductance approaches zero, *i.e.* for $-\delta_1 < \delta < \delta_1$, otherwise it is zero. This regime further coincides with $\langle n \rangle \sim 2$. Each dot thus becomes nearly singly occupied and spins on the two dots form a local singlet (S=0) due to effective exchange coupling $J_{\text{eff}} = 4t_d^2/U$.

In Fig. 4 we present Kondo temperatures T_K vs. δ extracted from the widths of Kondo peaks. Numerical results in the regime where $\langle n \rangle \sim 1$ and 3 fit the analytical expression obtained using the Schrieffer-Wolf transformation that projects out states with even electron occupancy and leads to an effective single quantum dot problem with renormalized parameters. We obtain

$$T_K = 0.182U \sqrt{\rho_0 J} \exp[-1/\rho_0 J] \tag{3}$$

with $\rho_0 J = \frac{2\Gamma}{\pi}(\alpha/|E_1| + \beta/|E_2 - E_1|)$, where $\alpha = 1/2$ and

$$\beta = \frac{\left(4t_d + U + \sqrt{16t_d^2 + U^2}\right)^2}{8\left(16t_d^2 + U\left(U + \sqrt{16t_d^2 + U^2}\right)\right)}. \tag{4}$$

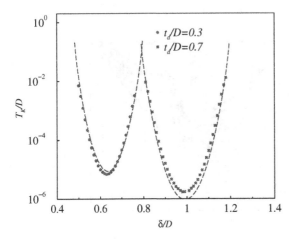

Figure 4. Kondo temperatures T_K vs. δ as measured from the widths of Kondo peaks obtained from NRG calculations (full circles and squares). Analytical estimate, Eq. 3, is shown using dashed lines. The rest of parameters are identical to those in Fig. 2.

The prefactor $0.182U$ in Eq. 3 is the effective bandwidth. The same effective bandwidth was used to obtain T_K of the Anderson model in the regime $U < D$ (Krishna-Murthy et al., 1980; Haldane, 1978).

4. Weak intra-dot coupling

We now turn to the *limit when $t_d \to 0$*. Unless otherwise specified, we choose the effective temperature T to be finite, *i.e.* $T \sim 10^{-9}D$, since calculations at much lower temperatures would be experimentally irrelevant. In this case one naively expects to obtain essentially identical conductance as in the single-dot case. As δ decreases below $\delta \sim U/2$, G/G_0 indeed follows result obtained for the single-dot case as shown in Fig. 5a. In the case of DQD, however, a sharp Fano resonance appears at $\delta = U/2$. This resonance coincides with the sudden jump in S, $\langle n \rangle$, as well as with the spike in Δn^2, as shown in Figs 5b,c, and d, respectively. The Fano resonance is a consequence of a sudden charging of the nearly decoupled dot a, as its level ϵ crosses the chemical potential of the leads, *i.e* at $\epsilon = 0$. Meanwhile, the electron density on the dot d remains a smooth function of δ, as seen from $\langle n_d \rangle$ in Fig 5c. With increasing t_d, the width of the resonance increases and at $t_d \gtrsim 0.1$, the resonance merges with the Kondo plateau and disappears (see Fig. 2a). The regime where the Fano resonance exists is also specified in the phase diagram, Fig. 3.

We now return to the description of the results presented in Fig. 5a in the regime where $\delta < U/2$. As δ further decreases, the system enters a regime of the two-stage Kondo effect (Cornaglia and Grempel, 2005). This region is defined by $J_{\text{eff}} < T_K$ (see also Fig. 5f), where T_K is the Kondo temperature, approximately given by the single quantum dot Kondo temperature, Eq. 3 with $\rho_0 J = \frac{2\Gamma}{\pi}(1/|\delta - U/2| + 1/|\delta + U/2|)$. This regime is also indicated in the

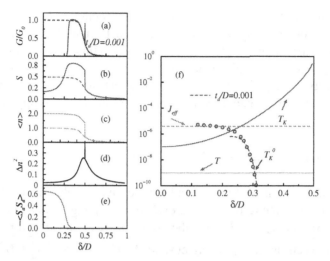

Figure 5. Conductance and correlation functions at $t_d/D = 0.001$ (a,...,e). Dashed lines represent in a) G/G_0 and in b) S of a single quantum dot with otherwise identical parameters. Dashed line in c) represents $\langle n_d \rangle$ of DQD. In f) a schematic plot of different temperatures and interactions is presented as explained in the text. NRG values of the gap in $\rho_d(\omega)$ at $\omega = 0$ and $T \ll T_K^0$ are presented with open circles. Values of J_{eff} and analytical results of T_K^0 are presented with dashed lines. For analytical estimates of T_K^0 $\alpha = 2.2$ was used. Other parameters of the model are $\Gamma/D = 0.03$, $U/D = 1$ and $J_{ad} = 0$.

phase diagram, presented Fig. 4, where condition $J_{\text{eff}} = T_K$ was used to separate two-stage Kondo from the spin-singlet regime. Just below $\delta < U/2$, T falls in the interval, given by $T_K^0 \ll T \ll T_K$, where $T_K^0 \sim T_K \exp(-\alpha T_K/J_{\text{eff}})$ denotes the lower Kondo temperature, corresponding to the gap in the spectral density $\rho_d(\omega)$ at $\omega = 0$ and α is of the order of one (Cornaglia and Grempel, 2005). Note that NRG values of the gap in $\rho_d(\omega)$ (open circles), calculated at $T \ll T_K^0$, follow analytical results for $T_K^0(\delta)$ when $J_{\text{eff}} < T_K$, see Fig. 5f, while in the opposite regime, *i.e.* for $J_{\text{eff}} > T_K$, they approach J_{eff}.

As shown in Fig. 5a for $0.3D \lesssim \delta < U/2$, G/G_0 calculated at $T = 10^{-9}D$ follows results obtained in the single quantum dot case and approaches value 1. The spin quantum number S in Fig. 5b reaches the value $S \sim 0.8$, consistent with the result obtained for a system of two decoupled spin-1/2 particles, where $\langle \hat{S}^2 \rangle = 3/2$. This result is also in agreement with $\langle n \rangle \sim 2$ and the small value of the spin-spin correlation function $\langle S_a \cdot S_d \rangle$, presented in Fig. 5c and 5e respectively.

With further decreasing of δ, G/G_0 suddenly drops to zero at $\delta \lesssim 0.3D$. This sudden drop is approximately given by $T \lesssim T_K^0(\delta)$, see Figs. 5a and f. At this point the Kondo hole opens in $\rho_d(\omega)$ at $\omega = 0$, which in turn leads to a drop in the conductivity. The position of this sudden drop in terms of δ is rather insensitive to the chosen T, as apparent from Fig. 5f.

Below $\delta \lesssim 0.25D$, which corresponds to the condition $J_{\text{eff}} \sim T_K(\delta)$, also presented in Fig. 5f, the system crosses over from the two stage Kondo regime to a regime where spins on DQD form a singlet. In this case S decreases and $\langle \mathbf{S}_a \cdot \mathbf{S}_d \rangle$ shows strong anti-ferromagnetic correlations, Figs. 5b and e. The lowest energy scale in the system is J_{eff}, which is supported by the observation that the size of the gap in $\rho(\omega)$ (open circles in Fig. 5f) is approximately given by J_{eff}.

5. Ferromagnetic coupling

In Fig. 6 we present results of conductance and correlation functions for different values of the ferromagnetic exchange coupling J_{ad}, $t_d/D = 0.3$ and $U/D = 1$. At finite t_d and $J_{ad} = 0$ the ground state of an isolated DQD containing two electrons is a spin singlet state. At $J_{ad}^c/D = 2(\sqrt{4t_d^2 + U^2} - U) \sim$ 0.33 spin singlet and triplet states become degenerate. With further increasing $J_{ad} > J_{ad}^c$ we expect formation of a $S = 1$ state in the regime when $\langle n \rangle \sim 2$. Since in this case $2S > K$ where $K = 1$ is the number of channels, this systems falls into a class where the spin on the DQD is not fully compensated by the conduction electrons (Vojta et al., 2002b; Hofstetter and Schoeller, 2003). Results at finite J_{ad}, presented in Fig. 6, can be divided into two groups. For $J_{ad} < J_{ad}^c$ the main effect of increasing J_{ad} is seen as expansion of the $S = 1/2$ Kondo regime where $\langle n \rangle \sim 1$. For $J_{ad} > J_{ad}^c$, two Kondo plateaus appear, one associated with $S = 1/2$ Kondo regime and the other with $S = 1$ Kondo regime. In both cases conductance reaches unitary limit, while in the transition regime it drops to zero.

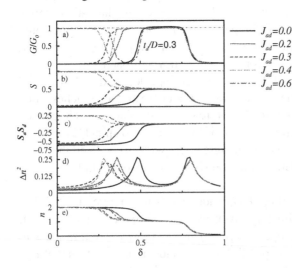

Figure 6. Conductance and correlation functions of DQD vs. δ for various values of J_{ad} and $t_d/D = 0.3$. Other parameters of the model are identical to those in Fig. 2. Temperature T is chosen to be far below T_K, i.e. $T \ll T_K$.

6. Conclusions

In this work we have explored different regimes of the side-coupled DQD. Conclusions can be summarized as follows: a) when quantum dots are *strongly coupled*, wide regions of enhanced, nearly unitary conductance exist due to the underlying Kondo physics. Analytical estimates for their positions, widths, as well as for the corresponding Kondo temperatures, are given and numerically verified. When two electrons occupy DQD, conductance is zero due to formation of the spin-singlet state which is effectively decoupled from the leads. b) In the limit when quantum dots are *weakly coupled* the Fano resonance appears in the valence fluctuation regime. Its width is enhanced as a consequence of interactions which should facilitate experimental observation. Unitary conductance exists when two electrons occupy DQD due to a two-stage Kondo effect as long as the temperature of the system is well below T_K and above T_K^0. The experimental signature of the two-stage Kondo effect in weakly coupled regime should materialize through the inter-dot coupling sensitive width of the enhanced conductance vs. gate voltage.

When intra-dot ferromagnetic coupling exceeds a critical value, there is a phase transition from a spin-singlet, non-conducting regime to $S = 1$ Kondo regime where the spin on the DQD is under-screened but nevertheless the conductivity reaches the unitary limit.

Acknowledgements

Authors acknowledge useful discussions with A. Ramšak. We also acknowledge the financial support of the Slovenian Research Agency under grant P1-0044.

References

Apel, V. M., Davidovich, M. A., Anda, E. V., Chiappe, G., and Busser, C. A. (2004) Effect of topology on the transport properties of two interacting dots, *Eur. Phys. J. B* **40**, 365.

Bulka, B. R. and Stefanski, P. (2001) Fano and Kondo resonance in electronic current through nanodevices, *Phys. Rev. Lett.* **86**, 5128.

Chen, J. C., Chang, A. M., and Melloch, M. R. (2004) Transition between Quantum States in a Parallel-Coupled Double Quantum Dot, *Phys. Rev. Lett.* **92**, 176801.

Cornaglia, P. S. and Grempel, D. R. (2005) Strongly correlated regimes in a double quantum dot device, *Phys. Rev. B* **71**, 075305.

Costi, T. A. (2001) Magnetotransport through a strongly interacting quantum dot, *Phys. Rev. B* **64**, 241310.

Costi, T. A., Hewson, A. C., and Zlatic, V. (1994) Transport coefficients of the Anderson model via the numerical renormalization group, *J. Phys.: Condens. Matter* **6**, 2519.

Craig, N. J., Taylor, J. M., Lester, E. A., Marcus, C. M., Hanson, M. P., and Gossard, A. C. (2004) Tunable Nonlocal Spin Control in a Coupled-Quantum Dot System, *Science* **304**, 565.

Haldane, F. (1978) Theory of the Atomic Limit of the Anderson Model: I. Perturbation expansion re-examined, *J. Phys. C: Solid State Phys.* **11**, 5015.

Hofstetter, W. (2000) Generalized Numerical Renormalization Group for Dynamical Quantities, *Phys. Rev. Lett.* **85**, 1508.

Hofstetter, W. and Schoeller, H. (2003) Quantum Phase Transition in a Multilevel Dot, *Phys. Rev. Lett.* **88**, 016803.

Holleitner, A. W., Blick, R. H., Hüttel, A. K., Eberl, K., and Kotthaus, J. P. (2002) Probing and Controlling the Bonds of an Artificial Molecule, *Science* **297**, 70.

Jeong, H., Chang, A. M., and Melloch, M. R. (2001) The Kondo Effect in an Artificial Quantum Dot Molecules, *Science* **293**, 2221.

Jones, B. A. and Varma, C. M. (1987) Study of Two Magnetic Impurities in a Fermi Gas, *Phys. Rev. Lett.* **58**, 843.

Kang, K., Cho, S. Y., Kim, J.-J., and Shin, S.-C. (2001) Anti-Kondo resonance in transport through a quantum wire with a side-coupled quantum dot, *Phys. Rev. B* **63**, 113304.

Kim, T.-S. and Hershfield, S. (2001) Suppression of current in transport through parallel double quantum dot, *Phys. Rev. B* **63**, 245326.

Kobayashi, K., Aikawa, H., Katsumoto, S., and Iye, Y. (2002) Tuning of the Fano Effect through a Quantum Dot in an Aharonov-Bohm Interferometer, *Phys. Rev. Lett.* **88**, 256806.

Kobayashi, K., Aikawa, H., Sano, A., Katsumoto, S., and Iye, Y. (2004) Fano resonance in a quantum wire with a side-coupled quantum dot, *Phys. Rev. B* **70**, 035319.

Krishna-Murthy, H. R., Wilkins, J. W., and Wilson, K. G. (1980) Renormalization-group approach to the Anderson model of dilute magnetic alloys. I. Statis properties for the symmetric case, *Phys. Rev. B* **21**, 1003.

Lara, G. A., Orellana, P. A., Yanez, J. M., and Anda, E. V. (2004) Kondo effect in side coupled double quantum-dot molecule, cond-mat/0411661.

Meir, Y. and Wingreen, N. S. (1992) Landauer formula for the current through an interacting electron region, *Phys. Rev. Lett.* **68**, 2512.

Stefanski, P., Tagliacozzo, A., and Bulka, B. R. (2004) Fano versus Kondo Resonances in a Multilevel "Semiopen" Quantum Dot, *Phys. Rev. Lett.* **93**, 186805.

van der Wiel, W. G., Franceschi, S. D., Elzerman, J. M., Tarucha, S., Kouwenhoven, L. P., Motohisa, J., Nakajima, F., and Fukui, T. (2002) Two-Stage Kondo Effect in a Quantum Dot at a High Magnetic Field, *Phys. Rev. Lett.* **88**, 126803.

Vojta, M., Bulla, R., and Hofstetter, W. (2002a) Quantum phase transitions in models of coupled magnetic impurities, *Phys. Rev. B* **65**, 140405.

Vojta, M., Bulla, R., and Hofstetter, W. (2002b) Quantum phase transitions in models of coupled magnetic impurities, *Phys. Rev. B* **65**, 140405(R).

Wilson, K. G. (1975) The renormalization group: Critical phenomena and the Kondo problem, *Rev. Mod. Phys.* **47**, 773.

SPIN-ORBITAL ORDERING AND GIANT MAGNETORESISTANCE IN COBALT OXIDES: INTRINSIC MAGNETIC-FIELD-EFFECT TRANSISTOR

A. N. Lavrov (lavrov@che.nsk.su)
Institute of Inorganic Chemistry, Lavrentyeva-3, Novosibirsk-630090, Russia

A. A. Taskin and Yoichi Ando
Central Research Institute of Electric Power Industry, Komae, Tokyo 201-8511, Japan

Abstract. Layered cobalt oxides $RBaCo_2O_{5+x}$ (R is a rare-earth element) possess very rich phase diagrams owing to the competition of various spin-charge-orbital ordered states in the doped square-lattice CoO_2 planes. To clarify the mechanism of the giant magnetoresistance (GMR) that accompanies magnetic transformations in these compounds we have prepared and studied $GdBaCo_2O_{5+x}$ single crystals with precisely tuned doping levels. We find that the GMR is observed even in the parent composition ($x = 0.50$) with all cobalt ions in the Co^{3+} state, which has neither phase segregation nor charge ordering that are responsible for the MR in manganites. A new MR mechanism is suggested for these cobalt oxides where the charge carrier generation in conducting channels is controlled by the magnetic state of their local environment.

Key words: Magnetoresistance, Phase Diagram, Cobalt Oxide

1. Introduction

Since the discovery of the colossal magnetoresistance (CMR) in manganites, a great deal of experimental and theoretical efforts have been made to clarify the nature of this phenomenon. It has soon been realized that the observed resistivity changes are way too large to be accounted for within a homogeneous-medium picture, and it should be a nanoscopic phase separation or some other kind of electron self organization that lies behind this novel phenomenon; for a review, see (Nagaev, 1996; Dagotto et al., 2001). By now several MR mechanisms operating in manganites have been qualitatively identified, namely: (i) A so-called "spin valve" where the conduction is governed by tunneling between ferromagnetic (FM) metallic regions with completely spin-polarized carriers, such as metallic layers, grains in ceram-

K. Scharnberg and S. Kruchinin (eds.),
Electron Correlation in New Materials and Nanosystems, 381–391.
© 2007 *Springer.*

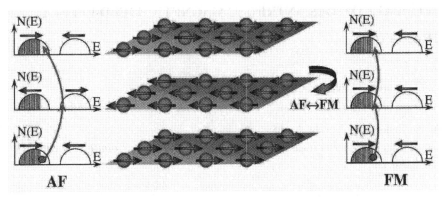

Figure 1. A "spin valve" formed by a stack of ferromagnetically ordered layers. In each layer, the charge carriers are completely polarized and an energy gap separates subbands with different spins orientations. For the AF arrangement of layers the spin valve is "closed", since carriers should be activated over the gap to move from one layer to another (left). However, the valve is opened in the FM state where carriers can propagate between the layers, staying within the same subband (right).

Figure 2. An insulating two-phase state composed of FM metallic droplets embedded into a non-conducting AF matrix (left). Magnetic field can change the size and topology of the FM droplets, causing percolation through the metallic phase and thus an insulator-metal transition (right).

ics, or metallic domains in heterogeneous systems (Kimura et al., 1996). The magnetic field aligns all the magnetic moments, thus allowing carriers to tunnel between metallic regions and facilitating the charge transport (Fig. 1). (ii) A nanoscopic phase separation into FM metallic and antiferromagnetic (AF) insulating phases (Fig. 2). The "spin-valve" mechanism is also operating in such phase mixture, but even more important is that the magnetic field can modify the scale and topology of FM domains, thus causing a macroscopic metallic percolation (Nagaev, 1996; Dagotto et al., 2001). (iii) A field-induced melting of the charge order usually observed in half-doped manganites (Tomioka et al., 1995).

Recently, a giant MR with the resistivity change exceeding an order of magnitude has been found in another family of compounds – layered cobalt oxides $RBaCo_2O_{5+x}$ (where R is a rare-earth element) (Martin et al., 1997; Respaud et al., 2001); the ensuing single-crystal studies by Taskin et al. (2003; 2005) have confirmed the bulk nature of the observed GMR. What

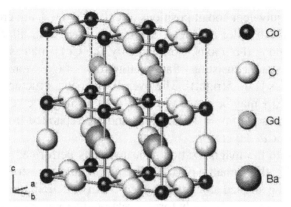

Figure 3. Schematic picture of GdBaCo$_2$O$_{5.50}$. For $x = 0.5$, the oxygen ions in GdO$_x$ layers order into alternating filled and empty rows running along the a axis.

distinguishes RBaCo$_2$O$_{5+x}$ from the CMR manganites is an exceptionally strong uniaxial anisotropy of the Co spins, which dramatically simplifies the possible spin arrangements, turning it into a model system (Taskin et al., 2003; Taskin et al., 2005). A particularly interesting region of compositions is around $x = 0.5$, where the valence state of Co-ions approaches 3+, so that the charge ordering and electronic phase separation typical for mixed-valence systems become irrelevant. Moreover, for $x \approx 0.5$, the oxygen ions in RO_x layers order into alternating filled and empty rows running along the a axis (Fig. 3), thus allowing one to deal with almost perfect crystal structure (Frontera et al., 2002).

To clarify the mechanism of the giant magnetoresistance (GMR) in these layered cobalt oxides, we have grown GdBaCo$_2$O$_{5+x}$ single crystals by the floating-zone method, as described in detail by Taskin et al. (2005), have tuned precisely their doping levels in the range around $x \approx 0.5$, and studied their transport, magnetic, and structural properties.

2. Results and discussion

GdBaCo$_2$O$_{5+x}$ crystals of the parent $x = 0.5$ composition behave as a metal at high temperatures, yet abruptly switch into a semiconducting state below the metal-insulator transition at $T_{MIT} \approx 360$ K (Fig. 4). The gap opening at the Fermi level is associated with a cooperative spin-state transition, which involves one half of the cobalt ions that turn into the low-spin $S = 0$ state, while the other half keeps its spin state $S = 1$ unchanged (Frontera et al., 2002; Taskin et al., 2005). Upon further cooling, ordering of the $S = 1$ cobalt spins brings about successive paramagnetic-ferromagnetic-antiferromagnetic transitions [Figs. 4 and 5(left)].

Two non-equivalent cobalt positions in $GdBaCo_2O_{5.50}$ are created by the oxygen ions which order in $GdO_{0.5}$ layers into alternating filled and empty rows running along the a axis. Consequently, the CoO_2 planes also develop a spin-ordered state, consisting of alternating rows of Co^{3+} ions in the $S = 1$ and $S = 0$ states [Fig. 5(right)]. The overall magnetic structure of GdBaCo $_2O_{5.50}$ below T_{MIT} may be conceived of as a set of magnetic "2-leg ladders", which are composed of $S = 1$ Co^{3+} ions and are separated from each other by non-magnetic ac layers.

According to the magnetization measurements performed on detwinned single crystals, the cobalt spins exhibit a remarkably strong Ising-like anisotropy, being aligned along the oxygen-chain direction; thus the FM ladders in Gd $BaCo_2O_{5.50}$ can *only* form a relative ferromagnetic or antiferromagnetic arrangement (Fig. 5). Owing to the weak coupling between ladders, their relative magnetic order can easily change, as actually happens at the Néel temperature. In the AF state, the magnetic cell can be doubled along the b or c axis, or both.

Quite naturally, magnetic fields applied to such a stack of FM ladders (along their easy-axis) stabilize their relative FM order and shift the FM-AF transition to lower temperatures (Fig. 6). We have found that the FM state can similarly be stabilized by doping, though the rôle of doping is more complex. Indeed, whenever the oxygen concentration deviates from

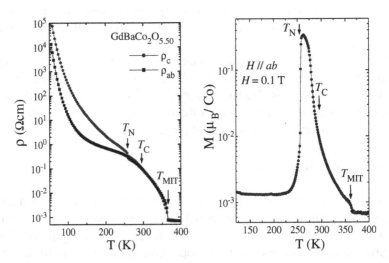

Figure 4. (Left) Resistivity of a $GdBaCo_2O_{5.50}$ crystal measured along the c axis and along the ab plane. (Right) Magnetization of $GdBaCo_2O_{5.50}$ (the paramagnetic contribution of Gd^{3+} ions is subtracted); the data were taken on cooling in a magnetic field of 0.1 T applied along the ab plane. Note that the metal-insulator transition at $T_{MIT} \approx 360\,K$ is accompanied by a spin-state transition.

the parent composition $x = 0.500$, be it on the electron-doped or the hole-doped side, the FM phase becomes more stable *in comparison* with the AF one and the FM→AF transition quickly shifts to lower temperatures (Fig. 7). However the FM ordering itself (at $T_C \approx 300$ K) is either unaffected by doping ($x < 0.500$), or even slightly suppressed for $x > 0.500$. It is worth noting that the double-exchange mechanism is irrelevant to this AF-FM competition, since no increase in conductivity was found with the electron doping (Taskin et al., 2005).

The composition dependence of the transition temperatures is summarized in the phase diagram in Fig. 8. While outside of the central region GdBaCo$_2$O$_{5+x}$ exhibits a nanoscopic or even macroscopic (hatched region) phase separation (Taskin et al., 2005), in the range limited by 5% electron or hole doping ($0.45 < x < 0.55$) it stays fairly homogeneous and the picture of magnetic transformations remains very transparent. The parent composition GdBaCo$_2$O$_{5.50}$, possessing neither structural disorder nor nanoscopic phase segregation, provides therefore a good ground for understanding the GMR mechanism in layered cobalt oxides.

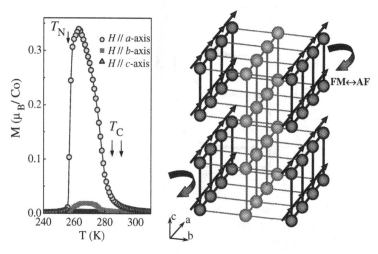

Figure 5. (Left) Magnetization of a detwinned GdBaCo$_2$O$_{5.50}$ crystal measured in $H = 0.1$ T applied along one of the crystal axes (the contribution of Gd^{3+} ions has been subtracted). The b-axis magnetization comes mostly from remaining misaligned domains. (Right) A sketch of the magnetic structure, where only Co ions are shown. At $T < T_{MIT} \approx 360$ K, the Co^{3+} ions in octahedral environment acquire a low-spin ($S = 0$) state and form non-magnetic ac layers, while those in pyramidal environment have an intermediate ($S = 1$) state. Upon cooling below T_C, the $S = 1$ Co spins (indicated by arrows) order into ferromagnetic ladders, running along the a axis and only weakly interacting with each other. At T_N, half of the FM ladders flip their moments into the opposite direction; note that the huge magnetocrystalline anisotropy allows only two possible directions for the ladders' moments, namely along the a axis.

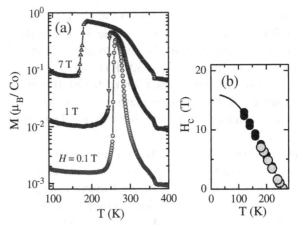

Figure 6. (a) *a*-axis magnetization of GdBaCo$_2$O$_{5.50}$ measured in different magnetic fields. (b) Critical field of the AF-FM phase transition determined from the magnetization (O) and magnetoresistance (●) measurements.

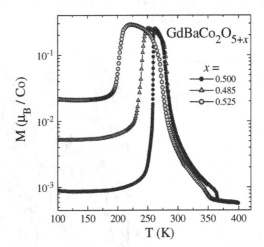

Figure 7. Evolution of the magnetization behavior in GdBaCo$_2$O$_{5+x}$ with doping. Even a slight deviation in the oxygen content from $x = 0.50$ noticeably shifts T_N down and induces an increase in the low-T magnetization.

We have found that GdBaCo$_2$O$_{5+x}$ with the oxygen composition precisely tuned to $x = 0.500$ still show GMR not much different from the $x = 0.4$ samples studied by Martin et al. (1997). As an example, Fig. 9 shows the *c*-axis resistivity of a GdBaCo$_2$O$_{5.500}$ crystal measured at $H = 0$ and 14 T. In zero field, the FM→AF transition at $T_N \approx 260$ K brings about a step-like increase of the resistivity. The 14-T field applied along the *ab* plane shifts the magnetic transition towards lower temperatures and wipes out the resistivity

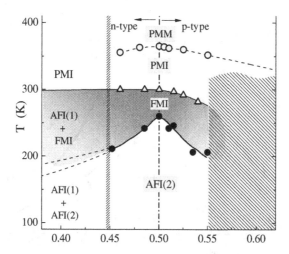

Figure 8. Phase diagram of GdBaCo$_2$O$_{5+x}$, including regions of a paramagnetic metal (PMM), paramagnetic insulator (PMI), ferromagnetic insulator (FMI), and antiferromagnetic insulator (AFI). Apart from the fairly homogeneous region around the parent composition $x = 0.50$ $(0.45 < x < 0.55)$, GdBaCo$_2$O$_{5+x}$ exhibits macroscopic (hatched) or nanoscopic phase separation.

increase, thus causing the resistivity to drop by several times. As can be seen in Fig. 9(a), ρ_c grows roughly exponentially [$\propto \exp(E_a/2k_BT)$] upon cooling below ~ 230 K regardless of the applied field, yet the gap values differ by $\Delta E_a \approx 30$ meV. The charge carriers, therefore, have different activation energies in the FM and AF states, and the MR originates from an increase in the number of carriers, rather than from an enhancement of their mobility. As soon as the magnetic field becomes insufficient to maintain the FM order, the system switches into the AF state and the resistivity jumps to its zero-field value.

Abrupt changes in resistivity taking place together with the field-induced AF-FM transition are also well seen upon sweeping the magnetic field at fixed temperatures (Fig. 10). One can see that the transition field H_c increases in accord with the dependence in Fig. 6(b) and gets over 14 T below ≈ 130 K.

Following the decrease in E_a at the AF-FM transition, the $\rho_c(0)/\rho_c(H)$ ratio grows exponentially with cooling [$\rho_c(0)/\rho_c(H) \approx \exp(\Delta E_a/2k_BT)$], reaching a value of 3 at $T = 150$ K [Fig. 10(b)]; it would keep growing to much larger values if fields sufficient to maintain the FM state down to low temperatures ($\gtrsim 20$ T) were applied. Indeed, a resistivity change by two orders of magnitude was observed by Respaud et al. (2001) when a 30-T field was applied to a ceramic sample at 4.2 K.

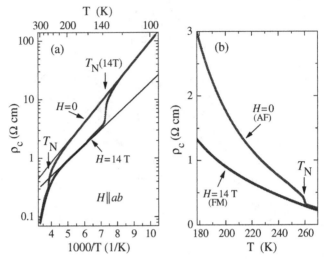

Figure 9. (a) Temperature dependence of ρ_c measured at $H = 0$ and 14 T applied along the *ab* plane. The dashed lines show simple activation fits $\rho_c \propto \exp(E_a/2k_B T)$ for both the AF and FM states, and T_N indicates the temperature of the FM-AF transition that is shifted to lower temperatures by the applied field. (b) A linear-scale view of the high temperature region, illustrating a step-like increase of ρ_c at T_N.

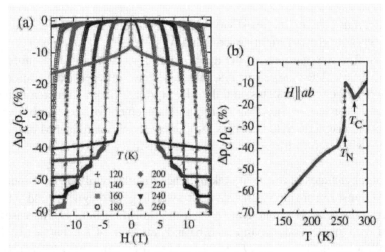

Figure 10. *c*-axis magnetoresistance of a GdBaCo$_2$O$_{5.50}$ crystal measured for $\mathbf{H} \| ab$. (a) Field dependences of $\Delta\rho_c/\rho_c$ measured at several temperatures, and (b) the temperature dependence of $\Delta\rho_c/\rho_c$ at $H = 14$ T.

What can be the mechanism of GMR in GdBaCo$_2$O$_{5.50}$, where neither the charge ordering nor the nanoscopic phase separation discussed in the Introduction can take place? One might speculate that the MR originates from

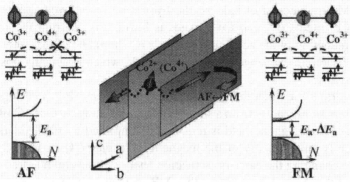

Figure 11. A schematic picture illustrating the generation of charge carriers (Co^{2+} or Co^{4+} states) in nominally nonmagnetic CoO_2 *ac* layers of $GdBaCo_2O_{5.50}$ and the origin of giant MR. For the AF arrangement of neighboring FM ladders, the energy of Co^{2+} and Co^{4+} states excited in the CoO_2 *ac* layer is lifted up, because the exchange interaction with ladders is inevitably frustrated (left panel). In contrast, the FM arranged ladders remove this frustration, thus reducing the required activation energy for the carrier generation by ΔE_a (right panel).

a "spin-valve" effect (see Fig. 1), that is, from the spin-dependent tunneling between the FM ladders. This explanation, however, does not seem likely. First, along the c axis the ladders are separated from each other by rows of oxygen vacancies, while the *ab* planes are intact, and thus very different tunneling probabilities can be expected along the c and *ab* directions. The measured conductivity is nevertheless virtually isotropic in the FM phase [Fig. 4(left)], indicating that the charges move within fairly isotropic nonmagnetic *ac* layers instead of tunneling between the FM ladders. Second and more important is that the "spin-valve" model can provide explanations neither to the temperature-induced FM↔AF transition nor to its doping dependence.

To account for the MR behavior in $GdBaCo_2O_{5.50}$ we suggest an alternative explanation, namely, that the generation of charge carriers in octahedral *ac* CoO_2 planes is significantly influenced by the relative orientation of the magnetic moments in neighboring FM ladders. At first sight, it might look strange that reorientation of weakly coupled FM ladders may affect the charge transport, bringing about a giant magnetoresistance, particularly in the nominally non-magnetic CoO_2 planes. One should remember, though, that carriers are generated in semiconducting $GdBaCo_2O_{5.50}$ through the formation of electron-hole pairs thus creating Co^{2+} and Co^{4+} ions which are *inevitably magnetic* (Fig. 11). Clearly, the energy of these excited states must depend on the surrounding magnetic order (Taskin et al., 2003; Taskin et al., 2005). For the AF arrangement of neighboring FM ladders, the exchange interaction of magnetic Co^{2+} and Co^{4+} states with ladders is inevitably frustrated, and their energy includes the penalty for this frustration [Fig. 11(left)]. However, for the FM arrangement of ladders, the frustration is removed and thus the energy of excited Co^{2+} and Co^{4+} states is lower [Fig. 11(right)]; presumably, the reduction in the activation energy ΔE_a should be of the same order as the inter-atomic exchange energy $J \sim 25$ meV, which is actually observed experimentally.

It is worth noting that the gap-reduction mechanism (Fig. 11) suggested above differs significantly from a simple Zeeman splitting. In $GdBaCo_2O_{5.50}$, the effect of the magnetic field is remarkably amplified by FM ladders, and an apparently small energy of the magnetic field $g\mu_B H/k_B \sim 1$ K becomes capable of changing the carriers' activation energy by several hundred K. To distinguish this MR scheme from other MR mechanisms, we have coined a term "magnetic field-effect transistor", based on a hypothetic structure where the charge-carrier injection into a 2D semiconducting channel is controlled not by an electric field, but by a magnetic state of neighboring "ligands".

Interestingly, the gap reduction suggested above gives a natural explanation for both the temperature-induced AF→FM transition and for the doping dependences of T_N. Indeed, in the ground state, the FM ladders are coupled antiferromagnetically with an energy estimated as $J_{l-l} \sim 1.5$ meV (Taskin et al., 2003). Upon heating, carriers are generated in the valence and conduction bands [Fig. 11(left)], which will sooner or later make the AF state unstable with respect to the FM one, since the AF→FM transition will reduce the energy of doped carriers by $\Delta E_a/2 \sim 15$ meV [Fig. 11(right)]. A rough estimate shows that the AF→FM transition should occur when the density of carriers reaches $\sim 10\%$ (this value may be lower if the inter-ladder coupling weakens with increasing temperature). The same should hold for the chemical doping, be it with holes or electrons, which qualitatively explains the doping dependence of T_N in the phase diagram.

Acknowledgements

A.N.L. gratefully acknowledges support from the Russian Foundation for Basic Research (grant 05-02-16973).

References

Dagotto, E., Hotta, T., and Moreo, A. (2001) Colossal magnetoresistant materials: the key role of phase separation, *Phys. Rep.* **344**, 1–153.

Frontera, C., García-Muñoz, J. L., Llobet, A., and Aranda, M. A. G. (2002) Selective spin-state switch and metal-insulator transition in $GdBaCo_2O_{5.5}$, *Phys. Rev. B* **65**, 180405-1– 180405-4.

Kimura, T., Tomioka, Y., Kuwahara, H., Asamitsu, A., Tamura, M., and Tokura, Y. (1996) Interplane Tunneling Magnetoresistance in a Layered Manganite Crystal, *Science* **274**, 1698–1701.

Martin, C., Maignan, A., Pelloquin, D., Nguyen, N., and Raveau, B. (1997) Magnetoresistance in the oxygen deficient $LnBaCo_2O_{5.4}$ (Ln = Eu, Gd) phases, *Appl. Phys. Lett.* **71**, 1421–1423.

Nagaev, E. L. (1996) Lanthanum manganites and other giant-magnetoresistance magnetic conductors, *Phys. Usp.* **39**, 781–805.

Respaud, M., Frontera, C., García-Muñoz, J. L., Aranda, M. A. G., Raquet, B., Broto, J. M., Rakoto, H., Goiran, M., Llobet, A., and Rodríguez-Carvajal, J. (2001) Magnetic and magnetotransport properties of $GdBaCo_2O_{5+\delta}$: A high magnetic-field study, *Phys. Rev. B* **64**, 214401-1–214401-7.

Taskin, A. A., Lavrov, A. N., and Ando, Y. (2003) Ising-Like Spin Anisotropy and Competing Antiferromagnetic-Ferromagnetic Orders in $GdBaCo_2O_{5.5}$ Single Crystals, *Phys. Rev. Lett.* **90**, 227201-1–227201-4.

Taskin, A. A., Lavrov, A. N., and Ando, Y. (2005) Transport and magnetic properties of $GdBaCo_2O_{5+x}$ single crystals: A cobalt oxide with square-lattice CoO_2 planes over a wide range of electron and hole doping, *Phys. Rev. B* **71**, 134414-1–134414-28.

Tomioka, Y., Asamitsu, A., Moritomo, Y., Kuwahara, H., and Tokura, Y. (1995) Collapse of a Charge-Ordered State under a Magnetic Field in $Pr_{1/2}Sr_{1/2}MnO_3$, *Phys. Rev. Lett.* **74**, 5108–5111.

FERRIMAGNETIC DOUBLE PEROVSKITES AS SPINTRONIC MATERIALS *

Lambert Alff (alff@oxide.tu-darmstadt.de)
Institute of Materials Science, Darmstadt University of Technology, Petersenstr. 23, 64287 Darmstadt, Germany

Abstract. It is a formidable task of material science to find materials suitable for spintronics applications. Candidates within the large group of perovskites are the ferrimagnetic double perovskites which combine high Curie-temperatures with half-metallic behavior. Research in this field is focussed particularly on strategies for tayloring ferrimagnetic double perovskites with still higher Curie-temperatures and new properties such as antiferromagnetic half-metallicity.

Key words: Spintronic, Double Perovskites, Half Metals

1. Introduction

Materials suitable for spintronic applications still have to be identified. Half-metallic materials that can supply fully spin polarized charge carriers at the Fermi level are of particular interest. Such half-metals hold the promise of being ideal candidates for injection devices and magnetic memory elements. There are several directions in which the search for such materials is being pursued. The first and perhaps most obvious is the search for ferromagnetic semiconductors like $Ga_{1-x}Mn_xAs$ and $Ga_{1-x}Mn_xN$ which would ideally fit into a technology based solely on semiconductors. The challenge here is to produce materials with Curie temperatures well above room temperature (see Hori *et al.*, these proceedings). Within the intensively discussed wide band-gap semiconductors, ZnO is a material that might also be suited for spintronics applications. Combined with a (not yet achieved) *p*-doping, an energy saving oxide electronic technology could be developed. Within the class of simple oxides several other interesting candidates do exist: While for CrO_2 the Curie-temperature is quite low, Fe_3O_4 with $T_C \approx 850\,K$ is a good candidate. Another promising group is formed by the Heusler-alloys where critical temperatures above $1000\,K$ can be achieved. In the group of

---

* Dedicated to H. Fuess on the occasion of his 65th birthday.

K. Scharnberg and S. Kruchinin (eds.),
Electron Correlation in New Materials and Nanosystems, 393–400.
© 2007 *Springer.*

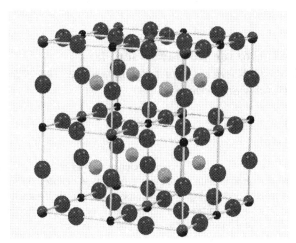

Figure 1. Cubic crystal structure of a typical double perovskite. Two magnetic ions are separated by two oxygen ions and a non-magnetic ion.

perovskites, the simple perovskites like the manganites all have critical temperatures below 400 K. An interesting subgroup within the perovskites is made up of ferrimagnetic double perovskites which form a rich playground for designing materials.

Material design, of course, has to start from the accessible combinations of elements and crystal structures employing suitable techniques. Here it is important to mention that, using thin film synthesis like oxide molecular beam epitaxy, phases out of thermodynamic equilibrium and metastable crystal structures can be produced. Techniques like this one will become increasingly important for materials science.

Ferrimagnetic double perovskites came into the focus of research after the discovery of high magneto-resistance in Sr_2FeMoO_6 (Kobayashi *et al.*, 1998). In the group of the double perovskites, the highest Curie-temperature so far is 635 K for Sr_2CrReO_6 (Kato *et al.*, 2002). It is astonishing that the double perovskites (see crystal structure in Fig. 1) have critical temperatures higher than the simple perovskites since the distance between the magnetic ions is doubled. However, this can be understood, at least qualitatively, within the framework of a kinetic energy driven exchange model.

2. Kinetic energy driven exchange coupling

The basic idea of the kinetic energy driven exchange model has been developed by Sarma *et al.* and others (Sarma *et al.*, 2000; Fang *et al.*, 2001; Kanamori and Terakura, 2001). We illustrate the model using the example

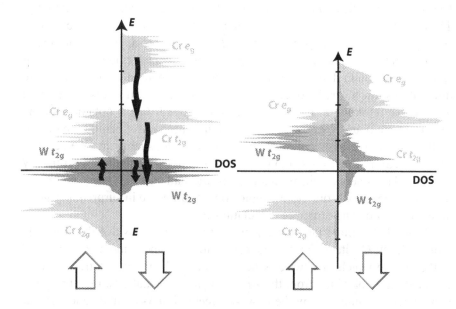

Figure 2. Kinetic energy driven exchange model for double perovskites. Left side: Before hybridization. The black arrows indicate the shift of orbitals. Right side: The resulting band structure after hybridization clearly shows the spin-majority carrier gap at the Fermi level.

of Sr_2CrWO_6 (Philipp *et al.*, 2001 and Philipp *et al.*, 2003). The Cr^{3+} ion is in the $3d^3$ state. The crystal field splitting shifts the e_g-levels above the Fermi edge, leaving the fully occupied t_{2g} below the Fermi edge. The strong Hund's splitting shifts all spin-minority bands above the Fermi level. As a result charge carriers have to be supplied from the W^{5+} sites, which is in the $5d^1$ state. For W (and Mo, Re etc.) it is known that there is usually no spin splitting. Thus, only the crystal field splitting is active leaving unpolarized charge carriers in the t_{2g}-levels. This situation is shown in Fig. 2, left side. Now we have to take into account the hybridization of orbitals of same symmetry. The main feature here is that the Cr t_{2g} and W t_{2g} levels hybridize and shift towards the Fermi level as indicated by the black arrows in Fig. 2, left side. Consequently, the other bands also shift: the main results are that Cr t_{2g} spin-minority levels now contribute to the charge transport, while the tungsten t_{2g} spin-majority bands are shifted above the Fermi-level. This situation is shown in Fig. 2, right side. In other words: all electrons contributing to the charge transport are of spin-minority type, i.e. a half-metal is obtained. This is a key property of ferrimagnetic double perovskites: the mechanism leading to magnetic coupling is intrinsically associated with half-metallicity. The full spin polarization, however, can be reduced by the presence of strong spin-orbit-coupling (Vaitheeswaran *et al.*, 2005). This applies theoretically to Sr_2CrReO_6. As there is still no good experimental evidence for a full spin-polarization in double perovskites (and other materials), it is even more difficult to show a small reduction of spin polarization. Still, it is fair enough to state that a totally or almost half-metallic ground state is obtained for the ferrimagnetic double perovskites. The driving mechanism behind the orbital shifts is, of course, the kinetic energy gained by delocalizing the W electrons. These electrons, however, can only delocalize if they are in the spin-minority channel. This energy gain is maximal in the half-metallic state. Note that the results obtained from these simple consideration are in surprisingly good agreement with band-structure calculations.

A simple but important consequence of this scenario is that the magnetic moment at the Cr site is slightly reduced, while a magnetic moment is *induced* at the W site which otherwise is non-magnetic. As the magnetic moments can be directly obtained from the band-structure calculations, measuring the local magnetic moments by the powerful method of x-ray magnetic circular dichroism (XMCD) (see Rogalev *et al.*, 2001) or that of neutron scattering is an important test of the kinetic energy driven exchange model. As a matter of fact, the first observation of the induced spin magnetic on Mo (of about $0.25 \mu_B$) in the compound Sr_2FeMoO_6 has been made by XMCD (Besse *et al.*, 2002). Note that using XMCD, spin and orbital magnetic moments can be obtained separately.

3.   Increasing T_C by electron doping

As is well known for example from manganites, the properties of complex oxides can be strongly affected by carrier doping. A first obvious way of doing that is to replace Sr^{2+} with ions of similar radii but different valencies. Electron doping can be achieved by using La^{3+}, following the example of high-temperature superconductors and manganites. These experiments have shown that the critical temperature can be further increased by increasing the band width in $La_xSr_{2-x}FeMoO_6$ (Navarro *et al.*, 2001; Serrate *et al.*, 2002; Frontera *et al.*, 2003; and Navarro *et al.*, 2003). This has been confirmed by the finding that the same behavior is also present in the CrW-system (Geprägs *et al.*, 2005). Within the above model, the increase in critical temperature is a direct consequence of the increased density of states at the Fermi-level. However, La doping also poses serious problems. First of all, the amount of Sr^{2+} that can be replaced by La is limited, and the growth of parasitic phases is difficult to avoid. It is not possible to introduce an amount of La^{3+}

(inducing electrons) sufficient to observe a reduction in T_C. Second, doping with La can also change the number of antisites and induce lattice distortions. These effects also strongly affect T_C, so that it is difficult to disentangle the various effects doping can have. It has been found that for almost all ferri-magnetic double perovskites the maximum T_C is obtained in the undistorted cubic structure, corresponding to a tolerance factor close to one (Philipp *et al.*, 2003). The only exception to this rule is found in the FeRe-system, where the orthorhombic structure has the highest Curie-temperature.

A more efficient way to change the doping is the direct replacement of the $5d$ ion. This leads to a simple explanation for the fact that Sr_2CrReO_6 has a critical temperature more than 100 K higher than Sr_2CrWO_6: Re corresponds to electron doping comparable to the replacement of one Sr by La per unit cell. As the highest observed value for the increased T_C by La-doping is also around 100 K, the effect of Re is a simple increase in the density of states at the Fermi-level. Unfortunately, the element to do this job for the FeMo-system would be Technetium which is radioactive. It would certainly be one of the most interesting developments for the double perovskites to test this idea by using ions supplying even more electrons. Again, for the FeMo-system a problem would be the use of Ru which will most likely not favor a high-spin state which, however, is necessary to invoke the kinetic energy driven exchange energy gain mechanism to increase T_C.

Finally, let us note the possibility that the high T_C in the Re-compounds might be due to an enhancement effect of an intrinsic magnetic moment of Re^{5+}. In order to elucidate this possibility in more detail, double perovskites would be of great interest where Re is not paired with a magnetic ion, for example one could think of Mg as non-magnetic ion. In that case, how-ever, one compares Re^{5+} and Re^{6+}, i.e. different valency states. In this case one would expect no *induced* magnetic moment. More ideal would be the combination with a trivalent transition metal without magnetic moment as scandium. Again, XMCD would be the method of choice to determine the element specific magnetic moments in such compounds.

4. Scaling laws

As discussed above, the kinetic energy driven exchange is associated with the induction of a magnetic moment at the non-magnetic site. Majewski *et al.* have found that a simple scaling law holds: the critical temperature scales with the induced magnetic moment as shown in Fig. 3. The scaling law is valid for all ferrimagnetic double perovskites studied so far. It is remarkable that the measured values are in really good agreement with band-structure calculations (Vaitheeswaran *et al.*, 2005; Jeng *et al.*, 2003).

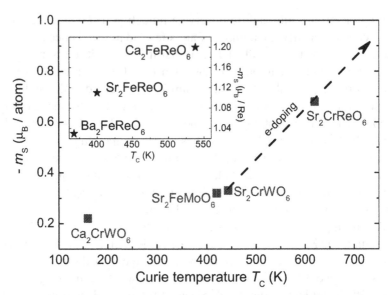

Figure 3. Scaling law between the critical temperature and the induced spin magnetic moment at the non-magnetic site of ferrimagnetic double perovskites as measured by XMCD. The data of the main plot is from Majewski *et al.*, 2005, the data in the inset is from Sikora *et al.*, 2005. Note that the absolute values of the spin magnetic moments are not comparable.

For the Co_2-based Heusler alloys a similar scaling law has been found. However, here it is the magnetic moment of the *magnetic* ion that scales with T_C (Fecher *et al.*, 2005). For the compound Co_2FeSi with $6\,\mu_B$ per unit cell, a Curie-temperature as high as $1100\,K$ has been observed. In contrast to the Heusler-alloys, for the ferrimagnetic double perovskites the increased induced magnetic moments at the non-magnetic sites correspond to a *reduced* over all magnetic moment per formula unit. It is therefore possible within the ferrimagnetic double perovskites to search for materials which come close to half-metallic antiferromagnets. Such compounds would be of great interest, as the full spin-polarization is combined with the absence of a net magnetic moment, which seems to be a contradiction in itself. Such materials have been theoretically predicted, but so far experimentally not yet observed (van Leuken and de Groot, 1995; Pickett, 1998). For example in the CrOs- or VOs-systems ($LaAVOsO_6$ with A = Ca, Sr, Ba has been predicted to be a half-metallic antiferromagnet by Wang and Guo, 2006), which of course are extremely difficult to synthesize, there could be a quasi half-metal with vanishingly small magnetic moment, i.e almost antiferromagnetic order. Note that as a competing ground state there can also be antiferromagnetic *insulating* order which is mediated by a more conventional superexchange mechanism. However, such compounds are likely to have low Néel-temperatures.

5. Summary

In this paper, the potential of ferrimagnetic double perovskites as spintronic materials has been highlighted. In view of the vast possibilities to increase the critical temperatures and to design these materials to become for example even (close to) half-metallic antiferromagnets, these oxides are promising spintronics candidates. In particular, the double perovskites are suited for fully epitaxial thin film structures based on perovskite materials. Possible applications could be spin injection devices or magnetic memory storage.

Acknowledgements

This work was supported by the ESRF and ILL (Grenoble).

References

M. Besse, V. Cros, A. Barthélémy, H. Jaffrès, J. Vogel, F. Petroff, A. Mirone, A. Tagliaferri, P. Bencok, P. Decorse, P. Berthet, Z. Szotek, W.M. Temmerman, S.S. Dhesi, N.B. Brookes, A. Rogalev, and A. Fert. *Experimental evidence of the ferrimagnetic ground state of Sr_2FeMoO_6 probed by X-ray magnetic circular dichroism.* Europhys. Lett. **60**, 608 (2002).

Z. Fang, K. Terakura, and J. Kanamori. *Strong ferromagnetism and weak antiferromagnetism in double perovskites: Sr_2FeMO_6 (M = Mo, W, and Re).* Phys. Rev. B **63**, R180407 (2001).

G. H. Fecher, H. C. Kandpal, S. Wurmehl, C. Felser, and G. Schnhense. *Slater-Pauling rule and Curie-temperature of Co_2-based Heusler compounds.* cond-mat/0510210.

C. Frontera, D. Rubi, J. Navarro, J. L. García-Muñoz, J. Fontcuberta, and C. Ritter. *Effect of band-filling and structural distortions on the Curie temperature of Fe-Mo double perovkites.* Phys. Rev. B **68**, 012412 (2003).

S. Geprägs, L. Alff, P. Majewski, Y. Krockenberger, A. Erb, R. Gross, C. Ritter, J. Simon, and W. Mader. *Electron doping in the double perovskite $La_xA_{2-x}CrWO_6$ with A = Sr and Ca.* J. Appl. Phys. **99**, 08J102 (2006).

Horng-Tay Jeng, and G. Y. Guo. *First-principles investigations of orbital magnetic moments and electronic structures of the double perovskites Sr_2FeMoO_6, Sr_2FeReO_6, and Sr_2CrWO_6.* Phys. Rev. B **67**, 094438 (2003).

J. Kanamori and K. Terakura. *A general mechanism underlying ferromagnetism in transition metal compounds.* J. Phys. Soc. Jpn. **70**, 1433 (2001).

H. Kato, T. Okuda, Y. Okimoto, Y. Tomioka, Y. Takenoya, A. Ohkubo, M. Kawasaki, and Y. Tokura. *Metallic ordered double-perovskite Sr_2CrReO_6 with maximal Curie temperature of 635 K.* Appl. Phys. Lett. **81**, 328 (2002).

K.-I. Kobayashi, T. Kimura, H. Sawada, K. Terakura, and Y. Tokura. *Room-temperature magnetoresistance in an oxide material with an ordered double-perovskite structure.* Nature **395**, 677 (1998).

P. Majewski, S. Geprägs, O. Sanganas, M. Opel, R. Gross, F. Wilhelm, A. Rogalev, and L. Alff. *X-ray magnetic circular dichroism study of Re 5d magnetism in Sr_2CrReO_6.* Appl. Phys. Lett. **87**, 202503 (2005).

P. Majewski, S. Geprägs, A. Boger, M. Opel, A. Erb, R. Gross, G. Vaitheeswaran, V. Kanchana, A. Delin, F. Wilhelm, A. Rogalev, and L. Alff. *Magnetic moments of W 5d in Ca_2CrWO_6 and Sr_2CrWO_6 double perovskites.* Phys. Rev. B **72**, 132402 (2005).

J. Navarro, C. Frontera, Ll. Balcells, B. Martínez, and J. Fontcuberta. *Raising the Curie temperature in Sr_2FeMoO_6 double perovskites by electron doping.* Phys. Rev. B **64**, 092411 (2001).

J. Navarro, J. Nogués, J. S. Muñoz, and J. Fontcuberta. *Antisites and electron-doping effects on the magnetic transition of Sr_2FeMoO_6 double perovskite.* Phys. Rev. B **67**, 174416 (2003).

J. B. Philipp, J. Klein, D. Reisinger, M. Schonecke, A. Marx, A. Erb, L. Alff, and R. Gross. *Spin-dependent transport in the double-perovskite Sr_2CrWO_6.* Appl. Phys. Lett. **79**, 3654 (2001).

J. B. Philipp, P. Majewski, L. Alff, A. Erb, R. Gross, T. Graf, M. S. Brandt, J. Simon, T. Walther, W. Mader, D. Topwal, and D. D. Sarma. *Structural and doping effects in the half-metallic double perovskite A_2CrWO_6 (A = Sr, Ba, and Ca).* Phys. Rev. B. **68**, 144431 (2003).

W. E. Pickett. *Spin-density-functional-based search for half-metallic antiferromagnets.* Phys. Rev. B **57**, 10613 (1998).

A. Rogalev, J. Goulon, Ch. Goulon-Ginet, and C. Malgrange. *Instrumentation developments for polarization dependent X-ray spectroscopies* in *Magnetism and Synchrotron Radiation*, E. Beaurepaire *et al.* (Eds.), LNP vol. 565 (Springer, 2001).

D. D. Sarma, P. Mahadevan, T. Saha-Dasgupta, S. Ray, and A. Kumar. *Electronic Structure of Sr_2FeMoO_6.* Phys. Rev. Lett. **85**, 2549 (2000); see also D. D. Sarma. *A new class of magnetic materials: Sr_2FeMoO_6 and related compounds.* Curr. Opinion in Solid State Mat. Sci. **5**, 261 (2001).

D. Serrate, J. M. De Teresa, J. Blasco, M. R. Ibarra, L. Morellón, and C. Ritter. *Large low-field magnetoresistance and T_C in polycrystalline $(Ba_{0.8}Sr_{0.2})_{2-x}La_xFeMoO_6$ double perovskites.* Appl. Phys. Lett. **80**, 4573 (2002).

M. Sikora, D. Zajac, Cz. Kapusta, M. Borowiec, C. J. Oates, V. Prochazka, D. Rybicki, J. M. De Teresa, C. Marquina, and M. R. Ibarra. *Direct evidence of the orbital contribution to the magnetic moment in $AA'FeReO_6$ double perovskites.* cond-mat/0503358.

G. Vaitheeswaran, V. Kanchana, and A. Delin. *Pseudo-half-metallicity in the double perovskite Sr_2CrReO_6 from density-functional calculations.* Appl. Phys. Lett. **86**, 032513 (2005).

H. van Leuken and R. A. de Groot. *Half-metallic antiferromagnets.* Phys. Rev. Lett. **74**, 1171 (1995).

Y. K. Wang and G. Y. Guo. *Robust half-metallic antiferromagnets $LaAVOsO_6$ and $LaAMoYO_6$ (A = Ca, Sr, Ba; Y = Re, Tc) from first-principles calculations.* Phys. Rev. B, in print; cond-mat/0601468 (2006).

A POSSIBLE MODEL FOR HIGH-T_C FERROMAGNETISM IN GALLIUM MANGANESE NITRIDES BASED ON RESONATION PROPERTIES OF IMPURITIES IN SEMICONDUCTORS

Hidenobu Hori (h-hori@jaist.ac.jp), and Yoshiyuki Yamamoto
School Materials Science, Japan Advanced Institute of Science and Technology (JAIST), 1-1 Asahidai, Nomi, Ishikawa 923-1292, Japan

Saki Sonoda
School of Engineering Science, Osaka University, 1-3 Machikaneyama, Toyonaka 560-8531, Japan

Abstract. High-T_C ferromagnetism in (Ga,Mn)N was observed and almost all results are found to be similar to the experimental results in (Ga,Mn)As, except for the value of T_C. Though all standard experiments on magnetism clearly support the results, the value is unexpectedly high. This work presents and discusses possible origins of high-T_C ferromagnetism after a brief review of the experimental results. The key speculation on the Bosonization method in three dimensions is related to the problem of Anderson localization.

Key words: High T_C ferromagnetism, Dilute magnetic semiconductor, (Ga,Mn)N, Ordered phase, Boson representation, Carrier induced magnetism, Superparamagnetism, Laser, Resonator

1. Introduction

Ferromagnetism in Dilute Magnetic Semiconductors (DMS) has come to be the center of attention for the researchers because of the potential for new functionalities in the main stream of application to electronics. Besides the potential of such applications, it is especially to be noted that the problems of dilute impurity systems or the low carrier density in semiconductors seem to provide us with common interesting physical problems. For example, high efficiency in the thermo-electric transformation effect in room temperature and high-T_C superconductivity are considered to be the typical phenomena.

BiTe[1] with its highly efficient thermo-electric transformation and high-T_C superconductors have carrier concentrations of less than $\sim 10^{27}$ per m^3

K. Scharnberg and S. Kruchinin (eds.),
Electron Correlation in New Materials and Nanosystems, 401–416.
© 2007 *Springer.*

($\sim 10^{21}$ 1/cc) and impurity concentrations of several percent. High-T_C ferromagnetism occurs in semiconductors which have carrier concentrations in a similar range of $10^{25} \sim 10^{28}$ per m^3. It should be noted that these effects appear at relatively high temperatures, although the impurities are randomly distributed and in low concentration. These results lead us to infer that some unrecognized common property may exist in the dilute impurity systems in some semiconductors. The effects also appear under these conditions as the "unthinkable phenomena" in relatively high temperature range. In the present work, we will try to look for the physical key concept through the detailed investigation of high-T_C ferromagnetism in (Ga,Mn)N.

The first observation of ferromagnetism in semiconductors with dilute magnetic impurities (DMS) was reported for $Ga_{1-x}Mn_xAs$ ($1 \geq x \geq 0$) with T_C of 110 K. (the family of the Mn-doped crystals $Ga_{1-x}Mn_xAs$ is simply written as (Ga,Mn)As and similar abbreviations are used for any other doped crystals) [2]. The value of $T_C = 110$ K is relatively high for the Mn impurities with a concentration of 3%. After this work, a number of observations of DMS-ferromagnetism have been reported for various materials. Recently, our group reported the extremely high T_C value of 940 K in (Ga,Mn)N[3,4]. Besides these reports, ferromagnetism of $Ca_{1-x}La_xB_6$ has specially attracted our attention, because all ions in the crystal are non-magnetic. Moreover, the estimated T_C is about 600 K[5]. However, some careful experiments have shown the existence of iron impurities with a concentration of about 100ppm[6] and the Auger experiment shows that the thickness of the Fe impurity layer is greater than $1\mu m$[7]. Here, the question of "what is the origin of the ferromagnetism for the extremely low concentration of iron ions?" has emerged. Several models have been proposed to answer this question, but the most likely one in our opinion is the model of DMS-ferromagnetism.

At first, we experimentally confirm the ferromagnetism and the unbelievably high T_C value of (Ga,Mn)N by comparing with possible explanations in terms of superparamagnetism. Although these experiments are quite standard, some theorists have strongly discounted the possibility of such a high T_C value. Therefore, as experimentalists we will raise some naive questions with regards to their calculations, based on physical pictures and, moreover, we want to present a somewhat speculative model to explain the experimental results of high-T_C ferromagnetism in (Ga,Mn)N. The important clue to the problem of high-T_C ferromagnetism comes from ideas relating to the lasing effect. The most remarkable property of the lasing effect is that the lasing action can be considered as a kind of phase transition at high temperatures[8]. The important point is that the lasing action appears at high temperatures and is produced by a low concentration of randomly distributed impurities. In the present work, the possibility of high-T_C ferromagnetism in (Ga,Mn)N is discussed referring to the conditions of lasing action. After that, we will briefly argue in defence of the adequacy and applicability of the key concepts used in the model.

1.   The problem of ferromagnetism in DMS materials of (Ga,Mn)N

For GaN based DMS, Dietl et al. suggested that Mn doped GaN would have high T_C's, exceeding room temperature[3]. Recently, the present authors reported the successful growth of III-V based wurtzite (Ga,Mn)N films using Molecular Beam Epitaxy (MBE) and measuring their spontaneous magnetization in 0.1T for the first time. The details of the magnetic properties are discussed in [4] and the results are summarized in Fig. 1.

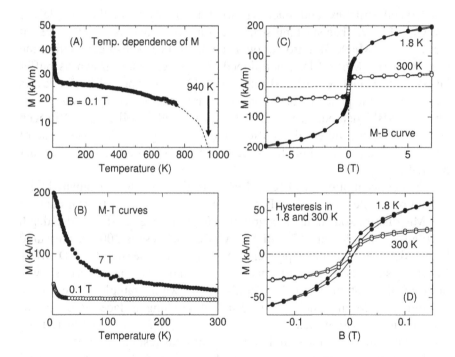

Figure 1. Measurements of Ferromagnetism in (Ga,Mn)N. **(A)** Temperature dependence of magnetization of (Ga,Mn)N (M-T curve) in 0.1 Tesla(T) observed up to 750K to estimate the value of T_C. The value of T_C is about 940K on the theoretical curve obtained from the mean field theory. **(B)** M-T curves in low (0.1 T) and high fields(7T). **(C)** Field dependence of magnetization M-B curve. **(D)** Hysteresis curves of the ferromagnetism in liquid He and room temperatures.

The Curie temperature is roughly estimated by fitting the spontaneous magnetization curve (derived from the molecular field model using Brillouin function with S = 5/2) to the temperature dependence of magnetization. The steep drop of magnetization around 10 K is not included in the fitting curve. Data were taken up to 750 K, as shown in Fig. 1(A), and the estimated value of T_C is 940 K. The method of sample preparation is well established now. It has been discussed and described in detail elsewhere[3,4].

These results make us feel quite confident of the anomalously high T_C value and this is the central point of this work. To confirm these results, the experimental procedures are reviewed in successive subsections.

1.1. EXPERIMENTAL PROCEDURES AND THE RESULTS FOR (Ga,Mn)N

The crystal structure and stacking structure of layers in the sample are shown in Fig. 2. The stacking structure is a typical one used in the present work for (Ga,Mn)N film growth on a sapphire(0001) substrate: Before the growth of (Ga,Mn)N films, non-doped GaN layers were grown as buffer layers with a thickness of 200 nm (2000 Å). Two types of buffer layers were used in the present work. For one sample, the buffer layer was grown by NH_3-MBE method, in which NH_3 gas is used in MBE as a Nitrogen source. For the other, the buffer layers were grown by rf-plasma of Nitrogen gas and the sample films are grown in a different vacuum chamber of MBE.

Mn concentrations x in $(Ga_{1-x}Mn_xN)$ of the films were estimated with the aid of an Electron Probe Micro Analyzer (EPMA). As typical data for the Mn distribution along the growth direction, Fig. 2 shows the depth profile of a sample with ~7% of Mn whose thickness is 2000 Å measured by EPMA. The depth dependence in Fig. 2(C) measured by D-SIMS (dynamic secondary ion mass spectroscopy) and these data clearly show a uniform distribution of the Mn atoms in the whole sample space with a resolution of 20 nm. The uniform distribution along surface direction was also confirmed within the resolution of 100 nm. This means that there is no grain-like structure in the sample space within this resolution.

The X-ray diffraction is clearly observed and the Wurtzite crystal structure is confirmed. Although a broadening of the diffraction spots is observed in Mn doped GaN, the basic structure remains unchanged. The experiments of XAFS and CAISIS are observed for the (Ga,Mn)N. The measurements of EXAFS on (Ga,Mn)N were carried out at the BL38B1 beam line of SPring-8. The experimental results give us information of substitutional doping by comparing the coherent effect of X-ray radiation from the ions surrounding Mn and Ga. The results clearly show substitutional doping of Mn ions onto Ga sites of the wurtzite structure[9]. Therefore, the segregation of Mn atoms or the formation of other Mn alloys can be ruled out.

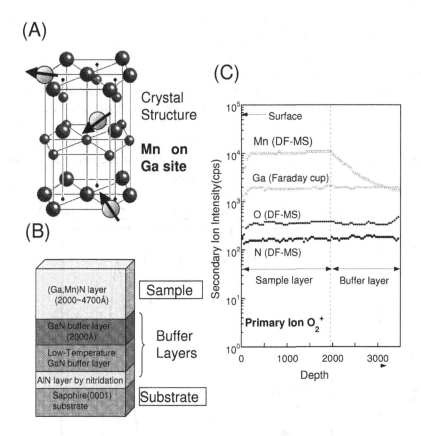

Figure 2. Structure of (Ga,Mn)N sample. **(A)** Crystal structure. Arrows mean spin moments on Mn impurities. **(B)** Stacking Layer structure of the sample. The top layer is (Ga,Mn)N. **(C)** Distribution of constituent atoms observed by D-SIMS.

1.2. MAGNETIC MEASUREMENTS

As is easily understood from the data in Fig.1, the value of T_C is unbelievably high. The hysteresis curve clearly indicates typical properties of ferromagnetism. The remnant moment indicates a spontaneous spin-polarization and the existence of domain. The domain size is given by the energy balance between exchange energy and the sum of spin dipole energies within the domain. The usual domain sizes are larger than several hundreds of nanometer. This means that the ferromagnetic particles should be larger than several hundreds of nanometers, if the ferromagnetism of (Ga,Mn)N originated from an assembly of ferromagnetic grains. However, the experiments of EPMA and SIMS clearly rule out such a grain-like structure.

Figures 1(B) and (C) are explained by a coexistence model of ferro- and para- (or super-para-) magnetisms[4] and the ferromagnetic part is about 20% of saturation moment in the sample with 8%-Mn concentration. The concentration dependence is large around 3% Mn, but the dependence is not large above the concentration of 6%. The clear ferromagnetic part is observed in the samples with the concentration above 3%-Mn. Quite recently, we observed the steep decrease of the remnant moment at zero field with decreasing temperature below 10K, while the remnant moment is approximately constant from 10 K to 400 K[10]. Such a phenomenon is quite anomalous in the research field of magnetism.

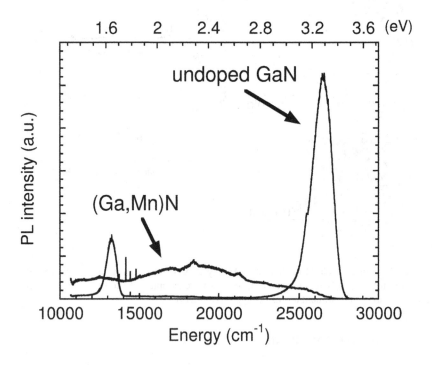

Figure 3. Emission spectra of GaN and (Ga,Mn)N with 6% Mn. The sharp peak near 13000 cm^{-1} is considered to be a second reflection by the grating in the spectrometer corresponding to strong spectrum at 26500 cm^{-1}. No sharp spectrum corresponding to these sharp lines in GaN is not found in the spectra in (Ga,Mn)N. This demonstrates the large change in electronic structure by Mn doping.

1.3.   OPTICAL, TRANSPORT AND MAGNETIC PROPERTIES

Because the ferromagnetism in (Ga,Mn)N originates from the properties of DMS materials, investigations of transport and magnetic properties may provide us with important clues about the origin of the magnetism. Optical measurements can also give us useful information on the electronic states, because 6% Mn impurities cause a large change to the spectra and any peak in the sharp and strong emission spectrum of pure GaN is no longer observed at the concentration of 94%. This means that there is a large change in the electronic structure of the whole host crystal and that the possibility of a mixed crystal, consisting of pure GaN (94 at%) and some segregated Mn compound (6 at%), can be ruled out.

Takeyama et al. observed the hydrogenic exciton absorption spectra superimposed on the band absorption edge in the ultraviolet range[11]. This demonstrates the existence of the hydrogen like atomic exciton in (Ga,Mn)N.

As (Ga,Mn)N is a DMS materials, the origin of the ferromagnetism is closely related to the properties of the carriers. In fact, Fig. 4 (C) and (D) clearly show this relation. The temperature dependence to the carrier density in Fig. 4 was obtained from the data of Hall resistance. It is noticable that for the Mn-concentration higher than 3% in (Ga,Mn)N, carriers are p-type, while pure GaN grown by NH$_3$-MBE method has n-type conduction. The GaN film has n-type carrier with the density of 10^{27}m^{-3} (10^{21} cm^{-3}), but in the case of a (Ga,Mn)N sample with Mn-concentration of 6.8%, the hole carrier density is 10^{26} m^{-3} (10^{20} cm^{-3}).

The GaN produced by NH$_3$-method has electron conduction and metallic temperature characteristics. A most likely model for the electron conduction in GaN invokes some lattice defects, such as an anti-site structure. As the concentration of Mn impurities increases, the carrier type changes from n-type to p-type and the ferromagnetism in (Ga,Mn)N appears only above the Mn concentration of 3% . Although the hole density becomes higher with increasing Mn concentration, the ratio of the ferromagnetic part to the total magnetization does not increase correspondingly.

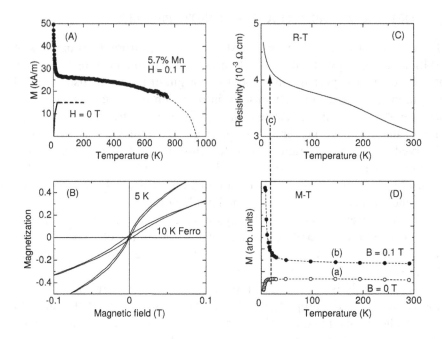

Figure 4. Review of M-T curves and their relation with the R-T curve. **(A)**Temperature dependence of magnetization in 0.1 T and schematic curve of temperature dependence of zero field magnetization of the sample with Mn-concentrations of 8%. **(B)** Hysteresis curves. It is noticed that coercive field of 10 K is much larger than that of 5K. **(C)** Temperature dependence of resistance (R-T curve). The temperature at (c) is the threshold temperature below which the carrier trapping becomes markedly large. **(D)** The remnant magnetization at zero field (a) and magnetization at the field just above the hysteresis loop (b).

The temperature dependence of the carrier density can be obtained from the Hall resistance data. It is shown in Fig. 5. As can be seen from Fig. 5, the increase of the resistance at low temperatures strongly depends on the decrease of carrier density. The hole conduction can be explained by the electron hopping motion from Mn^{2+} to Mn^{3+}. The existence of the mixed valence state of Mn ions in (Ga,Mn)N has recently been reported by our group[10]. From both data sets shown in Fig. 5 it follows that the main reason for the increase of resistivity can be attributed to the decrease in carrier density. Furthermore, the decrease of carrier concentration also coincides with the decrease of the spontaneous magnetization, as is seen in Fig. 4(D).

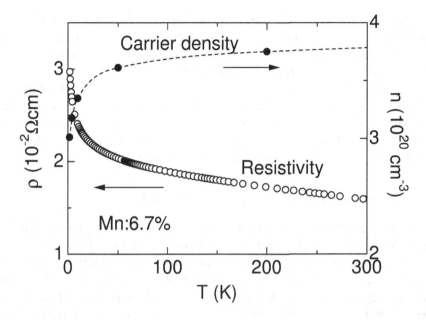

Figure 5. Temperature dependence of carrier density and resistivity on (Ga,Mn)N. Because of the clear correlation with the carrier density, the main reason for the increase in the resistivity can be attributed to the decrease of the carrier density and furthermore, the decrease of magnetization is also correlated with the decrease of the spontaneous magnetic polarization.

As discussed in [4], the double exchange mechanism explains the experimental results of the ferromagnetism in (Ga,Mn)N. The double exchange model was first proposed by the present authors in their previous work[4] and, more recently, the model has also been discussed and supported theoretically. Electron hopping conduction between Mn atoms decreases with decreasing temperature. Therefore, the spontaneous magnetization arising from the double exchange mechanism may decrease at low temperatures. The dotted arrow in Fig. 4 (C) and (D) indicates that the the threshold temperature below which the carrier trapping becomes markedly large corresponds to the temperature below which the spontaneous magnetization shows the reduction.

From these experimental results, the following physical picture evolves: some Mn^{3+} ions in (Ga,Mn)N compensate the electrons in GaN and Mn^{2+} is produced near the Fermi level. Such Mn^{2+} in (Ga,Mn)N splits into two virtual bound states. These virtual bound states have two opposite spin polarizations for the spin on Mn^{2+}. The spin polarizations are produced by an s-d Hamiltonian. The virtual bound states are responsible for the high density of states near the Fermi level. The two densities of states have opposite spin polarizations near the Fermi level. The physical picture of the double exchange mechanism can be described as the collective motion of the electrons, hopping in the spin-polarized bands.

This mechanism can explain the change in carrier type in (Ga,Mn)N at the critical concentration of about 3%. This model requires the coexistence of Mn^{2+} and Mn^{3+} ions. The experimental results of EXAFS support this model[10]. In fact, the XANES spectra are consistent with neither Mn^{2+} nor Mn^{3+} and a mixed valence state has been suggested[9].

1.4. REVIEW OF THE RESULTS AND DISCUSSION

The magnetization measurements show typical ferromagnetism and the possibility of segregation or phase separation is experimentally ruled out by the hysteresis curve, the existence of zero field magnetization and its temperature dependence and uniform distribution of Mn ions in the whole crystal region. We can also experimentally rule out segregation by the experiments on ferromagnetic powder samples. A step-like field dependence of the magnetization is shown by the powder sample with a concentration higher than 90%. Even in such a high density ferromagnetic powder sample, hysteresis is not observed at room temperature. These observations clearly show that the segregated superparamagnetism with the 6% Mn impurities cannot reproduce the magnetization data of (Ga,Mn)N.

Mixed valence states of Mn^{2+} and Mn^{3+} and double exchange mechanism based on the hopping conduction is supported by experiments of EXAFS and XANES spectra, transport properties and the strong coincidence in the anomalous region between transport and magnetic properties. In particular, it is quite noticeable that the zero-field spin polarization in high-T_C DMS sharply decreases below 10K. This effect proves directly that the conduction electrons produce the ferromagnetism, because the temperature range is clearly consistent with decreasing region of the hole carriers. Thus, the problem is "what is the origin of the high-T_C ferromagnetism in (Ga,Mn)N with several % of Mn impurities".

2.   A possible model and the application to other problems

2.1. QUESTION TO THE THEORETICAL CALCULATION

Though ferromagnetism in (Ga,Mn)N is experimentally proven, some theorists raise objections to the experiments based on their calculation[9]. According to the theory, the band structure of (Ga,Mn)N is calculated by the so-called KKR-CPA method. The calculational method of the band structure is a kind of APW-methods and the space of the crystal is divided into free electron- and atomic sphere-regions (or muffin-tin type potential regions). In this method, the atomic potential region is restricted to within a sphere having a charac-teristic radius R such that it fits into the Wigner-Seitz cell.

We would question this procedure of segmentation into Wigner-Seitz cells! Because the shielding lengths in low carrier density DMS materials are much longer than those in metals, it might be totally inadequate. In the case of metals, the electron density is so high that the size of the shielding cloud around the cation is small compared with the size of Wigner-Seitz cell. The shielding length around an imperfection in (Ga,Mn)N is simply estimated with the help of the Thomas-Fermi approximation[12]. According to this approximation, the shielding length r_{shield} is obtained from the carrier density and other fundamental constants. The role of atomic potential is relatively important in the imperfect shielding region. In fact, the hydrogenic exciton spectra in (Ga,Mn)N have already been observed in the ultraviolet region[11]. Applying the Thomas-Fermi approximation to the case of (Ga,Mn)N, the estimated value of the shielding size (called "Thomas-Fermi diameter" in this work) is longer than 7 nm and thus much longer than the longest lattice constant of c = 0.517nm. In contrast, the Muffin-tin radii used in the theory[13] are 0.1016nm for cations and 0.9252 nm for anions. Although the free electron region is reduced as much as possible in the APW method and there is a smooth connection between core and free electron wave functions, the spin state is fairly restricted by the properties of plane wave functions. The free electron part exhibits the spin singlet state for every energy level. Such a singlet spin state is given by the resultant contribution of Coulomb- and exchange- integrals based on the plane wave functions. Thus, the Thomas-Fermi effect might cause some severe problems for the numerical estimates. Because the long range effect in the Coulomb interaction is more important in the Thomas-Fermi region, it can be considered that all magnetic ions of (Ga,Mn)N in the Thomas-Fermi region directly contribute to the magnetic order.

2.2. A MODEL FOR HIGH T$_C$ FERROMAGNETISM IN (Ga,Mn)N

Although ferromagnetism in (Ga,Mn)N is experimentally proven, the question remains "why such a high T$_C$ occurs in a system with low carrier density and a low concentration of randomly distributed Mn ion". In our model, the origin of the ferromagnetic order is attributed to the double exchange mechanism based on virtual bound states coupled by spin correlated electron hopping. The spin correlated transfer is generated by the s-d Hamiltonian. This model is presented schematically in Fig. 6.

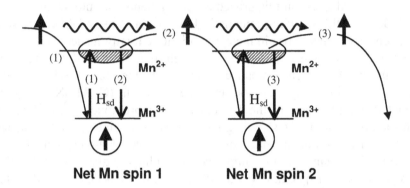

Figure 6. Schematic view of Double Exchange Model. Only one virtual bound state near the Fermi level is shown in this figure. Process (1) is electron trapping and processes (2) and (3) are spin correlated transfer process. An electron is trapped to Mn^{3+} and the Mn^{3+} ion changes to Mn^{2+} near the Fermi level. The Mn^{2+} is in a virtual bound state and the spin state is restricted to the net Mn spins because the transition is generated by spin-correlated interaction of s-d Hamiltonian. This is the elementally process of electron hopping.

The condition of low carrier density follows from the experimental results shown in Fig. 4(D). These conditions are very disadvantageous for the magnetic order in usual materials. As noted in previous sections, the important clue to the source of high T_C ferromagnetism comes from the idea of the lasing effect. The most remarkable point is that the lasing action can be considered as a kind of phase transition at high temperatures[8] and that the lasing action takes place at high temperatures, even though the lasing media operate under the conditions of randomly distributed and low concentration lasing atoms. In the present work, the possibility of the high T_C ferromagnetism in (Ga,Mn)N is discussed with reference to the conditions of lasing action.

We begin with the one-dimensional case for simplicity and then discuss the three-dimensional one. We note that the Mn impurities can act as scattering centers for the conduction electrons and the ions on substitutional sites can form resonators for the conduction electrons, because some electron states can form standing waves in the space between two Mn ions. Thus, if we assume that the role of laser light corresponds to the role of electron waves in the conduction band, a pair of Mn impurities can be considered as   one-dimensional resonator. For the time being we shall call a resonator made up of Mn impurities "impurity resonator" . The impurity resonator for the electron waves corresponds to the resonator in laser action. On this basis we propose a lasing-like mechanism for high T$_C$ ferromagnetism in (Ga,Mn)N by using the following correspondences:

Table I. Corresponding elements between Lasing Action and High T$_C$ (Ga,Mn)N.

	Laser	High T$_C$
1)	Lasing light	Electron wave in conduction band
2)	Excited state E$_2$	Electron trapped Acceptor of Mn^{2+}
3)	Lower state E$_1$	Lower state on Ga site of Mn^{3+}
4)	Induced optical transition: E$_2 \rightarrow$E$_1$	Electron emission : Mn$^{2+} \rightarrow$Mn^{3+}
5)	Photon absorption process : E$_1 \rightarrow$E$_2$	Electron absorption: Mn$^{3+} \rightarrow$Mn^{2+}
6)	Inversion population	Hole dominated : existence of Mn^{2+}
7)	Induced optical transition	Quantum transitions: electron emission
8)	Ratio of responsive atoms ~1%	Concentration of Mn impurities 3~7%
9)	Electric dipole transition	Transition by s-d interaction
10)	Resonator	"Impurity resonator"
11)	Bose statistics of light	Fermi statistics of electron wave

Looking at the correspondence of conditions for Laser action and High T$_C$ ferromagnetism, we can understand the similarity between them, except for the statistics in #11) of Table I. Correspondence #6) is deduced from the following experimental results on (Ga,Mn)N. a) The existence of mixed ion states of Mn^{2+} and Mn^{3+} together with hole conduction in the ground state (the concentration dependence shows that there exists both n- and p-type conduction, but the p-type is the majority in the ferromagnetic state with Mn-concentration higher than 3%, as was discussed in [4]. b) GaN crystals without Mn ions made by the ammonia-method show n-type conductivity. c) The appearance of ferromagnetism is strongly related to the conductivity as discussed in the text and in [4]. These characteristics originate from the properties of the ground state in this crystal. For these results in (Ga,Mn)N we may assume that the hole conduction is produced by hopping conduction between Mn^{2+} and Mn^{3+} states. This model just follows the "double exchange mechanism". This directly means that the ferromagnetic ground state is connected to the hopping conduction. If the ferromagnetism is represented by the electron states in the conduction band and two Mn ion states (mixed valence state) on Ga-sites of GaN, the ferromagnetic state looks like some excited state. In the case of (Ga,Mn)N, a lot of holes of Mn^{2+} already exists as Mn^{3+} ions (which is the excited state generated by the ionization of Mn^{2+}). The population inversion of holes is realized in the ground state when the population of Mn^{3+} is higher than that of holes in virtual bound states of Mn^{2+}, which is across the Fermi level. The observation of hole conduction supports the existence of such mechanism. The structure is produced in the process of crystal growth of (Ga,Mn)N.

The correspondence #11) is satisfied in the one-dimensional case because the scattering phenomena near $\cdot_F$ in the electron system can be described by a process of bosonization. The basis of the Bosonization representation is given by standing wave states and the standing wave representation is produced by a Bogoliubov transformation of free electron states and the transformed creation annihilation operators satisfy the Bose commutation relations. This means that the one-dimensional system satisfies the condition of lasing action.   In the case of High T_C ferromagnetism, the "lasing" electron is in a standing wave state which can be considered as some kind of pairing state.

Against this   background, we present the following somewhat speculative model:

1) In the 3-dimensional case, not all electrons can be described in a bosonized representation, but the special electron modes may satisfy this condition. This possibility is similar to the problem of Anderson localization in highly condensed electron system. If this speculation is true, the coexistence of ferro- and para- (super para-) magnetisms becomes an intrinsic problem and the ratio of about 20% corresponds to the ratio of bosononic states to the total number of states.

2) The electronic states are given by double exchange electronic states coupled with the lasing standing waves formed in the conduction bands.

3) The origin of the high working temperature is explained by following physical picture: because of the low carrier density, the electron shielding is poor and the Thomas-Fermi shielding length is quite long, about 7 nm, and the Coulomb interaction between the Mn ions and the localized nature of the electron states becomes relatively important. Such a structure follows from the strong spin correlated interaction. The stability of inter Mn ion coupling depends on the stability of the standing waves in the conduction band.

4) The physical picture of high-T_C is explained by the stability of standing waves against the thermal fluctuation of the lattice. The point of the stability is in the long wavelength standing waves in the long distant impurity resonator. The long distant resonator in (Ga,Mn)N can be several ten times larger than the lattice constant. Such a resonator generally shows

high efficiency as a resonator because the effect of lattice fluctuation on the resonator is much lower than the resonator with the distance of lattice constant. Thus stored standing waves with long wavelengths are relatively stable in comparison to the band edge electron waves, which have large effect on the usual metallic ferro-magnetism.

Thus, the present model is believed to provide an explanation for the coexistence of ferro- and para-magnetisms and the high-T$_C$ value of (Ga,Mn)N.

2.3. DISCUSSION

This model can possibly be applied to other phenomena produced by substitutionally doping mixed valence ions with low carrier density into semiconductors. One of the characteristic properties in these phenomena are the high working temperatures. The characteristics are supported by some mode in the "impurity resonator" structure. In the three-dimensional crystal, a part of conduction mode is responsible up to high temperature. These effects should reflect the electronic state. Some papers have reported the existence of narrow spectra near · F even at high temperatures for working range. In some X-ray photoemission research, such spectra are called "coherent". The existence of coherent spectra has been reported in DMS ferromagnetic materials[14,15], high-T$_C$ superconductors[16] and other spin-correlated materials[17]. The existence of these sharp bands at high temperatures seems to support the model presented here. But the model depends on the speculations of the existence of an "impurity resonator" and the ratio of the standing wave in three dimensions. The present authors hope for theoretical support in addressing these problems.

Acknowledgements

The synchrotron radiation experiments were performed at SPring-8 BL-47XU with the approval of Japan Synchrotron Research Institute (JASRI) as Nanotechnology Support Project of the Ministry of Education, Culture, Sports, Science and Technology.

References

1. There are many paper and textbooks. One of the recent work of sophisticated measurement method is given by H. Iwasaki, M. Koyano, and H. Hori, *Jpn. J. Appl. Phys.* 41, 6606-6609 (2002).
2. H. Ohno, and F. Matsukura, *Solid State Comm.* 117, 179-186 (2001).
3. S. Sonoda, S. Shimizu, T. Sasaki, Y. Yamamoto, and H. Hori, *J. Crystal Growth* 237-239, 1358-1359 (2002) (Proceedings of the 13th Int. Conf. of Crystal Growth, Kyoto, 2001).

4. H. Hori, S. Sonoda, T. Sasaki, Y. Yamamoto, S. Shimizu, K. Suga, and K. Kindo, *Physica B* 324, 142-150 (2002).
5. D. P. Yang, D. Hall, M. E. Torelli, Z. Fisk, J. D. Thompson, H. R. Ott, S. B. Oserff, R. G. Goodrich, and R. Zysler, *Nature* 397, 412-414 (1999).
6. K. Matsubayasi, M. Maki, T. Moriwaka, T. Tsuzuki, T. Nishioka, C. H. Lee, A. Yamamoto, T.Ohta, and N. K. Sato, *J. Phys. Soc. Jpn.* 72 2097-2102 (2003).
7. C. Meegoda, M. Trenary, T. Mori, and S. Otani, *Phys. Rev. B* 67, 172410-1-17240-3 (2003).
8. H. Haken: *Synergetics-an introduction nonequiliblium phase transitions and selforganization in physics, chemistry and biology 2nd ed.*, Chapter 8, (Springer-Verlag, 1978).
9. M. Sato, H. Tanida, K. Kato, T. Sasaki, Y. Yamamoto, S. Sonoda, S. Shimizu, and H. Hori, *Jpn. J. Appl. Phys.* 41, 4513-4514 (2002).
10. S. Sonoda, I. Tanaka, H. Ikeno, T. Yamamoto, F. Oba, T. Araki, Y. Araki, Y. Yamamoto, K. Suga, N. Nakanishi, Y. Akasaka, K. Kindo, and H.Hori, *J. Phys. Cond. Mat.* 18, 4615-4621 (2006).
11. D. Innami, H. Mino, S.Takeyama, S.Sonoda, and S. Shimizu, "Meeting Abstract of the Phys. Soc. Jpn", 57, Part 4, (2002) 634 (57th Annual Meeting of Japanese. Phys. Society).
12. C. Kittel, *Introduction to Solid State Physics* (John Wiley &Sons, Inc., New York, 1986).
13. K. Sato, and H. Katayama-Yoshida, *Jpn. J. Appl. Phys.* 40, L485-L487 (2001).
14. Y. Takata, K. Tamasaku, M. Yabashi, K. Kobayashi, J. J. Kim, T. Yao, T. Yamamoto, M. Arita, and H. Namatate, *Appl. Phys. Lett.* 84, 4310-4312 (2004).
15. T. Takeuchi, M. Taguchi, Y. Harada, T. Tokushima, Y. Tanaka, A. Chainani, J. J. Kim, H. Makino, T. Yao, T. Tsukamoto, S. Shin, and K. Kobayashi, *Jpn. J. Appl. Phys.* 44, L153-L155 (2005).
16. S. Sugai *et al.*, to be published in Proc. of LT'24 (2005, Ontario, Florida).
17. K. Horiba, M. Taguchi, A. Chainai, Y. Takata, E. Ikenaga, A. Takeuchi, M. Yabashi, H. Namatame, M. Taniguchi, H. Kumi, H. Konuma, K. Kobayashi, T. Ishi, *Phys. Rev. Lett.* 93, 236401-1-236401-4 (2004).

LARGE MAGNETORESISTANCE EFFECTS IN NOVEL LAYERED RARE EARTH HALIDES

R. K. Kremer, M. Ryazanov, and A. Simon

Max-Planck Institut für Festkörperforschung, D-70569 Stuttgart, Germany

Abstract. We give a survey of the structures, electric, magnetic and magnetoresistance properties of the two novel low dimensional rare-earth halide systems, GdI_2 and $GdIH_y$ ($2/3 < y \leq 1$). The large magnetoresistance effect observed for GdI_2 can be understood on the basis of a conventional spin disorder scattering mechanism, however, strongly magnified by the structural anisotropy and the special topology of the Fermi surface. Bound magnetic polarons are formed in $GdIH_y$ leading to a metal insulator transition below ≈ 30 K. The mobility of the magnetic polarons can be effectively modified by external magnetic fields resulting in the large experimentally found magnetoresistance.

Key words: Magnetoresistance, Magnetic polarons, Rare earth halides, Layered systems

1. Introduction

High-T_C superconductivity in oxocuprates and large magnetoresistance effects in mixed valence transition metal oxides are the phenomena in solid state physics which have attracted broadest attention over the past 20 years. It is widely accepted that the 'colossal response' and the 'complexity' of these systems is induced by electronic inhomogeneities on nano- and micrometer scales which occur as a consequence of an electronic phase separation[1, 2, 3, 4]. Currently there is increasing evidence that such a micro-phase-separation may occur by self-organization of the electronic state per-se, by introducing quenched disorder, and/or by material design like structuring on a nanometer scale. This electronic complexity frequently leads to entangled multiphase diagrams the subtleties of which are under ongoing scrutiny and far from being understood in all detail at present [5, 6, 7, 8, 9].

The large magnetoresistance in mixed valence transition metal oxides is a well known phenomenon which had been found already in the 1950's while searching for magnetically highly permeable insulators for high-frequency applications [10]. With respect to applications in micro-devices and high-capacity data storage, the investigation of these systems has gained significant pace since high quality thin films could successfully be prepared by the

K. Scharnberg and S. Kruchinin (eds.),
Electron Correlation in New Materials and Nanosystems, 417–436.
© 2007 *Springer.*

pulsed-laser deposition method which has been developed and refined for the high-T_C oxocuprates [11, 12].

Large and unusual magnetoresistance effects and concomitant metal-insulator transitions in *rare earth systems* were a topic of intensive research in the 1960's and 1970's, a field which since then has almost fallen into oblivion. In particular, rare earth chalcogenides, as e.g. EuX (X=O, S, Se, Te) which crystallize in the NaCl structure or the RE_3S_4 system (Th_4P_3 structure) have been of interest in those days [13]. Interestingly, a number of concepts discussed today to explain the large magnetoresistance in mixed valence transition metal oxides have been developed and discussed already for the rare earth materials, even though there are marked differences in the characteristics of the electronic structure and the relevant energy scales for the electronic interactions. The properties of the transition metal oxides are essentially determined by the competition of a number of interactions, all on an energy scale of typically ≈ 1 eV, namely the Mott-Hubbard and charge transfer interaction, on-site exchange, crystal field splitting, width of the the $3d$ electronic band. Very interesting physics arises from the strong electron correlations in the narrow d bands. Unlike in the transition metal oxides with the $3d$ levels exhibiting a substantial bandwidth, the tightly bound $4f$ levels are only slightly broadened by overlap and hybridization and the $4f$ levels in general do not participate in charge transport in rare earth systems. The itinerant electrons rather have $5d$ and $6s$ character. Crystal field splitting effects and - in general - also the exchange interactions are smaller than in transition metal oxides and magnetic ordering phenomena appear at lower temperatures. Except for Gd^{3+} and Eu^{2+}, which have a half-filled $4f$ shell with a spin-only moment, the rare-earth magnetic moments comprise a substantial orbital contribution. The large magnetoresistance effects in rare-earth systems were attributed to spin disorder scattering [14, 15] with the largest amplitude close to the magnetic ordering temperatures. However, it soon became evident that magnetic polarons i.e. magnetically polarized clusters arising from exchange interaction between bound carriers and the localized $4f$ spins also need to be considered to explain some of the observed effects [15, 16, 17, 18].

In this contribution we will discuss two novel low-dimensional rare-earth systems that exhibit large magnetoresistance effects. Both systems have a layered crystal structure and contain Gd in its usual oxidation state 3+ with a spin-only $S = 7/2$ magnetic moment. In the first system, GdI_2, one can understand the large magnetoresistance near room-temperature within the scope of spin-disorder scattering in a ferromagnetic metal. The second system, $GdXH_y$ ($2/3 < y \leq 1$) with X being the halogens Br or I, shows a metal-insulator transition below $\approx 30K$ if y approaches the lower limit of its homogeneity range. A concomitant large negative magnetoresistance effect is seen which we attribute to the formation of magnetic bound polarons the effective mass of which can be tuned by an external magnetic field.

2.   Spin-disorder scattering in the ferromagnetic metal GdI$_2$

GdI$_2$ crystallizes in the 2H-MoS$_2$ structure with triangular metal atom layers (Fig. 1) sandwiched between close- packed I atom double layers[19, 20].

GdI$_2$ is metallic and shows spontaneous magnetization due to ferromagnetic ordering of the 4f electrons with a remarkably high Curie temperature, $\approx$ 280K (Fig.1(b))[20, 21, 22, 23, 24]. Spin-polarized electronic structure calculations predict almost full polarization of the 5d electrons, in addition to the polarization of the f electrons, and a total saturation magnetic moment of ~ 7.3 μ_B which is, in fact, observed in the experiments [22].

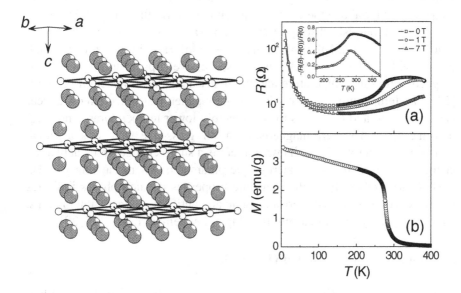

Figure 1. (left) Crystal structure of GdI$_2$ with the layers of the Gd atoms outlined. Gd and I are represented by circles with increasing diameter. (a) Resistance of GdI$_2$ in zero-field, 1T and 7T. Inset: Magnetoresistance at the indicated fields. (b) Magnetization of GdI$_2$ measured in a field of 10 mT. (after [22, 23]).

The ferromagnetic ordering is accompanied by an edge in the electrical resistance near the Curie temperature and a plateau above [20]. An external magnetic field strongly decreases the resistance in the plateau regime, and a very large negative magnetoresistance results [22, 23]. The magnitude of the magnetoresistance effect is similar to or even larger than the magnetoresistance observed in mixed valence oxomanganates but is shifted to a temperature regime around room temperature (Fig.1(a)) in contrast to most of these mixed valence systems.

A quantitative theoretical explanation for the large negative magnetoresistance effect in GdI_2 has recently been proposed by Eremin et al. [25]. According to their theory the magnetoresistance is due to strong spin-disorder scattering of the conduction electrons with essentially $5d_{z^2}$ character by the localized $4f^7$ electrons with their spin-only moment of $S = 7/2$ coupled via s - f exchange [14]. The effect is enhanced with respect to spin-flip scattering in usual three dimensional ferromagnets because of the special topology of the Fermi surface and a Fermi wavevector which lies near the Γ point of the Brillouin zone and thus couples effectively to the ferromagnetic critical fluctuations. The two-dimensional character of the spin fluctuations broaden this effect to a regime significantly above room-temperature. Recent ESR measurement on GdI_2 carried out to investigate the spin fluctuations at temperatures far above the Curie temperature confirm this suggestion [26].

GdI_2 continuously absorbs hydrogen up to a maximum metal-to-hydrogen ratio of 1:1, i.e. until the insulating phase GdI_2H_y with $y = 1$ is reached. Phases with $0.28 < y < 0.34$ show a successive suppression of the ferromagnetic ordering temperature and thermally activated conduction behavior [20]. The temperature regime where large magnetoresistance effects appear is smeared out and decreases to considerably lower temperatures. The magnetism changes from ferromagnetism to a spin-glass-like behavior at a critical hydrogen content $y \approx 1/3$. Likewise, there is a sharp change in the activation energy as well as in other macroscopic parameters when reaching $y \approx 1/3$ [27]. A detailed study of the magnetic properties of such phases indicates the presence of competing ferromagnetic and antiferromagnetic interactions, which eventually result in a spin freezing at low temperatures. [27, 28]

3. Magnetic polarons in $GdXH_y$ (X=Br, I)

The hydride halides $REXH_y$ ($2/3 < y \leq 1$); X = Cl, Br, I) of the trivalent lanthanides constitute a class of layered metal-rich compounds in which close-packed metal atom bilayers are sandwiched between halogen atoms to form the elementary structural building unit RE_2X_2 [29]. In the crystal structure such X-RE-RE-X slabs are linked to each other via van der Waals forces (Fig. 2). Hydrogen atoms occupy tetrahedral voids within the metal atom bilayers. Filling of all the tetrahedral interstices corresponds to the upper limit REXH [30, 31]. Hydrogen atoms can be removed and the phases exist with a homogeneity range $2/3 < y \leq 1$.

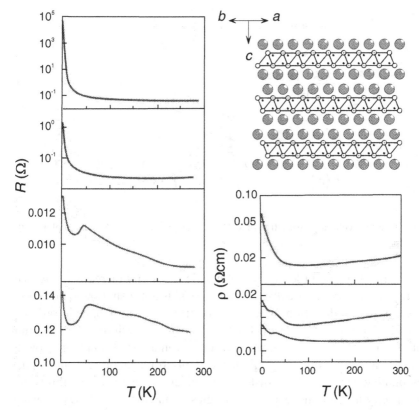

Figure 2. (left) Electrical resistivities of GdBrH$_y$=0.7, 0.8, 0.9, 1, from top to bottom respectively. (right top) Crystal structure of REXH displayed in a projection along [110]. X, RE, and H atoms are represented by circles with decreasing diameter. (right bottom) Electrical resistivities of TbBrH$_y$=0.7, 0.8, 0.9, from top to bottom respectively. (after [32, 33]).

The phases GdXH$_y$ (X=Br and I) show a strong correlation between the hydrogen content y and temperature dependence of the electrical resistivity [29, 32] Below ~30 K, a huge increase in resistivity by several orders of magnitude has been observed for GdBrH$_y$ samples as y approaches the lower limit of the homogeneity range, while for samples with $y \geq 0.8$ the resistivity varies only weakly with temperature yet showing an anomaly at the Néel temperature T_N. Samples of GdBrH$_y$ show antiferromagnetic ordering with T_N decreasing from 53 K to 35 K if y is reduced from 1 to 0.67, respectively [33]. A magnetic field suppresses this resistivity increase considerably and leads to a large negative magnetoresistance effect [32]. The isostructural phases TbBrD$_y$ also show metallic behavior with anomalies associated with

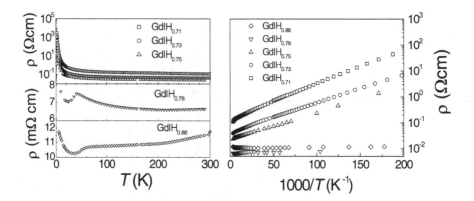

Figure 3. (left) Electrical resistivities of GdIH$_y$ (after [35]). (right) Arrhenius plot of the resistivities.

the antiferromagnetic ordering. The dramatic resistivity increase at low temperatures, as observed for GdBrH$_y$ for $y \rightarrow 0.67$ is not seen. Low-temperature neutron powder diffraction patterns of TbBrD$_y$ indicate magnetic ordering below 22 K for x=0.81 and 0.88 with ferromagnetic alignment of the moments within a metal atom layer and antiferromagnetic coupling between adjacent planes in a bilayer, while for x=0.69 no magnetic reflections have been found down to 2 K [33]. Instead, ac- and dc-magnetic susceptibility and heat capacity measurements reveal spin-glass behavior for TbBrD$_{0.69}$, which has been ascribed to random single ion anisotropies of the Tb ions induced by a statistical distribution of the vacancies in the D substructure generating random crystalline electric fields [34].

As already described by Bauhofer *et al.* [32] for two representative H concentrations and recently investigated in greater detail by Ryazanov *et al.* the phases GdIH$_y$ ($y \rightarrow 0.67$) show a similar resistivity increase at low temperatures (see Fig. 3(a)) as GdBrH$_y$ also indicating a metal-insulator transition for the lower end of the H homogeneity range. [35] GdIH$_y$ samples with y=0.86 and 0.78 show a metallic characteristic with room-temperature resistivities of ≈ 10 mΩcm and anomalies at the Néel temperatures. The resistivities of samples with a lower hydrogen content increase by up to 5 orders of magnitude below about 25 - 30K. Anomalies associated with the magnetic transition identified in the heat capacity and magnetization measurements (see below) at about 50 K are not visible any more.

An Arrhenius plot (Fig. 3(b)) reveals that the growth of the resistivities follows an activated behavior with the activation $E_a(y)$ showing a continuous increase from 0.8 meV for y=0.78 to a saturated value of about 3.5 meV

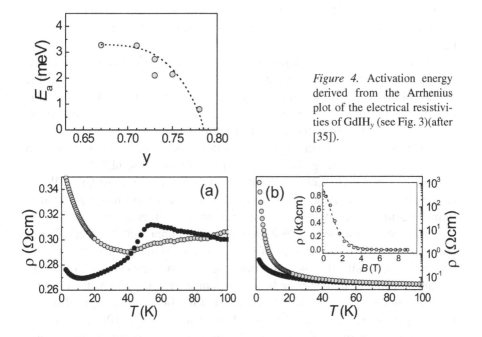

Figure 4. Activation energy derived from the Arrhenius plot of the electrical resistivities of GdIH$_y$ (see Fig. 3)(after [35]).

Figure 5. Magnetic field dependence of the electrical resitivities of GdIH$_y$. (a) $y=0.86$; (b) $y=0.73$. Filled symbols correspond to the zero-field measurement, hollow symbols to the measurement in a magnetic field of 9 T. ((b) inset) Magnetic field dependence of the resistivity of GdIH$_{0.73}$ at 2K. (after [35]).

for ($y \rightarrow 0.67$) (Fig. 4). The critical hydrogen concentration, y_{crit}, below which the phases are semiconductors is determined as $y \approx 0.78(1)$. Above this concentration the phases are metallic.

An activation energy of a few meV corresponds to a Zeeman splitting which can be induced already by a few Tesla magnetic induction. Consequently, one can expect that the low-temperature resistivity increase can be influenced very effectively by an external field of this magnitude. This is in fact the case, as demonstrated in Fig. 5 by comparing the resistivities measured in zero-field and in a field of 9 Tesla for a sample GdIH$_{0.73}$. The low temperature resistivity increase amounts to less than an order of magnitude. A magnetic field of 9 Tesla is sufficient to effectively suppress the metal-insulator transition. Samples with higher H contents, in contrast, show only little variation of the resistance and with a magnetic field rather exhibit a smearing of the anomaly in the vicinity of the Néel temperature at ≈ 50K.

An understanding of the large negative magnetoresistance effects in the phases GdIH$_y$ ($y \rightarrow 2/3$) was gained from a detailed investigation of the temperature and field dependent magnetizations and the thermodynamic properties.

Figure 6 shows the magnetic susceptibilities of GdIH$_y$ measured after the samples had been cooled in zero-field (zfc) or in an applied (fc) field of 0.01T. All samples undergo an antiferromagnetic transition with the Néel temperature decreasing from $\approx$ 50 K to 33 K and 25 K for x = 0.86, 0.73 and 0.69, respectively. Apparent is the low-temperature increase of the magnetic susceptibilities with decreasing hydrogen content, indicating a relative increase of ferromagnetic exchange interaction. The analysis of the high-temperature Curie-Weiss behavior shows a gradual change from a negative Curie-Weiss temperature Θ_{CW} corresponding to predominant antiferromagnetic interaction to positive Curie-Weiss temperatures with the intersection close to the critical H concentration y_{crit}. Within a mean-field approach, the Curie-Weiss temperature represents the sum of the exchange interactions J_i of a particular moment to all other neighboring magnetic moments, $\Theta_{CW}(y) = \Sigma_i J_i$. A vanishing Θ_{CW} therefore indicates a compensation point with a Curie-type susceptibility. The zero-field cooled and field-cooled susceptibilities of the sample GdIH$_{0.69}$ show a thermal hysteresis below $\approx$ 10K reminiscent of a spin-glass or magnetic cluster-glass behavior. Frequency dependent ac-susceptibility measurements substantiate this observation [35, 36].

A very similar compensation point for $\Theta_{CW}(y)$ has been seen for the system TbBrH$_y$, a system which exhibits spin-glass behavior for $y \rightarrow 0.7$ but no resistivity increase [33, 34].

The emergence of a ferromagnetic component in the magnetizations of GdIH$_y$ induced by the reduction of the hydrogen content y is clearly visible from field-dependent measurements (see Fig. 7). The magnetizations for $y=0.73$ and 0.69 show a nonlinearity with increasing field which is reminiscent of a ferromagnetic component. However, the magnetizations do not level off towards high fields, but rather continue to increase linearly due to a paramagnetic or antiferromagnetic component. The field dependence of the magnetizations can successfully be fitted to a superposition of a Brillouin function assuming a saturation with increasing fields of spin clusters. The concentration and the magnitude of the cluster moments increase with decreasing H content and temperature [35].

Indication for an emerging ferromagnetic component with decreasing H content is also found in heat capacity measurements. Figure 8 shows the magnetic contributions C_{mag} to the heat capacities of GdIH$_y$. With decreasing y a downshift of the λ-anomaly associated to long-range antiferromagnetic ordering is clearly seen. The λ-anomaly deforms to an edge for all samples

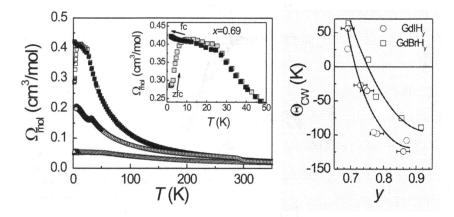

Figure 6. (left) Field cooled (fc, filled symbols) and zero-field cooled (zfc, open symbols) magnetic susceptibilities of GdIH$_y$ measured in a field of 0.01T of GdIH$_{0.69}$, GdIH$_{0.73}$ and GdIH$_{0.86}$ from top to bottom, respectively. The inset shows the low temperature regime for GdIH$_{0.69}$ in an enlarged scale (after [35]). (right) Paramagnetic Curie temperatures Θ_{CW} as derived from a Curie-Weiss analysis of the high temperature susceptbilities of GdXH$_y$ (X=Br, I) vs. H content y.

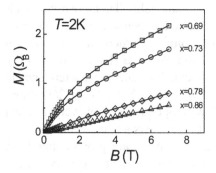

Figure 7. (Magnetization of GdIH$_y$ at 2K. The solid lines result from a fit assuming a superposition of a saturating contribution and an antiferromagnetic or paramagnetic contribution $\propto B$ (after [35]).

with $y \leq 0.75$. This edge is further rounded with decreasing H content. Concomitantly, there is a significant increase of the magnetic contribution to the heat capacity below 15 K. For an H content $y=0.66$, a sharp kink develops at 12 K (see inset of Fig. 8). Below this anomaly, the magnetic heat capacity follows a $T^{3/2}$ power law, clearly different from a T^n ($n=2, 3$) dependence, as expected for magnon contributions to the heat capacity of a 2-dim or 3-dim antiferromagnetic system, respectively. Integration of C_{mag}/T yields a magnetic entropy of R×ln 8, with R being the molar gas constant, in good agreement with a value expected for a $S = 7/2$ spin system. The heat capacity study reveals a gradual shift of the magnetic excitation away from a coherent long-range order to a magnetic state with low-energy excitations. The $T^{3/2}$

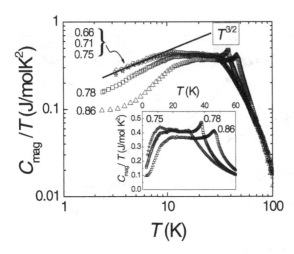

Figure 8. Magnetic contribution to the heat capacity for GdIH$_y$. The solid line indicates a $T^{3/2}$ power law (after [35]).

power-law dependence seen for $y \rightarrow 2/3$ indicates the advent of magnetic excitations with quadratic dispersion relation reminiscent of magnon excitations in a ferromagnetically polarized cluster.

Two essential ingredients have to be taken into consideration to understand the observed metal-insulator transition and the extreme sensitivity of the resulting insulating state to external magnetic fields. This sensitivity leads to the large magnetoresistance observed for GdIH$_y$ below the critical H concentration y_{crit}=0.78. Firstly, there is concomitant structural disorder generated when H atoms are randomly removed from the H substructure [37]. Random vacancies will lead to potential fluctuations and cause a tendency to an Anderson-type localization. In fact, diamagnetic YIH$_y$ ($y \rightarrow 2/3$) shows a weak increase of the resistivity at low temperatures, however, no clear indications of an insulating state [35]. TbBrHy shows a somewhat more pronounced increase of the resitivity but also no insulating ground state [33]. In particular, from the described magnetization and heat capacity experiments it is, secondly, evident that in the 'mixed magnetic phase' below y_{crit} emerging ferromagnetically polarized clusters embedded in an antiferromagnetic background are decisive for the observed electronic transition. These ferromagnetically polarized clusters ('ferromagnetic polarons' or 'ferrons' [15, 17, 18]) created by doped charge carriers in an antiferromagnet matrix have a substantial binding energy due to their large magnetic susceptibility, and they can be stabilized by Coulomb attraction to the charged doping center, viz. the H vacancies in GdIH$_y$. Kasuya coined such a stable localized configuration a 'bound' polaron [15]. Bound polarons gain their magnetic contribution to localization from the energy difference between the antiferromagnetic order of the host and the ferromagnetic polarization within the polaron. Delocalization of the polarons may be achieved if this energy difference is decreased, e.g. by polarizing the antiferromagnetic host by an external magnetic field. For GdIH$_y$ it appears that the spin-only moment of the $4f^7$ electronic configuration with very small crystal field anisotropy is an additional prerequisite to form such polarons. In contrast, in TbBrH$_y$ with a large orbital contribution to the magnetic moment, crystal field anisotropy prevents easy polarization of the moments and suppresses polaron formation such that a metal-insulator transition cannot take place.

4. Summary

Layered metal-rich rare earth halide systems provide a rich zoo to study the complex interplay of magnetic ordering and electronic phenomena in systems close to metal-insulator transitions. For both systems described in this paper, large magnetoresistance effects have been established. GdI$_2$ can be understood on the basis of conventional spin disorder scattering mechanism, however, strongly magnified by the structural anisotropy and the special topology of the Fermi surface. Bound magnetic polarons are formed in GdIH$_y$. Their effective mass is highly susceptible to external magnetic fields giving rise to the observed large magnetoresistance effect.

References

1 Müller, K. A., and Benedek, G. (Eds.) "Proceedings of the Workshop on Phase Separation in Cuprate Superconductors" (World Scientific 1993)
2 Sigmund, E. and Müller, K. A. (Eds.) "Phase Separation in Cuprate Superconductors" (Springer 1994)
3 Nagaev, E. L., (1996) Lanthanum Manganites and Other Giant-Magnetoresistance Magnetic Conductors, *Phys. Usp.* **39**, 781 - 805.
4 Nagaev, E. L., (1996) Magnetoimpurity Theory of Resistivity and Magnetoresistance for Degenerate Ferromagnetic Semiconductors of the LaMnO$_3$ Type, *Phys. Rev.* **B 54**, 16 608-16617.
5 Ramirez, A. P. (1997) Colossal magnetoresistance, *J. Phys.: Condens. Matter* **9**, 8171-8199.
6 Rao C. N. R. and Raveau, B. (Eds.) "Colossal Magnetoresistance, Charge Ordering, and Related Properties of Manganese Oxides" (World Scientific 1998).
7 Coey, J. M. D., Viret, M., and von Molnár, S. (1999) Mixed-valence Manganites, *Adv. Phys.* **48**, 167-293.
8 Tokura, Y., and Tomioka, Y. (1999) Colossal Magnetoresistive Manganites, *J. Magn. Magn. Mater.* **200**, 1-23.
9 Dagotto, E., "Nanoscale Phase Separation and Colossal Magnetoresistance" (Springer, 2002)

10 Volger, J., (1954) Further Experimental Investigations on some Ferromagnetic Oxidic Compounds of Manganese with Perovskite Structure, *Physica* **20**, 49-66.

11 von Helmholt, R., Wecker, J., Holzapfel, B., Schultz, L., and Samwer, K., (1993) Giant Negative Magnetoresistance in Perovskitelike $La_{2/3}Ba_{1/3}MnO_x$ Ferromagnetic Films, *Phys. Rev. Lett.* **71**, 2331-2333.

12 Chahara, K., Ohno, T., Kasai, M., and Kozono, Y., (1993) Magnetoresistance in Magnetic Manganese Oxide with Intrinsic Antiferromagnetic Spin Structure, *Appl. Phys. Lett.* **63**, 1990-1992.

13 for a review see e. g. ref. [7].

14 deGennes, P. G., and Friedel, J. (1958) Anomalies de Resistivite dans certains Metaux Magnetiques, *J. Phys. Chem. Solids* **4**, 71-77.

15 Kasuya, T. (1970) Mobility of Antiferromagnetic Large Polaron *Solid State Commun.* **8**, 1635-1638.

16 von Molnar, S. and Kasuya, T., in "Proceedings of the 10th international conference on the physics of semiconductors", p.233-241, edited by Keller, S. P., Hencal, J. C., and Stern, F., US Atomic Energy Commission, Washington 1970)

17 Nagaev, E. L., (1974) Spin Poalron Theory for Magnetic Semiconductors with Narrow Bands, *Phys. Stat. sol.* **b 65** 11 - 60.

18 Nagaev, E. L., (1992) Self-trapped States of Charge carriers in Magnetic Semiconductors, *J. of Magn. magn. Mater.* **110**, 39 - 60.

19 Bärnighausen, H., Ungewöhnliche Verbindungen der Seltenermetall-Halogen-Systeme und deren besondere Eigenschaften, Hauptvortrag auf der Hauptversammlung der GdCh in München 1977 (Verlag Chemie 1977).

20 Michaelis, C., Bauhofer, W., Buchkremer-Hermanns, H., Kremer, R. K., Simon, A., and Miller G. J. (1992) $LnHal_2H_n$ - Neue Phasen in den ternären Systemen Ln/Hal/H (Ln = Lanthanoid, Hal = Br,I) - III. Physikalische Eigenschaften, *Z. anorg. allg. Chem.* **618**, 98-106.

21 Kasten, A., Müller, P. H., and Schienle M. (1984) Magnetic-Ordering in GdI_2, *Solid State Comm.* **51**, 919-921.

22 Felser, C., Ahn, K., Kremer, R. K., Seshadri, R., and Simon, A. (1999) Giant Magnetoresistance in GdI_2: Prediction and Realization, *J. Solid State Chem.* **147**, 19-25.

23 Ahn, K., Felser, C., Seshadri, R., Kremer, R. K., and Simon, A. (2000) Giant Magnetoresistance in GdI_2, *J. Alloys Compd.* **303-304**, 252-256.

24 Ryazanov M., (2004) Elektrische und Magnetische Eigenschaften metallreicher Seltenerdmetallhalogenide, PhD Thesis, Universität Stuttgart.

25 Eremin, I., Thalmeier, P., Fulde, P., Kremer, R. K., Ahn, K., and Simon, A. (2001) Large Magnetoresistance and Critical Spin Fluctuations in GdI_2, *Phys. Rev. B* **64**, 64425-1 - 64425-6.

26 Deisenhofer, J., Krug von Nidda, H.-A., Loidl, A., Ahn, K., Kremer, R. K., and Simon, A. (2004) Spin Fluctuations in the Quasi-two-dimensional Heisenberg Ferromagnet GdI_2 studied by Electron Spin Resonance, *Phys. Rev. B* **69**, 104407-1 - 104407-5.

27 Ryazanov, M., Kremer, R. K., Mattausch, Hj., and Simon, A. (2005) Influence of Hydrogen on the Magnetic and Electrical Properties of GdI_2H_x (0 < x < 1), *J. Solid State Chem.* **178**, 2339-2345.

28 Ryazanov, M., Kremer, R. K., Simon, A., and Mattausch, Hj. (2005) Large negative Magnetoresistance in the Hydride Halides GdI_2H_x: A System with competing Magnetic Interactions, *Phys. Rev. B* **72**, 092408-1 - 092408-4.

29 Simon, A., Mattausch, Hj., Miller, G. J., Bauhofer, W., and Kremer, R. K. in "Handbook on the Physics and Chemistry of Rare Earths, Vol. 15" edited by Gschneidner, K. A. ,Jr., and Eyring, L. (Elsevier, 1991).

30 Mattausch, Hj., Schramm, W., Eger, R., and Simon, A. (1985) Metallreiche Gadoliniumhydridhalogenide, *Z. anorg. allg. Chem.* **530**, 43-59.

31 Ueno, F., Ziebeck, K. R. A., Mattausch, Hj., and Simon, A. (1984) The Crystal-Structure of $TbClD_{0.8}$, *Rev. Chim. Min.* **21**, 804-808.

32 Bauhofer, W., Joss, W., Kremer, R. K., Mattausch, Hj., and Simon, A. (1992) Origin of the Resistivity Increase in Gadolinium Hydride Halides: $GdXH(D)_y$ (X=Cl, Br, I; $0.67 < y < 1.0$), *J. Magn. Magn. Mater.* **104-107**, 1243-1244.

33 Cockcroft, J. K., Bauhofer, W., Mattausch, Hj., and Simon, A. (1989) Electrical Resistivity and Magnetic Ordering of Gadolinium and Terbium Bromide Deuterides, $LnBrD_x$ ($2/3 < x \leq 1$), *J. Less Comm. Met.* **152**, 227-238.

34 Kremer, R. K., Bauhofer, W., Mattausch, Hj., Brill, W., and Simon, A. (1990) $TbBrD_{0.7}$: Spin Glass Behavior Introduced by Disorder in the Nonmagnetic Substructure, *Solid State Commun.* **73**, 281-284.

35 Ryazanov, M., Kremer, R. K., Simon, A., and Mattausch, Hj. (2006) Metal-Nonmetal Transition and Colossal Negative Magnetoresistance in Gadolinium Hydride Halides $GdIH_x$ ($0.67 < x < 1$), *Phys. Rev.* **B 73**, 035114-1 - 035114-11.

36 Ryazanov, M., Kremer, R. K., Simon, A., and Mattausch, Hj., unpublished results.

37 A superstructure formation due to ordering of the vacancies in the H substructure as discussed in more detail in Ref. [33] has not been confirmed yet. However, first rapid thermal quenching of samples from 900 °C to room-temperature revealed marked effects on the electrical properties and support this conjecture [38].

38 Mattausch, Hj., unpublished results.

AUTHOR INDEX